SDGs and Textiles

Series Editor

Hafeezullah Memon, International Institute of Silk, Zhejiang Sci-Tech University, Hangzhou, China

The book series “SDGs and Textiles” addresses the strategies to achieve sustainable development goals (SDGs) in the present, past, and future. It presents books about the present and future policies of textile ministries of different countries, and books related to sustainability education around different parts of the world in the textile sector. Moreover, it would welcome the conference proceeding related to SDGs and Textiles. The series would cover books comparing the sustainability and SDGs of different institutions and countries. The individual book volumes in the series are thematic. The goal of each book is to give readers a comprehensive overview of a different area of sustainability in the textile sector. As a collection, the series provides valuable resources to a broad audience in academia, the research community, industry, and anyone looking to expand their knowledge of SDGs and Textiles.

Textiles and life are together – life cannot be separated from textiles as it is the most important need for human beings after food. In 2015, the United Nations General Assembly proposed 17 interlinked global goals to be achieved by 2030. Since then, academia and industry have paid much attention to achieving these goals. Textile found its close relation with almost all of these 17 goals.

SDG 1 - No Poverty: Poverty would never be overcome by a charity only; it is essential to develop people’s skills to have a better and wealthy life. Thus, the textile can be considered an excellent discipline to achieve this goal by creating jobs and small and medium businesses.

SDG 2 - Zero Hunger: Through the effective utilization of advanced application of Agrotech Textiles, it is possible to have higher crop yields and save crops from rough weather, unexpected rains, floods, insects, etc.; thus, geotextiles play an essential in achieving this goal of sustainable development.

SDG 3 - Good Health & well-being: There has been much health consciousness after Covid19, and medical textiles assist in getting good health and well-being.

SDG 4 - Learning & Education: Textile or fashion has remained a significant discipline for societies for ages, and there has always remained much to explore in this field. Textile-related universities may play a vital role by offering free access to their education resources, training and spreading information among the locals.

SDG 5 - Gender Equality: The textile sector is one of the industrial sectors that accepted gender equality long ago; in particular, the garment sector has more females than males. Thus, the textile sector has been doing gender equality. Moreover, there has been a recent trend for Gender Neutral Clothing, which need worth studying and may further assist gender equality.

SDG 6 - Clean Water & Sanitation: Textiles could be achieved through filtration, and of course, textile is one of the critical materials for filtration.

SDG 7 - Affordable & Clean Energy: With the recent advancement in material science and engineering, the textile sector has come on the front for, not only by using this clean energy during textile production but also by assisting the production of this clean energy, either in the form of wind turbines blades made of textile composites or by energy harvesting from T-Shirts, etc.

SDG 8 - Decent Work: Recently, there has been much attention that the textile workers are not paid well, labor rights are not cared about, etc.

SDG 9 - Industry and innovation: Textile Industry always follows innovation; the textile companies that do not chase innovation cannot survive in the market.

SDG 10 - Reduced Inequalities: Getting better life and well-being would help reduce inequalities in the textile industry.

SDG 11 - Sustainable Cities: Sustainable Textile Cities through Buildtech and transport textiles.

SDG 12 - Consumption and Production: Textile and garment consumption and production all come under.

SDG 13 - Climate Action: Oekotech or Ecotech Textile, waste management of textiles are upfront to achieve this goal of sustainable development.

SDG 14 - Life Below Water: Mitigating microfiber waste in rivers and oceans may come under the context of it. There has been much attention on this subject after passing the bill at the parliament level of the UK.

SDG 15 - Life on Land: Geotech or Geotextiles studies life on land.

SDG 16 - Peace, Justice, and Strong Institutions: Protective textiles are doing their best to achieve peace, justice, and strong institutions.

SDG 17 - Partnerships for the Goals: The application of textiles to achieve sustainable development goals is only an example. In all textiles sectors, combined efforts of all the goals are essential to achieve true sustainability.

Subramanian Sundaramoorthy ·
S. Kubera Sampath Kumar · Md. Vaseem Chavhan
Editors

Textile Materials for Good Health and Wellbeing

Design and Applications

Springer

Editors
Subramanian Sundaramoorthy
Department of Textile Technology
A.C.Tech, Anna University
Chennai, India

S. Kubera Sampath Kumar
Department of Textile Technology
Vignan's Foundation for Science,
Technology and Research
Guntur, Andhra Pradesh, India

Md. Vaseem Chavhan
Department of Knitwear Design
National Institute of Fashion Technology
Hyderabad, India

ISSN 2948-1236 ISSN 2948-1244 (electronic)
SDGs and Textiles
ISBN 978-981-97-6495-2 ISBN 978-981-97-6496-9 (eBook)
https://doi.org/10.1007/978-981-97-6496-9

This Springer imprint is published by the registered company Springer Nature Singapore Pte Ltd.
The registered company address is: 152 Beach Road, #21-01/04 Gateway East, Singapore 189721, Singapore

Preface

Textile materials are fibers, yarns, and fabrics made from various natural and synthetic materials and they are the essential components in the manufacture of apparel and technical textile products. This book, titled "*Textile Materials for Good Health and Wellbeing*" comprises chapters contributed by experts in this field. The book is published under SDG3 for the book series on "SDGs and Textiles".

Textile materials play a vital role in the medical sector in various ways, from masks to extracorporeal devices. This book delves into the fascinating intersection of textiles and healthcare, exploring the innovative ways in which textile materials are revolutionizing the healthcare industry. Bringing together leading academicians, experts, and researchers, this book provides comprehensive insights into the latest developments, challenges, and applications of textile materials in healthcare. The chapters in this book cover a wide array of topics, including antimicrobial textiles, hygiene textiles, masks and respirators, personal protective gloves, thermal protective wears, filtration textiles for clean rooms and water purifiers, bandages, wound dressings, sutures, specially treated textiles for EMI shields, intelligent textiles, implantable medical textiles, and the scope of sustainable materials in medical textiles. This book is intended for students, researchers, healthcare professionals, and industry practitioners. We extend our deepest gratitude to you for choosing to read this book. We extend our sincere gratitude to all the contributors for their invaluable insights and expertise, which have made this book possible. We hope that the book will help the readers understand various applications of textiles in the healthcare sector and kindle for further research in this field.

Chennai, India — Dr. Subramanian Sundaramoorthy
Guntur, India — Dr. S. Kubera Sampath Kumar
Hyderabad, India — Dr. Md. Vaseem Chavhan

Contents

About the Editors

Dr. Subramanian Sundaramoorthy completed his M.Tech. textile engineering in the year 1997 at Indian Institute of Technology, New Delhi and his Ph.D. in the area of textile technology in the year 2005 at Anna University. He did his Postdoctoral fellowship at PTFE, Georgia Institute of Technology, Atlanta, USA. Presently he is working as a Professor at the Department of Textile Technology, Anna University, Chennai. His research area includes yarn spinning, technical textiles and management of textile industry. He has published more than 30 research articles in international journals, 10 research articles in international conferences and has 5 Indian patents. He has completed research projects and consultancy work worth of about Rs. 3 Million.

Dr. S. Kubera Sampath Kumar completed his Ph.D. in medical textiles at Anna University. He is working as an Associate Professor at Vignan's Foundation for Science, Technology and Research, Andhra Pradesh, India, and has published 32 research articles. He has 2 patents, in the field of medical textiles and new spinning technology; his research interests are technical textiles, medical textiles, clothing comfort and yarn engineering.

Dr. Md. Vaseem Chavhan completed his Ph.D. recently in 2022 in the domain of textile technology from VFSTR University, India. He is an Assistant Professor at the National Institute of Fashion Technology, Hyderabad. He is involved in the teaching of technical textiles, performance fabrics and wearable technology subjects. His research interest is textile modeling, natural fibres, nonwovens and fibrous filtration. He published more than 20 research papers in indexed peer-reviewed journals and presented 10 conference papers. Also, contributed book chapters to the edited books from international publishers.

Textile Materials and Structures for Health and Well-Being: An Overview

S. Subramanian and A. Muthumanickkam

1 Introduction

Textile materials in the form of fibers, yarns, and fabrics are being used extensively in our day-to-day life. They are primarily used for apparel and home furnishing applications; however, nowadays, applications in non-apparel areas are also growing. Technical textiles are the materials engineered for specialized functions. They are the textile materials and products manufactured primarily for their technical and performance properties rather than their aesthetic or decorative characteristics [1]. Technical textiles are categorized mainly into twelve, based on their function. Of the twelve categories, medical textile is one of the major areas covering hygiene textile products, health care products, and *invivo* medical devices. The global technical textiles market size in 2022 is about USD 188 billion and is expected to grow at a compound annual growth rate (CAGR) of 4.7% from 2023 to 2030. The global medical textiles market size is about USD 32.20 billion in 2022 and is expected to grow at a CAGR of 4.3% from 2023 to 2030. The rising awareness of improved healthcare services and effective medical treatments and the increased utilization of textile-based implantable devices like artificial tendons, ligaments, prosthetic valves, etc., are expected to contribute to market expansion globally (www.grandviewresearch.com). This chapter discusses the textile materials and structures used for healthcare applications.

S. Subramanian (✉)
Department of Textile Technology, A.C.Tech., Anna University, Chennai, India
e-mail: ssubbu.anna@gmail.com; ssubbu@annauniv.edu

A. Muthumanickkam
Institute of Textile Technology, Taramani Campus, Chennai, India

S. Sundaramoorthy et al. (eds.), *Textile Materials for Good Health and Wellbeing*, SDGs and Textiles, https://doi.org/10.1007/978-981-97-6496-9_1

2 Materials

2.1 Fibers

A fiber is a thin and long material with a certain level of tensile strength [2]. The length of fiber must be several hundred times its width or diameter and should be strong and flexible [3] to enable it to twist and make yarn. The major fibers used for the manufacture of textile products are (a) natural vegetable fibers like cotton, jute, flax, hemp, and natural animal fibers like silk and wool, (b) manmade fibers, (i) regenerated cellulose-based fibers like viscose, Lyocell, modal, and acetate, (ii) synthetic fibers like polyester, polypropylene, nylon, polyvinyl alcohol, acrylic, polypropylene, and carbon, (c) biopolymer fibers like polylactic acid, polyglycolic acid, chitosan, alginates, collagen, and gelatin, and (d) specialty fibers like high tenacity polyester, ultrahigh molecular weight polyethylene, aromatic fibers, glass, metallic fibers, etc.

2.2 Yarn

2.2.1 Filament Yarn

Two broad classifications of yarns are filament yarn and staple fiber yarn. Filament yarns are produced from polymers like polyester, nylon, viscose, acrylic, etc., by the fiber spinning process. Three major types of fiber spinning processes are (i) wet spinning, (ii) dry spinning, and (iii) melt spinning. In these methods, the polymer solution is spun from the spinneret, having many tiny orifices. In melt spinning, the polymer is first melted by heating, and then it is pumped through the spinneret. In wet and dry spinning, the polymer is first dissolved in a suitable solvent prior to extrusion. In dry spinning, the extrudate coming out of the spinneret is subjected to a hot atmosphere that removes the solvent by evaporation. In wet spinning, the spinneret is immersed in a solution or mixer of solutions that either diffuses the solvent or reacts with the fiber composition, precipitating it from the solution [4]. Filament yarns are made as monofilament or multifilament. In the case of monofilament yarn, a single filament is extruded from the spinneret and is used for applications like sewing threads, metallic threads, etc., after post-spinning operations. In the case of multifilament yarn, many filaments are extruded from the spinneret and twisted to form yarn. The filaments are texturized to increase bulkiness. Continuous filament yarns are obtained from natural protein silk cocoon also.

2.2.2 Staple Fiber Yarn

Staple fibers are of shorter length and require twisting together to produce a satisfactory length of yarn of suitable strength [5]. Staple fiber yarns are produced from natural fibers like cotton, jute, and flax by spinning machines. Synthetic fibers like polyester and viscose rayon are also made as staple fibers and then spun into yarn. The short-staple spinning system is used for spinning fibers like cotton, cotton-polyester blends, and cotton-polyester-viscose blends. The long-staple spinning system is used for spinning fibers like jute, flax, wool, blends of wool with polyester, and spinning of waste and cut filaments of silk.

2.2.3 Numbering of Fibers and Yarns

The fibers and yarns are expressed in terms of length of material per unit mass or mass of material per unit length; the first is the indirect system, and the second is the direct system of numbering. The fibers are mainly expressed in micronaire (micrograms per inch), tex (weight in grams per 1000 m length), or denier (weight in grams per 9000 m length) and the filaments are expressed in tex or denier. The widely used numbering system for yarn is tex, denier, and English count, Ne (length in 840 yards per 1 pound weight). The conversion for important yarn numbering systems is given below.

Tex = 590.5/Ne; Denier = 5314.5/Ne [6].

3 Material for Medical Textile Products

Medical textile is one of the growing sectors in technical textiles. Azam Ali and Shavandi [7] classified the areas within medical textiles as

- Implantable materials, e.g., sutures, grafts, scaffolds
- Non-implantable materials, e.g., wound dressings, pressure garments
- Healthcare/hygiene roles, e.g., clothing, disposable products such as wipes
- Extracorporeal devices, e.g., artificial kidney, liver, pancreas
- Intelligent medical and healthcare textiles

Materials used for apparels such as cotton, viscose rayon, polyester, nylon, polypropylene, etc., are used for hygiene and non-medical applications; for example, gloves-cotton, latex; medical gowns and hospital bed spreads—cotton; masks and respirators—polypropylene; coveralls-cotton, polypropylene.

Materials used for other medical applications are [8]:

Suture: silk, nylon, polyglycolic acid (PGA), polyester, polypropylene, ultrahigh molecular weight polyethylene.

Tissue engineering: silk, polylactic acid (PLA), Polycaprolactone (PCL), PGA, poly L(Lactic acid) PLLA.

Wound dressings: Polyurethane (PU), alginate, chitosan, collagen, silk, cotton.

Some important biopolymers used for medical textile applications are discussed below.

3.1 Biopolymers

Biodegradable polymers are solid polymeric materials and devices that break down due to macromolecular degradation with dispersion. They are attacked by biological elements, so that the integrity is affected, and they produce fragments or other degradation by-products. Such fragments can move away from their site of action but not necessarily from the body [9, 10]. Bio-resorbable polymers are solid polymeric materials and devices that show bulk degradation and further resorb *in vivo*. Bio-resorption is a concept that reflects the total elimination of initial foreign material and bulk degradation by-products with no residual effects. Bio-erodible polymers are solid polymeric materials or devices that show surface degradation *in vivo*. Bio-erosion is a concept that reflects the total elimination of the initial foreign material and surface degradation by-products with no residual side effects. Bio-absorbable polymers are solid polymeric materials or devices that can dissolve in body fluids without any polymer chain cleavage or molecular mass decrease. A bio-absorbable polymer can be bio-resorbable if the dispersed macromolecules are excreted [10–12]. Some important biopolymers are discussed below.

3.1.1 Chitin and Chitosan

Chitin and its de-acetylated derivative chitosan are natural polymers composed of randomly distributed β-(1-4)-linked D-glucosamine (de-acetylated unit) and *N*-acetyl-D-glucosamine (acetylated unit) [13]. Chitin is highly hydrophobic in nature and is insoluble in water and most of the organic solvents. It is soluble in hexafluoroiso-propanol, hexafluoro-acetone, and chloro alcohols in conjugation with aqueous solutions of mineral acids. Chitin is obtained basically from prawns and crab shells, and the chemical treatment of chitin produces chitosan. Chitin and chitosan possess excellent antibacterial, antifungal, biodegradability, biocompatibility, non-toxicity, and adsorption properties [14]. Chitosan contributes to faster wound healing processes, making it a suitable material for wound dressing applications. However, these scaffolds have low mechanical strength under physiological conditions, thus limiting their applicability [15].

3.1.2 Collagen

Collagen is an important biomaterial in medical applications due to its special characteristics, such as biodegradability and weak antigenicity [16, 17]. Collagen is the major constituent of connective and conjunctive tissues in the body and is the most abundant protein in the human body [18]. Collagen possesses a fibrous structure that can impart structure and strength to body tissues, such as tendons, ligaments, and skin. There are more than twenty disparate collagen types that exist in animal tissue. Type I collagen is the predominant fiber-forming collagen type and is found in bones, skin, teeth, and tendons. Type II collagen is the second most abundant, found in cartilaginous tissue, developing cornea, and vitreous humor [19, 20]. The major disadvantage is that collagen nanofibrous matrix may lose its structural integrity in an aqueous environment, such as in the human body. However, the combination of collagen with biocompatible polymers that enhance biological interactions with cells and speed up tissue regeneration could improve the dimensional stability of scaffolds [21].

3.1.3 Silk

Silk is popularly known in the textile industry for its luster and mechanical properties [22]. Silks are generally defined as protein polymers that are spun into fibers by some *Lepidoptera* larvae, such as silkworms, spiders, scorpions, mites, and flies [23]. Silk is a natural filament produced by the silkworm. Silk contains the fibrous protein 'fibroin' surrounded by the glue-like protein 'sericin' to bind the fibroins together. Fibroin is a highly insoluble protein containing up to 90% of the amino acids glycine, alanine, and serine, leading to anti-parallel β-pleated sheet formation in the filament [24]. After the removal of sericin from silk fibroin, it has been considered the starting raw material for biomedical, cosmetic, and biotechnological applications, such as surgical sutures, wound cover materials, controlled drug release carriers, tissue engineering scaffolds, and repair materials for skins, bones, ligaments, etc. [23]. The major silk varieties are, cultivated mulberry silk (*Bombyx mori*) and wild silks, eri (*Attacusricinii*), muga (*Antheraeassama),* and tassar (*Antheraeperni*).

3.1.4 Polycaprolactone

Polycaprolactone is an aliphatic polyester and hydrophobic polymer prepared by the ring-opening polymerization of caprolactone. The repeating molecular structure of the PCL homopolymer consists of five non-polar methylene groups and a single relatively polar ester group [25]. PCL has unique properties similar to those of polyolefin and is biodegradable. Polycaprolactone has slower biodegradation in *in vivo* than polylactic acid (PLA) due to its higher crystallinity and hydrophobic properties. It is used for soft and hard tissue engineering and drug delivery. The drawback of low cell attachment can be overcome by blending starch, cellulose, and PLA polymers with the PCL [26].

3.1.5 Poly (Dioxanone)

Poly (dioxanone) (PDO) is a polyether-ester aliphatic polyester synthesized by the ring-opening polymerization of p-dioxanone. Poly (dioxanone) exhibits a crystalline fraction of 55% and a glass transition temperature between 10 and 0 °C [27]. Poly (dioxanone) sutures provide higher flexibility, higher strength retention, slow absorption rates, lower inflammatory response rates, and easier knotting capacity when compared to poly (glycolic-co-lactic acid) and poly (glycolic acid) sutures. In addition, vascular prostheses made of PDO have been shown to be less thrombogenic than both PGA and polyester grafts [28]. PDO is a synthetic bio-resorbable polymer that offers several advantages over the more traditional bio-resorbable polymers like PLLA, PGA, and poly(lactic-co-glycolic acid) because it has a slower resorption rate and induces a lower inflammatory response [29].

3.1.6 Poly L(Lactic Acid)

PLLA is a homopolymer of the PLA family, which also includes the homopolymer PDLA and the copolymer PDLLA [30, 31]. Poly L (lactic acid) is a biocompatible, biodegradable, and low-toxic material. These bio-absorbable polymer materials can act as drug carriers in the application of drug delivery systems [32]. These biodegradable polymers are derived from renewable sources. These polymers have been considered the ideal biomaterials for biomedical applications [33]. PLA has extremely low moisture regain (0.4–0.6%), much lower than natural fibers and slightly higher than polyester, and PLA shows excellent wicking ability. PLA has fast water spreading and rapid drying capability, which gives the fiber a very positive inherent moisture management characteristic. It is used as baby wipes and diapers, feminine hygiene products, antimicrobial cleaners, etc. [11, 12, 34].

4 Structures

Different structures of textiles are used for textile product development: film, foam, and woven, knitted, braided fabrics, nonwoven, and electrospun mats.

4.1 Woven Fabrics

Woven fabrics are made by the interlacement of warp and weft threads. The warp is the yarn running along the length of the fabric, and the weft is inserted perpendicular to the warp. Woven fabrics are made with different weave structures, like plain, twill, satin and their derivatives, honeycomb, huck-a-back, etc. The weave structure defines the order of interlacement of warp and weft. The dimensions of woven fabric are warp count, weft count, end per cm (warp density), picks per cm (weft density), areal density (GSM), thickness, and width. The properties of the fabric, like tensile, tearing, bursting, bending strengths, abrasion, and pilling resistance, depend on the characteristics of the fiber, yarn, weave structure, and dimensions of the fabric. The features of woven fabric are dimensional stability in dry and wet conditions, higher strength, wash stability, easier dyeing and printing, and durability. Woven fabrics are mainly used for apparels, and in healthcare textiles, they are used for hospital bedspreads, medical gowns, caps, reusable masks, etc.

4.2 Knitted Fabrics

Knitting is the technique of constructing textile structures by forming a continuous length of yarn into columns of vertically intermeshed loops [5]. Knitting is broadly categorized into weft knitting and warp knitting. Weft knitting is otherwise known as circular knitting. In the weft knitting machine, yarn feeding and loop formation occur at each needle in succession across the needle bed during the same knitting cycle. Important structures of weft knits are plain, rib, interlock, and purl. In a warp knitting machine, there is simultaneous yarn-feeding and loop-forming action occurring at every needle in the needle bar during the same knitting cycle. As all needles receive their overlaps simultaneously, a guide under-lapping from one needle to another would pass from one knitting cycle or course to the next. Thus, the warp yarn passes from an overlap produced in one course to an overlap produced in the succeeding course [5]. The warp-knit fabrics are produced by Tricot and Raschel machines. The features of knitted fabrics are softness and extensibility. In apparel sector, they are extensively used for inner garments and casual wear. They find the following applications in the healthcare sector: compression and retention bandages, hand gloves, vascular grafts, reusable masks, etc. Chen et al. [35] produced knitted scaffold from silk filament, PLA, PCL, silk, etc., for tissue engineering applications. The knitted structure is the most adequate scaffolding for ligament tissue engineering [36].

4.3 Nonwovens

Nonwoven is a web of directionally or randomly oriented fibers, bonded by friction and/or cohesion and/or adhesion. Characteristics of a nonwoven fabric depend mainly on the type of fiber used, the preparation of the web, the method of bonding, and the parameters adopted in web preparation and bonding. The fibers can be natural or synthetic, organic or inorganic; the characteristic of a fiber is that it is longer than its thickness or diameter [37]. The nonwovens do not have an organized geometrical structure. They are essentially the result of the relationship between one single fiber and another [11]. The structure and properties of the nonwovens can be designed as per the requirements. Studies related to the application of nonwovens for tissue engineering have been reported by many researchers [38–40].

4.3.1 Manufacturing Processes

There are three major processes used for web forming: (i) dry-laid system with carding or air-laying, (ii) wet-laid system, and (iii) polymer-based system, such as spun-laying or specialized technologies like melt-blown or flash-spun fabrics, etc. The next process, consolidation, can be done by (i) the use of chemicals like binders, (ii) thermal means like cohesion bonding or thermal bonding, like the partial fusion of the constituting fibers or filaments; fusion can be achieved by calendaring or through air blowing or by ultrasonic impact and (iii) mechanical means (frictional bonding) like needling, stitching, water-jet entangling, or a combination of these various means [11].

4.3.2 Properties and Applications

Nonwovens can be made absorbent, breathable, drapeable, flame-resistant, heat-sealed, light, lint-free, soft, stable, stiff, tear-resistant, and water-repellent. Their uses are listed as follows [37]:

- Personal care and hygiene products like baby diapers, feminine hygiene products, adult incontinence items, dry and wet pads, nursing pads, or nasal strips
- Healthcare products like operation drapes, bed and table linens, gowns and packs, dressings and swabs, etc.
- Filtration: air and gas filters, HEPA and ULPA filters, face masks, and respirators

4.4 Braided Fabrics

Three or more yarns are used to form a three-dimensional integral braided structure oriented in various directions. The characteristics of individual yarns, the number of braids per unit length, and the tension at which they are braided determine the properties of braided yarn. The braided scaffolds are mostly fabricated for ligament growth due to their tensile strength. The advantage of the braided scaffold is the interconnected network of porous structure, which supports the transportation of oxygen and nutrients to the implant site [41].

4.5 Electrospun Fibrous Mats

Electrospinning is a unique process to produce fibers of diameter in the range of 100 nm–5 μm. Fibers produced by this approach are at least one or two orders of magnitude smaller in diameter than those produced by conventional production methods like melt or solution spinning [42, 43]. The nanofibrous mat has many favorable characteristics, such as high porosity, a wide range of pore size distribution, and a high surface area-to-volume ratio, which are favorable parameters for cell attachment, growth, and proliferation [44–46]. Electrospinning offers a good opportunity for manipulating the structural and mechanical properties [47] and also enabling the use of different polymers' fibrous assemblies as the scaffold for cell culture. Muthumanickam [48–51] produced electrospun mats from mulberry and eri silk and studied their properties related to biomedical applications. Sampath et al. [11] produced mats from the blend of eri silk fibroin and poly *L*-lactic acid polymer by the electrospinning process and evaluated the properties required for the skin contact layer of wound dressing systems. Many research have been carried out on biomedical applications of electrospun fibrous mats [52–56].

4.6 Films

The water-soluble salt particles are mixed into the biodegradable polymer–solvent solution. The mixture is then cast into the desired shape, and the solvent is removed by vacuum drying and lyophilization. The water-soluble salt particles are then leached out with water to leave a porous structure. The porosity of the material is determined by the amount of salt and the size of the particles added to the solution. In this method, the distribution of salt is not uniform within the polymer solution [57, 58]. The cast solvent and particulate method is used to produce a two-dimensional structure with a thickness between 500 and 2000 μm. In this method, it is very difficult to produce a scaffold with a thickness of more than 3000 μm [59]. The structures are made by the following methods also:

4.7 Melt Molding

Polymer is mixed with the leachable particles in the form of power and then loaded into a mold. Then heat and pressure are applied to the polymer; the heat causes the polymer to melt, and the compression maximizes the packing of the mixture. Then the molded polymer-porogen composite is immersed in a solvent that selectively dissolves the porogen agent. The drawback of this method is the requirement of high temperature [60, 61].

4.8 Gas Foaming

Using this technique, a highly porous biodegradable scaffold for tissue engineering can be formed. Nam et al. [62] and Salerno et al. [63] casted a mold from a paste of PLLA-solvent gel containing dispersed ammonium bicarbonate salt particles. It was submerged in a hot water solution to allow ammonia and carbon dioxide in the polymer matrix to gradually evolve. Consequently, a well-connected macroporous scaffold with mean pore sizes of around 300–400 μm was produced. The principal drawback of this methodology is its inability to achieve consistent porosity within the scaffold.

4.9 Membrane Lamination

Membrane lamination is achieved by solvent casting, particle leaching processes, and introducing peptides and proteins layer by layer during the fabrication process. Suitably shaped membranes were immersed in solvent and three-dimensional assembly featuring a continuous pore structure and morphology was produced [64]. The resulting 3D scaffolds had bulk characteristics that were same as each individual membrane. This method's drawbacks are (i) layering of porous sheets causes reduction in pore interconnectivity [9, 65], and (ii) the process takes more time because thin membranes can only be employed in this process [66].

4.10 Freeze Drying

Whang et al. [67] and Schoof et al. [68] developed scaffolds by this method. The scaffold with high porosity and interconnectivity was produced by dissolving polymer in a solvent and then lyophilization under high vacuum to eliminate the solvent [69, 70]. The scaffolds were prepared from the polymers such as PGA, PLLA, PLGA, silk proteins and blends of PLGA/PPF [23, 71]. The pH and freezing rate both affect pore size; a rapid freezing rate results in smaller pores. Schoof et al. [68] used controlled solidification in a single direction to build a homogenous 3D-pore structure. This technique's primary benefit is that it doesn't require high temperatures or separate leaching steps. Smaller pores, longer processing times, and poor pore structure uniformity are the disadvantages of this method [72, 73]. Silk protein porous scaffolds for tissue engineering applications have been created via this freeze-drying method [66, 69, 70, 74].

5 Conclusions

The materials used for healthcare applications and biomaterials for medical applications are presented in this chapter. Different types of textile structures and different methods of fabricating scaffolds for biomedical applications are described.

References

1. The Textile Institute (1994) Textile terms and definitions, 10th edn. Textile Institute, Manchester
2. Hongu T, Phillips GO, Takigami M (2005) New millennium fibers. Woodhead Publishing Limited, London, p 2
3. Cook GJ (2001) Handbook of textile fibres-natural fibres. Woodhead Publishing Limited, England
4. Gooch JW (2007) Encyclopedic dictionary of polymers. Springer, New York
5. Spencer DJ (2001) Knitting technology: a comprehensive handbook and practical guide, vol 3–7. Woodhead Publishing Limited, London, pp 48–49
6. Saville (2000) Physical testing of textiles. Woodhead Publishing Limited, Lodnon, pp 77–79
7. Azam Ali M, Shavandi A (2016). In performance testing of textiles, methods, technology, applications. In: Wang L (ed) Woodhead Publishing, England, p 132
8. Morris H, Murray R (2020) Medical textiles. Text Progr 2020:1
9. Hutmacher DW (2000) Scaffolds in tissue engineering bone and cartilage. Biomaterials 21:2529–2543
10. Vert M, Li SM, Spenlehauer G, Guerin P (1992) Bioresorbability and biocompatibility of aliphatic polyesters. J Mater Sci Mater Med 3:432–446
11. Kuberasampathkumar S (2015) Development of eri silk fibroin based multilayer wound dressing system. PhD thesis, Anna University, India
12. Muthumanickkam A (2012) Studies on electrospun eri silk fibroin scaffold for biomedical applications. PhD thesis, Anna University, India
13. RaviKumar MNV (2000) A review of chitin and chitosan applications. React Funct Polym 46:1–27
14. Dutta PK, Dutta J, Tripathi VS (2004) Chitin and chitosan: chemistry, properties and applications. J Sci Ind Res 63(20–31):2004
15. Ignatova M, Manolova N, Markova N, Rashkov I (2009) Electrospun non-woven nanofibrous hybrid mats based on chitosan and PLA for wound dressing applications. Macromol Biosci 9:102–111
16. Zhang Y, Ouyang H, Lim CT, Ramakrishna S, Huang ZM (2005) Electrospinning of gelatin fibers and gelatin/PCL composite fibrous scaffolds. J Biomed Mater Res B 72:156–165
17. Zhang Z, Li G, Shi BL (2005) Physicochemical properties of collagen, gelatin and collagen hydrolysate derived from bovine limed split wastes. J Soc Leath Technol Chem 90:23–28
18. Avila Rodríguez MI, Rodríguez Barroso LG, Sánchez ML (2018) Collagen: a review on its sources and potential cosmetic applications. J Cosmet 2018:1
19. Kadler KE, Holmes DF, Trotter JA, Chapman JA (1996) Collagen fibril formation. Biochem J 316:1–11
20. León-López A, Morales-Peñaloza A, Martínez-Juárez VM, Vargas-Torres A, Zeugolis DI, Aguirre-Álvarez G (2019) Hydrolyzed collagen-sources and applications. Molecules 24(22):4031. https://doi.org/10.3390/molecules24224031
21. Yeo IS, Oh JE, Jeong L, Lee TK, Lee SJ, Park WH, Min BM (2008) Collagen-based biomimetic nanofibrous scaffolds: preparation and characterization of collagen/silk fibroin bi-component nanofibrous structures. Bio Macromol 9:1106–1116

22. Kaplan DL, Mello SM, Arcidiacono S, Fossey S, Senecal KWM (1998). In: McGrath KKD (ed) Protein based materials. Birkhäuser, Boston, pp 103–131
23. Altman GH, Diaz F, Jakuba C, Calabro T, Horan RL, Chen JS, Lu H, Richmond J, Kaplan DL (2003) Silk-based biomaterials. Biomaterials 24:401–416
24. Shanmugam K, Sundaramoorthy S (2015) Development and characterization of an electrospun mat from Eri silk fibroin and PLA blends for wound dressing application. RSC Adv 5:31352–31364
25. Vroman I, Tighzert L (2009) Biodegradable polymers. Materials 2:307–344
26. Yu L, Dean K, Li L (2006) Polymer blends and composites from renewable resources. Prog Polym Sci 31:576–602
27. https://www.grandviewresearch.com/industry-analysis/technical-textiles-market. Viewed on 02.04.2024
28. Pillai CKS, Sharma CP (2010) Review paper: Absorbable polymeric surgical sutures: chemistry, production, properties, biodegradability, and performance. J Biomater Appl 25:291–366
29. Smith MJ, McClure MJ, Sell SA, Barnes CP, Walpoth BH, Simpson DG, Bowlin GL (2008) Suture-reinforced electrospun polydioxanone-elastin small-diameter tubes for use in vascular tissue engineering: a feasibility study. Acta Biomater 4:58–66
30. Capuana E, Lopresti F, Ceraulo M, La Carrubba V (2022) Poly-L-lactic acid (PLLA)-based biomaterials for regenerative medicine: a review on processing and applications. Polymers 14(6):1153. https://doi.org/10.3390/polym14061153
31. Casalini T, Rossi F, Castrovinci A, Perale G (2019) A perspective on polylactic acid-based polymers use for nanoparticles synthesis and applications. Front Bioeng Biotechnol 7:259. https://doi.org/10.3389/fbioe.2019.00259
32. Chen C, Gang L, Chao P, Min S, Chunhui W, Dadong G, Xuemei W, Chen B, Zhongze G (2007) Poly(lactic acid) (PLA) based nano composites—a novel way of drug-releasing. Biomed Mater 2:L1–L4
33. Antoniya T, Mariya S, Dilyana P, Nevena M, Iliya R (2013) Polylactide (PLA)-based electrospun fibrous materials containing ionic drugs as wound dressing materials: a review. Int J Polym Mater Polym Biomater 63:657–671
34. Blackburn R (2005) Biodegradable and sustainable fibres. Elsevier, London
35. Chen JL, Yin Z, Shen WL, Chen X, Heng BC, Zou XH, Ouyang HW (2010) Efficacy of hESC-MSCs in knitted silk-collagen scaffold for tendon tissue engineering and their roles. Biomaterials 31:9438–9451
36. Sahoo S, Cho-Hong JG, Siew-Lok T (2007) Development of hybrid polymer scaffolds for potential applications in ligament and tendon tissue engineering. Biomed Mater 3:169–173
37. Wilhelm A, Hilmar F, Walter K (2006) Nonwoven fabrics: raw materials, manufacture, applications, characteristics, testing processes. Wiley, London
38. Balan KK, Sundaramoorthy S (2019) Hydroentangled nonwoven eri silk fibroin scaffold for tissue engineering applications. J Ind Text 48(8):1291–1309. https://doi.org/10.1177/1528083718763779
39. Edwards SL, Russell SJ, Ingham E, Matthews JB, Mitchell W (2006) Nonwoven scaffolds of improved design for the tissue engineering of the anterior cruciate ligament. In: Anand SC, Kennedy JF, Miraftab M, Rajendran S (eds) Medical textiles and biomaterials for healthcare. Woodhead Publishing, London, pp 355–365
40. Trevisol TC, Langbehn RK, Battiston S, Immich APS (2019) Nonwoven membranes for tissue engineering: an overview of cartilage, epithelium, and bone regeneration. J Biomater Sci Polym 30(12):1026–1049. https://doi.org/10.1080/09205063.2019.1620592
41. Cooper JA, Lu HH, Ko FK, Freeman JW, Laurencin CT (2005) Fiber-based tissue-engineered scaffold for ligament replacement: design considerations and in vitro evaluation. Biomaterials 26:1523–1532
42. Srinivasan G, Reneker DH (1995) Structure and morphology of small diameter electrospun aramid fibers. Polym Int 36:195–201
43. Subbiah T, Bhat GS, Tock RW, Parameswaran S, Ramkumar SS (2005) Electrospinning of nanofibres. J Appl Polym Sci 96:557–569

44. Min BM, Lee G, Kim SH, Nam YS, Lee TS, Park WH (2004) Electrospinning of silk fibroin nanofibers and its effect on the adhesion and spreading of normal human keratinocytes and fibroblasts in vitro. Biomaterials 25:1289–1297
45. Nair LS, Bhattacharyya S, Laurencin CT (2004) Development of novel tissue engineering scaffolds via electrospinning. Expert Opin Biol Ther 4(5):659–668
46. Rim NG, Shin CS, Shin H (2013) Current approaches to electrospun nanofibers for tissue engineering. Biomed Mater 8(1):014102
47. Chan BP, Leong KW (2008) Scaffolding in tissue engineering: general approaches and tissue-specific considerations. Eur Spine J 17:467–479
48. Muthumanickam A, Subramanian S, Prasanna V, Natarajan TS, ShankerVerma R (2013) Characterization of electrospun fibrous scaffold produced from Indian eri silk fibroin. Int J Mater Res 104(5):498–506
49. Muthumanickkam A, Elankavi E, Gayathri R, KuberaSampathkumar S, Vijayakumar G, Muthukumar K, Subramanian S (2010) Tensile and in-vitro degradation study of electrospun fibrous mat produced from eri silk fibroin. Int J Mater Res 12:1548–1553
50. Muthumanickkam A, Subramanian S, Goweri M, SofiBeaula W, Ganesh V (2013) Comparison of eri silk and bombyx mori silk fibroin scaffolds for biomedical applications. Iranian Polym. J. 3:143–154
51. Muthumanickkam A, Subramanian S, Panda N, Gowri M, SofiBeaula W, Ganesh V, Krishna P (2013) Electrospun eri silk fibroin scaffold coated with hydroxyapatite for bone tissue engineering applications. Prog Biomater 2:1–11
52. Jin L, Wang T, Zhu M-L, Leach MK, Naim YI, Corey JM, Feng Z-Q, Jiang Q (2012) Electrospun fibers and tissue engineering. J Biomed Nano Tech 8(1):1–9
53. Khorshidi S, Solouk A, Mirzadeh H, Mazinani S, Lagarón JM, Sharifi S, Ramakrishna S (2015) A review of key challenges of electrospun scaffolds for tissue-engineering applications. J Tissue Eng Regen Med 10(9):715–738
54. Kitsara M, Agbulut O, Kontziampasis D, Chen Y, Menasché P (2017) Fibers for hearts: a critical review on electrospinning for cardiac tissue engineering. Acta Biomater 48:20–40
55. Maryam R, Mills DK, Urbanska AM, Saeb MR, Venugopal JR, Ramakrishna S, Mozafari M (2021) Electrospinning for tissue engineering applications. Progress Mat Sci 117:2021
56. Sill TJ, von Recum HA (2008) Electrospinning: applications in drug delivery and tissue engineering. Biomat 29(13):1989–2006
57. Mikos AG, Bao Y, Cima LG, Ingber DE, Vacanti JP, Langer R (1993) Preparation of poly(glycolic acid) bonded fiber structures for cell attachment and transplantation. J Biomed Mater Res 27:183–189
58. Mikos AG, Sarakinos G, Leite SM, Vacanti JP, Langer R (1993) Laminated three-dimensional biodegradable foams for use in tissue engineering. Biomaterials 14:323–330
59. Yang S, Leong KF, Du Z, Chua CK (2001) The design of scaffolds for use in tissue engineering. Part I. Traditional factors. Tissue Eng 7:679–689
60. Correlo VM, Boesel LF, Pinho E, Costa-Pinto AR, Alves da Silva ML, Bhattacharya M, Mano JF, Neves NM, Reis RL (2009) Melt-based compression-molded scaffolds from chitosan-polyester blends and composites: morphology and mechanical properties. J Biomed Mater Res A 91:489–504
61. Gomes ME, Godinho JS, Tchalamov D, Cunha AM, Reis RL (2002) Alternative tissue engineering scaffolds based on starch: processing methodologies, morphology, degradation and mechanical properties. Mater Sci Eng C 20:19–26
62. Nam YS, Yoon JJ, Park TG (2000) A novel fabrication method of macroporous biodegradable polymer scaffolds using gas foaming salt as a porogen additive. J Biomed Mater Res 53:1–7
63. Salerno A, Netti PA, Di Maio E, Iannace S (2009) Engineering of foamed structures for biomedical application. J Cell Plast 45:103–117
64. Maquet V, Jerome R (1997) Design of macroporous biodegradable polymer scaffolds for cell transplantation. Mater Sci Forum 250:15–42
65. Hutmacher DW (2001) Scaffold design and fabrication technologies for engineering tissues—state of the art and future perspectives. J Biomater Sci Polym Ed 12:107–124

66. Subia B, Kundu J, Kundu SC (2010) Biomaterial scaffold fabrication techniques for potential tissue engineering applications. In: Eberli D (ed) Tissue engineering. Intech Press, London
67. Whang K, Thomas CH, Healy KE (1995) A novel method to fabricate bioabsorbable scaffolds. Polymer 36:837–842
68. Schoof H, Apel J, Heschel I, Rau G (2001) Control of pore structure and size in freeze-dried collagen sponges. J Biomed Mater Res 58:352–357
69. Mandal BB, Kundu SC (2008) Non-bioengineered silk gland fibroin protein: characterization and evaluation of matrices for potential tissue engineering applications. Biotechnol Bioeng 100:1237–1250
70. Mandal BB, Kundu SC (2008) Non-bioengineered silk fibroin protein 3D scaffolds for potential biotechnological and tissue engineering applications. Macromol Biosci 8:807–818
71. Vepari C, Kaplan DL (2007) Silk as a biomaterial. Prog Polym Sci 32:991–1007
72. Boland ED, Espy PG, Bowlin GL (2004) Tissue engineering scaffolds. In: Wenk GE, Bowlin GL (eds) Encyclopedia of biomaterials and biomedical engineering. Taylor and Francis, London, pp 1633–1635
73. Hou Q, Grijpma DW, Feijen J (2003) Preparation of interconnected highly porous polymeric structures by a replication and freeze-drying process. Biomed Mater Res B 67(732–740):2003
74. Mandal BB, Kundu SC (2009) Osteogenic and adipogenic differentiation of rat bone marrow cells on non-mulberry and mulberry silk gland fibroin 3D scaffolds. Biomaterials 30:5019–5030

Antimicrobial Textiles

G. Krishnaraj, N. Gokarneshan, A. Jothimanikandan, P. Periyasamy, M. Ponmaniselvam, and K. Sridhar

Abstract In the realm of medical textiles, antimicrobial textiles play a prominent position and significant amounts of research work are carried out in this particular subject. Fabrics that are functionally active and have the ability to kill germs or inhibit their proliferation are referred to as antimicrobial textiles. The chapter provides an in-depth analysis of a number of noteworthy developments observed in the field of study. An emphasis has been placed on the various antimicrobial compounds such as copper, silver, zinc oxide, titanium, and silver-doped titanium, etc., imbedded in textiles used in healthcare services for imparting efficacy to lower the number of microbes that exist in such materials along with the rates of infection related to healthcare. It has also been discussed the physical characteristics of metal nanoparticles and possible mechanisms employed against bacteria. Further to the above the methods of manufacture as well as the effects of incorporating compounds that consist of metal-based nano particles and environmentally friendly antimicrobial materials have been highlighted. Antimicrobial fabric types such as antibacterial, antifungal, and antiviral have been reviewed. The improvement in nanostructured textile material through the application of sustainable finishing methods and the evaluation of its performance on wound treatment are discussed.

Keywords Anti-biofilm · Antifungal · Antimicrobial textiles · Health care · Microorganisms · Nano emulsion · Nanostructure

---

G. Krishnaraj
Department of Textile Design, National Institute of Fashion Technology, Chennai, Tamil Nadu, India

N. Gokarneshan (✉) · A. Jothimanikandan · P. Periyasamy · M. Ponmaniselvam · K. Sridhar
Department of Textile Chemistry, SSM College of Engineering, Komarapalayam, Tamil Nadu, India
e-mail: advaitcbe@rediffmail.com

S. Sundaramoorthy et al. (eds.), *Textile Materials for Good Health and Wellbeing*, SDGs and Textiles, https://doi.org/10.1007/978-981-97-6496-9_2

1 Introduction

The presence of pathogenic microbes may be originated from the patient's environment owing to several factors, including frequent physical contact and cross-contamination between health professionals, patients, and visitors during care activities. In the aforementioned framework, the textile materials present in healthcare services, including insentient surroundings, uniforms of professionals, and clothing and bed linen utilized by patients throughout their hospitalization are susceptible to contamination, proliferation, and dissemination by microorganisms [1, 2]. Earlier research highlights the extensive bacterial infection contamination that has been witnessed on privacy curtains utilized in clinical environments [3], and the white coats of healthcare personnel [4, 5]. Health associated infections (HAIs) and spread of infections during hospital care may also be related with bacterial infection in various textile materials utilized by patients and medical personnel, as well as in inanimate healthcare environments [6]. The application of textiles in the garment industry dates back a millennium where the fibres such as wool, silk, cotton, and nylon, among others are predominantly used. Textile materials are commonly prevalent and serve an important function in social surroundings like medical textiles, children's apparel, furniture upholstery, and even more [7]. Textile materials are categorized into three primary categories: animal, plant, and artificial origins. Cotton, bamboo, and viscose are a few examples of textiles derived from plants. Wool and silk serve as illustrations of products that fall within the animal-based category. Polyester, nylon, and acrylic are a few examples of textiles classified as synthetic. However, the susceptibility of textiles to microbial attack could be an important issue, particularly, those utilized in healthcare facilities, infant apparel, undergarments, and athletic apparel, which are in frequent contact with bacteria, may encourage the hostility of microorganisms due to their excellent moisture affinity and inherent properties [8]. The proliferation of pathogenic microorganisms on textiles, specifically *Escherichia coli*, *Staphylococcus aureus*, and *Klebsiella pneumoniae*, are to be restricted to prevent severe consequences like degradation in textile strength, dull appearance, and odors that endanger the wearer and contaminate the textile. In general, textile materials are affected by two microorganisms, namely assimilation and degradation. Indeed, their metabolites are found out as the cause of various issues, including the bubble generation on colored surfaces of textile products, causing breakage of bond in fibres, and entry into the cavities of natural fibres. Further to this, they have caused adverse impacts on mechanical strength, mass loss, and chemical properties.

The possibility that specific types of microorganisms reside in fabrics is explained by clothing microbiology, and the reaction between fabrics and the surface of skin microorganisms [2]. The application of natural antibacterial coatings onto textile materials can be traced back to antiquity, specifically to the Egyptians, who employed herbal and spice-infused coatings on clothing to construct mummy wraps. The use of antibiotics became prevalent during the period of the Second World War, coinciding with surge in demand for antimicrobials to prevent the deterioration of textiles. In order to enhance the longevity of the fabrics and prevent bacterial cluster formation, a number of military fabrics underwent treatment involving antimony ions, copper, and a combination of chlorinated waxes. In the last few years, medical practitioners are encountering a growing patient population suffering with a variety of skin ailments, like burns, and wounds, that are highly challenging to treat and recover from. The market for a diverse array of medical textiles has expanded at a rapid rate due to consumers' increased concern for hygiene and active lifestyles [9]. Subsequently, there has been a substantial surge in the quantity of functional textiles incorporating biopolymers and synthetic materials that possess antibacterial properties. Synthetic antimicrobial agents exhibit remarkable efficacy against a diverse array of microorganisms and impart a long-lasting impact on textiles. However, their utilization gives rise to apprehensions on account of the adverse effects on non-target microorganisms and water contamination. Therefore, there is significant interest in antimicrobial textiles formulated with non-toxic agents that not only effectively mitigate the negative consequences of microbial proliferation on textile materials but also adhere to regulatory mandates.

2 Study on Antimicrobial Textiles

Antimicrobial fabric has been classified based on the treatment and antimicrobial mechanism as leaching type or non-leaching type and biocide or biostatic, respectively. Further to the above, the antimicrobial fabric is categorized based on microorganisms such as fungi, bacteria, viruses, etc. It has been revealed from the findings that biostatic fabric is preferred owing to the protection of natural bacterial flora present in the skin and its protection whereas the biocidal fabric is used in healthcare and environmental uses [10]. Though quite a number of fabrics are possessing intrinsic antimicrobial properties, they are not active against specific microorganisms and result in narrow antimicrobial spectrum. However, the preference of antimicrobial agents can be carried out based on the fabric types and antimicrobial attribute requirements. Generally, the antimicrobial agents available in the market are designed to release the active components gradually in the presence of moisture. The above said additives may be introduced into the fibre in the even filament extrusion or conventional coating or application process like padding or exhaustion to textile surfaces [11]. A variety of techniques are employed in order to ascertain the antimicrobial characteristics of textile samples. The acquisition of antimicrobial properties for textiles by impregnation is of significance, as it serves to lessen microbial burden and contamination, thereby substantially reducing biological hazards. This process can be done in two main ways: either the antibacterial agent is mixed into the textile fiber matrix during the yarn manufacturing, or the antimicrobial agent is coated on the textile base in the stage of finishing [11]. Textile materials are treated with either organic or inorganic substances containing antimicrobial properties such as QAC, halamines, triclosan, chitosan, metal oxides, nanoparticles, etc. [12]. Active agents or chemical compositions can be evaluated independently in certain circumstances. The assessment of the active agent's antimicrobial function is conducted through various methodologies, as recommended by the Clinical and Laboratory Standards Institute [13]. The most accurate, efficient, and conventional broth micro dilution procedures using optical density measurement. The least inhibitory concentration (MIC) value denotes the minimum concentration at which the proliferation of the bacterial strains under investigation can be inhibited. While impregnating antibacterial bandages with Ag, ZnO, and mixed Ag/ZnO NPs, certain research utilized MIC value. Ag NPs exhibited the highest efficacy against *Pseudomonas aeruginosa* and *Acinetobacter baumanni* among these nanoparticles [14]. In order to impede bacterial growth, the MIC of the antimicrobial compound must be determined, this value finally undermines the cost of the material. The price-performance index for commercial silver nanoparticles and lignin extracts derived from sugarcane bagasse, which have been used as antibacterial agents in the development of antibacterial textiles, indicated that the implementation of bio-based agents might be cost-effective [15]. The bactericidal and bacteriostatic properties of the compounds can be verified by re-inoculating the clear broth that has been utilized to determine the minimum inhibitory concentration (MIC) of the antimicrobial compound. The Fabric Disc Diffusion Assay is utilized to ascertain the antimicrobial characteristic of the textile. A circular disc of sample fiber is mounted on the surface of the inoculated agar. The antibacterial function of the sample is assessed by the appearance of a zone of inhibition following incubation. For determining the methanolic extract of Helichrysumarenarium's antibacterial activity against *S. aureus* and *E. coli*, the disc diffusion technique is utilized. Wool fabrics that have been dyed using these natural dyestuffs exhibited noteworthy antimicrobial properties [16]. Certain antimicrobial agents can be selective in their targeting of particular bacteria. For instance, nano selenium brooms exhibited activity against only *B. Subtilis* when tested against *S. aureus, B. Subtilis, P. aeruginosa, E. coli*, and *Proteus vulgaris*. The minimum inhibitory and bactericidal concentrations for these brooms showed their effectiveness against only *B. Subtilis*, suggesting their potential use specifically in operating rooms to decrease the occurrence of Bacillus-associated bacteremia [8]. In order to ascertain the minimum bactericidal concentration (MBC), cells open to various concentrations of the antibacterial substance are sub-cultured in fresh growth medium. The MBC is calculated when no growth is observed [17]. An additional application of the tube dilution method is the determination of inhibitory concentrations (IC50) against microbes an essential indicator in evaluating the efficacy of antimicrobial compounds. The majority of the investigations have been dependent on the direct coating of such substances to ensure the maximum possible adhesion, as fabric of the leaching variety has been typically utilized.

2.1 Study on Antifungal Textile

Both disc diffusion and tube dilution can be employed to ascertain the textile's fungicidal and fungistatic functions. Filamentous fungi is considered as the most detrimental organisms to fabrics because of their capacity to proliferate on a wide variety of natural substrates. Fungi such as Cladosporium, Alternaria, and Aspergillus are frequently encountered [10]. A substantial inhibition in the proliferation of these fungi has been found on antimicrobial fabrics manufactured by incorporating Ag and C into modified silica [18]. Cotton fabric is treated with guanazole complexes containing zinc and silver ions which demonstrated a broad spectrum of antimicrobial activity against both fungal and bacterial pathogens, including *A. niger*, *Fusarium chlamydosporum*, and *Penicillium* sp. In addition to possessing antimicrobial properties, cotton fabrics coated with guanazole-metal complexes exhibited flame-retardant characteristics [19]. ZnO nanoparticles incorporated into cellulose fibers have exhibited anti-fungal properties to *Phanerochaete chrysosporium*, *Geotrichum candidum*, an opportunistic pathogenic yeast, *A. Niger*, and *P. Chrysosporium*, a wood degrading (cellulolytic) fungus [20].

2.2 Study on Anti-biofilm Textile

Biofilms formed by bacteria or fungi are directly associated with the majority of infectious diseases. Textiles with antimicrobial properties may inhibit the development of biofilms and reduce the incidence of infections. Textiles are currently undergoing the development of biocompatible nanoparticles (NPs) that are antibacterial, anti-biofilm, and anti-adhesive. These NPs may consist of various inorganic polymers (e.g., Cu, Zn, Ga, Si, or Ti) and biological components, such as functionalized lipids or antibacterial enzymes [21]. In order to evaluate the anti-biofilm activity, a fabric sample is incubated in the broth medium supplemented with the test bacterial strain for an entire night. The visual biofilm may be assessed by the instruments, such as confocal laser scanning microscopy, scanning electron microscopy, or high-power microscopy subsequent to cleansing with sterile water following incubation. However, the most traditional approach for assessing the anti-biofilm activity of nanoparticles is through the static biofilm assay employing violet crystal staining. This procedure can be quantified using a spectrophotometric device [22]. Beneficial anti-biofilm activity against *Pseudomonas aeruginosa* is noticed when cotton fabric is treated with propylene imine dendrimers and their Cu(II) complexes. SEM is used for comparing treated and untreated cotton fabrics for the adhesion and biofilm formation on cotton fabrics [23]. The fabrics containing gold nanoparticles have been adversely affected by *E. coli* and *S. aureus* where biofilm has been formed [24]. P1000-acid dendrimer-treated cotton fabric impeded the growth of *P. aeruginosa* biofilm [25]. By preventing bacterial cell adhesion to the fibers or eliminating cells that come into closer contact through different mechanisms, these anti-biofilm fabrics prevent the growth and development of biofilms by microorganisms.

2.3 Study on Antiviral Textile

The determination of antiviral activity in textiles can be accomplished in accordance with the specific standard procedure. Antiviral drugs are determined by treating the virus with the test chemicals and incubating viral sensitive cells for plaque assay. It has been revealed from earlier studies that the antiviral property of textile samples can be determined by passing or filtration of the viral suspension through the prepared textile and subsequently incubating the filtrate with virally susceptible cells [10]. Cotton fabrics treated with specified amount of sodium pentaborate pentahydrate, triclosan, and Glucapon have exhibited antimicrobial activity against a wide spectrum of microorganisms, including yeast, bacteria, and fungi. The cotton fabrics that underwent treatment have confirmed significant antiviral activity against adenovirus type 5 and poliovirus type 1 [8]. The sandwich test method has been used to ascertain the antiviral activity of nonwoven filters coated with 1-chloro-2,2,5,5-tetramethyl-4 imidazolidinone. The antiviral property of the filters has been confirmed by measuring the quantity of avian influenza viruses that survived in embryonated chicken eggs [26]. Radio chemically synthesized AgNPs have been fixed firmly on the cotton fabric surface. The fabric with a modified surface has exhibited antiviral properties against both Feline Calicivirus and Influenza A [27]. It has been recommended that the application of antiviral polysaccharides onto work surfaces, apparel, and masks could potentially impede the transmission of viruses. The spike proteins of the coronavirus can be affixed to these biocompatible and biodegradable polysaccharides. A material that has been nanocoated layer by layer functions as a decoy receptor, preventing the spike proteins from binding and inactivating the virus (Reddy et al. 2013). The COVID-19 pandemic has had a recent and positive impact on research pertaining to antiviral textiles. AgNPs have been proved to be efficacious against SARS-CoV-2 and are recommended for use in the lesser specified concentration range, with cytotoxic effects observed at specific higher concentrations. Inadequate disposal of these substances may also result in the disruption of ecological systems [28].

2.4 Study on Leaching and Non-leaching Type Antimicrobial Textile

Leaching of antimicrobial agents is tested in finished fabrics. The majority of textiles possessing inherent antimicrobial properties are typically non-leaching, whereas coated fabrics require antimicrobial compound leaching testing [10]. To conduct this examination, the final product is either rinsed with distilled water or a solution containing detergent and a non-ionic surfactant. To verify the extraction of the antimicrobial compound, the sample is subjected to multiple washes or treatments with various solvents. This can be verified through the application of various microorganisms to the washed sample. In the end, a decrease in the amount of antimicrobial agents may result in a decline in their antimicrobial activity. Using an agar well diffusion assay, the leaching and non-leaching characteristics of the antimicrobial textile can be distinguished by observing the active compound's diffusion into agar. A more extensive zone of inhibition encircling the fabric under examination signifies an increased capacity for leaching, whereas a more limited zone signifies reduced compound diffusion. A modest and consistent zone of inhibition after a number of laundering cycles verifies that the antimicrobial textile is non-leaching. A reduction in the magnitude of the zone of inhibition signifies the antimicrobial textile that is leaching. In general, it is suggested to use a non-leaching functional textile because it can maintain its activity for prolonged time period and is therefore deemed safe for direct skin contact. Biofilms formed by bacteria are regarded as a significant component of pathogenic microorganisms. Certain antimicrobial textiles have been developed with the purpose of inhibiting or even eliminating the biofilm formation. Leaching has been reduced when bioactive molecules, such as fluoroquinolone and eugenol derivatives, have been integrated with the cotton fabric using a triazine moiety. Cotton fabric treated with fluoroquinolone exhibited bacteriostatic properties with *S. aureus* against both planktonic and preformed biofilms. In a similar vein, fabric treated with eugenol effectively eliminated the pre-existing *P. aeruginosa* biofilms [29]. A quaternary ammonium salt-containing isocyanate group has been developed using a dipping–padding–drying procedure in order to explore its potential as a non-leaching type antimicrobial agent for cotton material. The treated cotton fabric has exhibited remarkable bactericidal efficacy against *S. aureus* and *E. coli*, even after undergoing specific number of washing cycles, without causing any detrimental effects on the epidermis [30]. By employing a sol–gel process, a transparent coating of quaternary ammonium-modified triethoxysilane has been applied to cotton textiles, showcasing their permanent, non-leaching antibacterial properties. Antimicrobial activity against *S. aureus* and *E. coli* continues to be significant even after 15 cycles of rinsing [31]. Certain cotton fabrics treated with antimicrobial agents like polyhexamethylene guanidine hydrochloride and polypropylene glycol diglycidyl ether, retained their antimicrobial efficacy even after undergoing laundering with a detergent solution [32]. The development of leaching and non-leaching fabrics is predominately determined by the cost of production and the accessibility of relevant technologies. The biocidal activity of leaching type fibers may surpass that of non-leaching type fibers, which, depending on the antimicrobial agent, production method, and fiber type, may exhibit biostatic or bio-repellent properties. In recent times, an assortment of textiles has been developed to combat various microorganisms, these fabrics include natural (derived from plants or animals, synthetic, and composite materials.

3 Study on Natural Fabric, Antimicrobial Dyes, and Pigments

Fabric staining has historically relied on natural pigments and dyes, which serve the dual purpose of imparting color and functioning as antimicrobial agents. Curcumin (1,7-bis4-hydroxy-3-methoxy)-1,6-heptadiene-3,5-dione), which is present in turmeric, is frequently employed as a natural dye that is non-toxic. Wool and cotton have been effectively stained with curcumin and their antimicrobial properties have been investigated. It is possible to provide color and antibacterial properties to wool and cotton by pad dyeing them with curcumin. It has been noticed that this treatment exhibited significantly greater resistance to common laundering practices than the untreated one. Natural compounds have been observed to impart exceptional antimicrobial properties to cotton and other natural fibers. Although silk is thought to possess inherent antimicrobial properties, it may not be effective against a broad spectrum of microbes. Conversely, a natural dye, namely curcumin, which can be extracted from plants including *Tectonagrandis*, *Artocarpus heterophyllus*, *Terminalia catappa*, and *Morinda citrofolia*, has effectively demonstrated substantial antimicrobial effect against the microorganisms [33]. Pretreated wool and silk fibers with neem oil have acquired more color while dyeing with chlorophyll, saffron crimson, and yellow natural dyes in comparison to untreated fibres [34]. In the same way, a silk fabric is treated with the natural mordant myrobalan and a natural dye derived from the fruit of Opuntia ficus-indica for improving its antimicrobial properties [35]. Further to synthetic or natural mordants, investigations utilizing natural dyes derived from Neem leaves and Green Walnut shells indicated that an elevated concentration of dyes also augmented the silk fabric's antimicrobial characteristic [36].

4 Study on Synthetic and Blended Based Antimicrobial Textile

Certain synthetic fibers, such as tetrafluoroethylene, acrylic, nylon, and polyester, may possess antimicrobial properties that can be enhanced by treating them with antimicrobial compounds [10, 37]. In order to enhance the antimicrobial properties of popular textiles such as nylon, polyester, acrylic, polypropylene, and natural fabrics, 3-Allyl-5,5-dimethylhydantoin (ADMH) has been applied onto them. These compounds have the capability of being introduced either directly into a fiber at the time of its production or the fiber surface during extrusion. Alternatively, they can be applied to fibers by methods like dip coating, spray application, polymer coating, or spin finishing. Antimicrobial agents can be integrated into nonwoven products during the bonding or finishing stage. Conversely, exhaust and pad-dry-cure methods are employed for knitted or woven textile treatment [38]. It has been revealed that graphene-modified performance textiles for personal apparel possess a significant degree of antimicrobial efficacy and can be both cost-effective and comfortable. This graphene-based textile comprises graphene 81 and graphene 82 nanomaterials. Such materials are proposed for the production of enhanced personal protective equipment (PPE) on the basis of their characteristics [39]. A substantial degree of antimicrobial activity has been observed in cotton/elastane garment fabrics that have been treated with quat-silane [40]. Copper nanoparticles (40–60 nm in size) have been physisorbed onto a modified surface of nylon 6 by means of mPEG grafting. The fabric that underwent treatment exhibited efficacy against pathogenic bacteria and additionally guaranteed cytocompatibility [41]. A sol–gel method has been employed to prepare hybrid coatings of tetra-ethoxysilane, glycidyloxy propyltrimethoxysilane, ZnO, and Ag:ZnO/chitosan. These coatings have been then applied to cotton and cotton/polyester (50/50) textiles utilizing the 'pad-dry-cure' method. The results of these coatings indicated a notable degree of antimicrobial activity [42].

5 Antimicrobial Compounds and Their Effects

The findings of earlier research reveal that textiles treated with antibacterial materials are dependent on lab studies, particularly on body surface, with few conducted in real healthcare conditions [43]. In these studies, the numerous combinations of independent and dependent variables are used in the methodological settings. The independent variables are textile types, antimicrobial chemicals, and the medical service application of impregnated/antimicrobial textiles. Healthcare-associated infection rates and textile microbial contamination analyses are the dependent variables. Studies on hospitalized patients have observed that textiles treated with heavy metals, such as copper, silver, zinc oxide, and titanium nanoparticles, either individually or in combination, have antibacterial properties and lesser bacterial accumulation and/or healthcare-associated infections rates after use by patient in comparison to traditional textiles [44–46]. Antimicrobial substance-treated textiles utilized by patients throughout their hospitalization have been subjected to characteristic analyses, including microbial burden analysis and healthcare-associated infection rates. Despite the absence of consistent evaluation or presentation of statistical significance, the methodological rigor of such studies has been rated as moderate to high [22, 47]. There have been no negative effects identified when treating the textiles with any one substance like copper, chlorhexidine, chitosan, and dimethylol dimethyl hydantoin (together) whereas itching, erythema, and rash have been noticed for the textiles treated with zinc oxide [48]. In the case of silver-treated textiles, no efficacy reduction in microbial load has been observed. But the textiles treated with quaternary ammonium have shown less microbial load reduction and similar effects have been observed with chlorhexidine, silver and copper together, alcohols and isothiazolone derivatives together, halamine, and chitosan and dimethylol dimethyl hydantoin together, quaternary ammonium, and halogens [49, 50]. Generally, antimicrobial textiles, except for those impregnated with silver alone, have exhibited lower microbial burdens in inanimate healthcare situations compared to normal textiles [51]. Heavy metals exert their antimicrobial effects on microbial cells via various mechanisms, including disruption of cell wall synthesis, depolarization of the plasma membrane's electrical potential, lysis of the plasma membrane, denaturation of proteins, and induction of oxidative stress. These mechanisms can also collaborate to amplify the antibacterial efficacy of heavy metals [36, 52]. The fabrication of functional fibers with potent antibacterial properties has been attained through the incorporation of N-halamine structure onto a coating composed of polydopamine. Polydopamine-based coatings functionalized with N-halamine have demonstrated antibacterial efficacy against both *S. aureus* and *E. coli* (Chien and Chiu 2020). Chlorophenols and various halogenated phenols are employed as antimicrobial coatings in textiles. Biguanide (polycationic amine) polymers can electrostatically kill bacteria [53]. By layer-by-layer assembly, cotton fabrics are treated with silicon and a nitrogen-containing compound, poly [3-(5,5-cyanuricacidpropyl-siloxane-co-trimethyl ammonium-propyl siloxane chloride], and phytic acid. The treated cotton fabrics lowered *E. coli* and *S. aureus* by one hundred percent during the first minute of contact [32]. A low-cost, chemical-free technique called plasma treatment has been suggested to change the surface properties of different textiles so as to add environmentally friendly rechargeable antimicrobial finishing. The antimicrobial efficacy of cotton fabric against *S. aureus* has been enhanced when it underwent nitrogen plasma treatment following padding with 5,5-dimethylhydantoin (pad-plasma-dry-cure) and chlorination with sodium hypochlorite (Zhou et al.). The application of silver nanoparticles onto a surface activated by plasma has demonstrated antimicrobial characteristics [54]. Likewise, polyamide 6,6 fabric treated with dielectric barrier discharge plasma has more Ag NPs [55].

6 Antimicrobial Textiles in Healthcare

The use of antimicrobial textiles in medical applications should be approached with extreme caution. This is because significant contradictions in laboratory studies concerning the reduction of microbial load, including drug-resistant microorganisms, and cytotoxicity are highlighted in the scientific literature [56, 57]. Furthermore, within authentic healthcare settings, clinical investigations utilizing various configurations of antimicrobial-impregnated textiles utilized by healthcare professionals or hospitalized patients yielded intriguing outcomes. Though the reduction of healthcare-linked infections is in satisfactory level, the bacterial reduction on health authorities' attire has been deemed unsatisfactory [46]. Additionally, it is crucial to consider the possibility that the substances applied on such materials could induce microbial resistance. This is especially true when the chosen antimicrobial and technique of application promotes controlled release mechanisms, which would result in significant leaching in a moist medium. Subsequent to this incremental and sustained release, the antimicrobial concentration inside the textile decreases to levels less than the Minimum Inhibitory Concentration (MIC), which signifies the threshold of antimicrobial effectiveness beyond which resistance to microorganisms shall be developed [12]. However, the definitive answers regarding the potential for specific materials applied on textiles to develop the choice of microorganisms resistant to pharmacological antimicrobials or other antibacterials are still elusive [58]. This raises concerns regarding the possibility of cross-resistance, co-resistance, and resistance by co-regulation to antimicrobials [59]. So, it is required to ascertain the antimicrobial compounds in hospital textiles that can reduce microbial load and healthcare-associated infection rates effectively compared to ordinary textiles.

In order to maximize biostatic or biocidal efficacy and reduce the risk of microbial resistance, various antimicrobials in textile impregnation have been considered. Still, most research studies using impregnated/antimicrobial textiles lack detailed information about the textile materials, antimicrobial substances, and impregnating method that restricts repetition, and systematic evaluation in identical situations. It has been noted that a deficit of details is observed about the hygiene settings of antimicrobial textiles in clinical trials, including products, techniques adopted, and sanitation procedures. Hence, there is no manner to precisely pinpoint the perfect states for sanitizing antibacterial textiles in actual medical environments to maximize antimicrobial endurance and reduce dirt, organic matter, and bacterial accumulation. With regard to the dependent variables under consideration, it is noted that various techniques can be employed to collect microbiological samples from textile surfaces. Moreover, analyses pertaining to microbial contamination can be conducted in a variety of ways: specific and nonspecific microbial load in terms of the type of microorganisms present; microbial load concentrated in specific regions of the textile's surface; and bacterial accumulation prior to and following utilization in healthcare activities. In relation to the rates of healthcare-associated infections, various approaches can be utilized to identify them. These approaches may involve the use of prior standards of medical facts or a theoretical background that is periodically revised. Furthermore, the analyses conducted may distinguish between specific and non-specific healthcare-associated infections based on the etiological agent and the affected tissues, organs, or physiological framework. Notwithstanding the fact that the intricacy of the epidemiological chain renders the emergence of each infection a multifactorial process, microbial burden invariably must be considered (Nortjie et al. 2022) [53]. Hence, investigations which utilize antibacterial textiles in actual healthcare situations and evaluate their effects on healthcare-associated infection rates and microbial load are invaluable. The scope of employing textiles applied by antibacterial materials are to be taken into consideration, as the patients and health authorities are in constant touch with such materials may experience adverse events primarily characterized by dermatological signs and symptoms. Though, there is no comprehensive examination for the untoward situations provided by the subjects of the study featured herein, all health hazards are not possible to be cast aside and demand additional consideration [43]. It becomes crucial to point out that irrespective of the antimicrobial agent incorporated into the textile material, the concentration of said substance should be considered a critical factor in achieving a harmonious equilibrium between biostatic or biocidal protective properties and user safety, specifically in terms of preventing toxicity and untoward situations [60]. Additionally, in clinical research without participant blinding, the nocebo effect may increase self-reported adverse events at notification rates. The environmental consequences should also be given appropriate consideration, as repeated decontamination processes involving the laundering of impregnated textiles can result in the leaching of specific antimicrobials and subsequent contamination of water ecosystems over extended, intermediate, or even short-term periods [61]. Therefore, for the purposes of biodegradability and environmental sustainability, natural bioactive materials having antibacterial characteristics derived primarily from medicinal plants and other organic bases can be utilized in the textile industry as an alternative to synthetic compounds [62], Nortjie et al. (2022). Additional strategies that can be mentioned include the application of passive coatings on textile surfaces that lack antimicrobial properties [43]. Although these coatings do not exhibit biocidal or biostatic activity, they effectively impede microbial adhesion and are readily decontaminable [63]. Antimicrobial textiles may be profitable economically, especially when compared to the costs of acquiring and applying other antimicrobial substances utilized in health activities to develop biosafety. Antimicrobial textiles have the capacity to prevent and control the occurrence of healthcare-related infections, which subsequently leads to significant financial investments in treatment and incalculable human fatalities. These potential yields comparable profitability. To assess the actual economic ratio of antibacterial textiles, economic modeling should be an integral part of future clinical trials in this field [64]. Though textiles with antimicrobials are resulting reduction of microbial load and healthcare-associated infection rates, the effect should not be overrated as it is imperative to standardize decontamination of these antimicrobial resources. Consequently, novel and pragmatic approaches that enable the detection of microbial contamination in textiles may serve as supplementary instruments for examining the antimicrobial effectiveness of impregnated textiles [65]. Moreover, they aid in the surveillance of microbial resistance and function as indicators of contamination by possible bacteria present on the textile surfaces, despite prolonged durations of exposure [43].

7 Antimicrobial Textile on Functional Perspective

During the initial stages of development, the adverse effects of antimicrobial compounds on human health and the ecosystem received more consideration than the side effects of these compounds [10]. An interdisciplinary effort has been conducted by ecologists, scientists, industrialists, and chemists to develop antibiotics which are favorable to the environment [66]. Antimicrobial textiles are of great assistance in modern hospitals, environments, and other hazardous microbe-infested areas. Patients, healthcare professionals, and physicians may be in contact with clothing that harbors an abundance of microbes that are readily transmitted between individuals. Antimicrobial fabrics offer abundant commercial prospects in any context involving the prevention of the transmission of pathogenic microorganisms [67]. Antimicrobial textiles are classified according to their specific activity against microorganisms, for instance, they may be antiviral, antibacterial, or antifungal. A number of antimicrobial textiles have the capability to concurrently combat bacteria, fungi, and viruses. Fabrics that are frequently encountered in public spaces such as hotels, restaurants, and railroads, such as towels used to wipe up liquids, curtains, and carpets, may potentially serve as sources of infection. Additionally, significant unmet demands exist in the realm of odor control, an additional area of research that is proliferating within this discipline. The textile might potentially accommodate numerous pathogenic microorganisms that can be transmitted from an infected individual to others. Continuous laundering of garments is the sole feasible and efficacious method for mitigating the microbial burden originating from textiles. However, this approach is impracticable in healthcare facilities characterized by continuous shifts. Thus, antimicrobial textiles are developed as an alternative approach to mitigate the risk of microbial transmission from one individual to another in sanitary-related occupations or effluent treatment, fields with a significant infection risk. Textile surface modification techniques, such as electrospinning, nanotechnology, plasma treatment, polymerization process, microencapsulation technique, and sol–gel methods, have been implemented in order to endow them with novel functional characteristics, including flame-retardant, antibacterial, and water-repellent properties [19]. The necessity of antimicrobial winter clothing is growing as it is seldom washed and exposed to sunshine. Owing to the extended storage duration of these garments, which may promote microbial proliferation, antimicrobial fabric may be a suitable substitute. In the same way, antimicrobial textiles could prove beneficial in locations that prohibit the use of plastic bags. Although food packaging composed primarily of biodegradable materials is environmentally friendly and does not compromise the quality of the food, it is crucial to incorporate antimicrobial coatings into these containers to prevent the proliferation of pathogens and food spoilage microorganisms. The following sectors require antimicrobial textiles primarily, in addition to the visually alluring combination of color, print, and design:

- Garments: winter wear, caps, jackets, sanitary napkins, sportswear, and undergarments
- Commercial: carpets, window coverings, seat, and vehicle coverings; tent and uniform, dusting linens; military fabric.
- Medical: bandages, earbuds, scrubs, masks, lab coats
- Household:Bedding, carpet, cover, draperies, mop, pillows, and towel.

8 Study on Antimicrobial Textile Application in Dermatology

Antimicrobial textile materials find application in various domains, such as first aid, clinical settings, and hygienic practices. Antibacterial textiles are utilized as bio-functional textiles in wearable drug delivery systems by integrating sophisticated pharmaceutical nanocarriers into conventional textiles. Bio-functional textiles have shown promise as potential products for enhancing dermal penetration while minimizing toxicity risks [68]. The antibacterial textiles are utilized as carriers for atopic dermatitis, analgesics, hormone therapy, melanoma and psoriasis treatments, and antimicrobial substances [69]. Lyocell and antimicrobial silver treated materials may have potential as treatments for common atopic dermatitis, according to a recent study (Fenton and Al-Salama 2021). The potential of Ag NPs to adhere to cotton fabric and penetrate it implies that they could serve as a wound healing dress material. Furthermore, this exceptionally effective antimicrobial cotton fabric has exhibited no cytotoxic impact on human cells [70]. A cotton wound dressing is devised as a therapeutic intervention for fungal infections of the skin which comprises of ketoconazole and β-cyclodextrin. It exhibits a regulated and gradual release of such antifungal substances for eliminating *C. Albicans* and *A. Niger*, which are skin fungi [71].

9 Study on Application of Antimicrobial Textile in Product Design

Antimicrobial fabrics are used in several industries, such as automotive textiles, HVAC (heating, ventilation, and air conditioning), hygiene, medical equipment, clothing, undergarments, food containers, and mechanical and thermal protection [10]. These antibacterial textiles are also used in mattresses, facial masks, stockings, baby diapers, and coverings, besides the protective applications for medical professionals. Antimicrobial clothes have been used to meet the needs of prolonged space travel. Though, the likelihood of microbiological growth in the spacesuit and space flight equipment is quite low, these textile materials are preferred owing to the limitation of laundry facility in space, and only a small number are carried on missions. When antimicrobial fabrics are utilized in space flight, laundry may be minimized or avoided [72]. The production and use of metal nanoparticle-based antibacterial textiles, such as gloves, facial masks, protective suits, shoes, and other items, has significantly increased because of COVID-19 epidemic. If such protective materials are not disposed of appropriately, it may have long-term adverse impacts on the environment [73]. Leaching of metals from a nano metal-based textile can be assessed to measure the amount of metal released at the decomposition stage of these textile products during landfill process [74]. Some of the brands that use antimicrobial fabrics for the applications such as apparels, curtains in healthcare sectors, anti-odor apparel, air mattresses, sleeping bag liners and window coverings, etc.

10 Study on Application of Antimicrobial Textile in Apparel Design

Antimicrobial apparel is a segment of the apparel category that is progressive, expanding, and fiercely competitive [10]. Antimicrobial textiles have been utilized in a novel manner to combat microbes, with an emphasis on aesthetics and fashion. However, utilization and awareness of these products might be restricted. An investigation into the buying patterns of antibacterial apparel, including infant clothes, socks, and underwear, in the Bangkok metropolitan area has unveiled that the majority of the consumer category consists of employed women between the ages of 32 and 38 [75]. The woman's apparel category placed significant demand on breathable, antiseptic, and antibacterial breast support bras and associated apparel. It comprises of a support component with an outer surface and an interior surface. The support member is composed of a minimum of one antimicrobial and one antibacterial material, in addition to one anti-absorbent material, or a combination of these substances [76]. In addition, antimicrobial designer undergarments for males are also utilized which is produced by using four 140D antibacterial nylon threads in the weaving technique. The graphic software has been utilized to create a graph and an overlapping function to decrease the density of the crotch. This resulted in the crotch, along with the stitch uplifting weave and the 140D antimicrobial fibers, being combined to form a bag [50]. Antimicrobial textiles are widely utilized in activewear and athletic apparel. Activewear or sportswear is experiencing a surge in prominence on both the international and Indian fashion markets. This change in fashion preferences could potentially be attributed to individuals' fascination with diverse ensembles for different sporting occasions, which are often influenced by celebrities or gym attire or casual activewear. Many celebrities have launched sports or active-wear fashion brands in recent years. Few examples for such brands are HRX, SKULT, Puma and so on. Therefore, these athletic garments incorporate specialized fibres, yarns, and fabric engineering designs, as well as a variety of chemical finishes. Additionally, there is a demand for athletic apparel that potentially inhibits microbial proliferation, perspiration odor and heat management to ensure proper thermo-physiological and tactile comfort. Sensory analysis on sportswear having odor controlling property showed that odor-control textile may smell less intense than comparable polyester samples [77]. A patent has been granted for a flexible antimicrobial sleeve that gets rid of any germs that are spread from the wearer to the sleeve by sneezing or coughing. This sleeve has been designed to be worn by infants on the elbow over their previous clothes. The construction permitted a tight fit that permitted the wearer easy to use and removal without restricting the motions [78]. While specific international brands, such as Diesel, assert that their novel denim possesses antimicrobial properties [79], sources indicate that it is capable of impeding more than 99 percent of viral activity within a two-hour period of contact. The fabric of dress shirts manufactured by Apposta, a company based in London, is claimed to prevent the growth of bacteria and viruses, and to decrease the probability and velocity of contamination and transmission by destroying such pathogens upon contact [80]. A product collection under the Lululemon activewear brand is referred to as 'Silverescent,' which utilizes silver threads to impart antimicrobial properties. Products manufactured by companies such as Dow Microbial Control, Microban, Lululemon, and Under Armour, among others, feature bacteria-killing and perspiration stain-free technologies. Lululemon employs a technology known as X-Static, which incorporates 99.9% pure silver into their fiber product. The price of these fabrics is approximately 15–20 percent greater than that of fabrics undamaged by the treatment. Aditya Birla group subsidiary Liva asserts antimicrobial protection is a unique viscose-based fabric that effectively inhibits the proliferation of microorganisms (including bacteria and viruses) on home textiles and apparel, achieving a 99% kill rate even after 50 washes [81].

11 Application of Nanostructure-Based Metal Compounds

Presently, the integration of nanotechnology into the fabrication of multifunctional fabrics is a subject of interest among scientists [7]. In general, nanotechnology is gathering attraction as a means of coating textiles with nanoparticles (NPs) to prevent bacterial proliferation and degradation and discoloration. In recent years, there has been considerable research focused on metal-based nanoparticles (NPs) with novel functionalities and the capacity to enhance pure textiles in physical, chemical, and biological aspects. The most frequently used metallic NPs in the development of antibacterial textiles are Ag, ZnO, Cu-based, and TiO_2. Additionally, investigations are carried out on NPs composed of other metals, including iron-based NPs such as CeO_2, Au, SiO_2, Ni, and Pt, with the aim of utilizing them in the field of biocidal textiles. In addition to their antibacterial properties, these metallic-based materials possess distinct characteristics that enhance textile finishing. Furthermore, the field of study encompasses the mechanisms by which antibacterial efficacy is enhanced and the variables that impact bacterial cell death. In order to augment the antibacterial activity of textiles, the metal-based nanoparticles (NPs) are functionalized in two ways: disseminated or incorporated into a polymer matrix [7].

11.1 Silver Nanoparticles

A wide range of application techniques for coating textile materials with Ag NPs have been developed. Several techniques have been used, including the pad-dry-cure method, dip and dry method, ex-situ immersion technique, direct sorption method, spray coating technique, hydrothermal method, microwave technique, γ-irradiation technique, sonochemical method, plasma treatment, and self-assembly technique [7, 78]. Generally, silver NPs are incorporated into textile through in situ immersion method where two steps, namely pre-activation and sorption are carried out. Silver nanoparticles are durable, more dyeable, and very toxic to a broad range of bacteria, but less harmful to human cells. Silver nanomaterials have gained significant recognition for their antimicrobial and self-cleaning properties [82]. AgNPs, which are less than 10 nm in diameter, have exhibited antiviral efficacy against SARS-CoV-2, the pathogen responsible for COVID-19 [74].

11.2 Zinc Oxide

Numerous coating techniques for zinc oxide (ZnO NPs have been investigated, including spraying and dipping, pad-dry-cure, hydrothermal, sonochemical, plasma treatment, and pulse laser ablation [7, 28, 72]. An amine-based material has been developed by using hexamethyl triethylenetetramine as the precursor. This material has facilitated the formation of zinc oxide nanoparticles (ZnO NPs) on cotton fabric via the pad-dry-cure method under alkaline conditions. The resulting ZnO NPs have exhibited antibacterial properties and UV protection. Through the pad-dry-cure method, poly-*N*-vinyl-2-pyrrolidone has been functionalized on the surface of cotton fiber, which exhibits enhanced antibacterial and reactive-dying properties. Similarly, zinc nanoparticles (Zn NPs) have been integrated into a silica matrix using two distinct methodologies and the results of the research are compared in terms of their impact on morphological characteristics, structural properties, thermal, and antibacterial characteristics. In the first method, the fabric has been submerged in the solution, subsequently dried, and cured [7]. Afterwards, it has been immersed in zinc nitrate solution. At last it has been treated with sodium hydroxide solution and dried. The second approach involved dissolving zinc nitrate in solution prior to submerging the fabric in the solution, the treated fabric is then desiccated and cured. Then, the fabric is desiccated after being combined with sodium hydroxide. The findings indicated that NP agglomeration has been found with the first technique, in contrast to the spherical NPs produced by the second method. Therefore, the fabric treated using the second technique exhibited greater antibacterial properties against *S. aureus* and *E. coli* in comparison to the first method, owing to the increased exposure of its surface area. These NPs have been then dip-coated onto cotton fabric to impart UV-protective, antibacterial, and long-lasting self-cleaning properties. The adhesion efficacy of zinc oxide (ZnO) to cotton fabric using the dipping and spraying methods has been compared. Consequently, the deposition of ZnO NPs is three times greater in the spraying procedure compared to the other method, despite the antibacterial reduction being remarkably at similar level [83], Shaheen et al. 2016, [7]. The cotton fabric has been treated with zinc oxide nanoparticles capped with seaweed (SW-ZnO NPs) utilizing the pad-dry-cure method [84]. The fabric has exhibited antibacterial properties against both Gram-positive bacteria (*Staphylococcus aureus* and *Streptococcus pyogenes*) and Gram-negative bacteria (*Escherichia coli* and *Klebsiella aerogenes*). In a similar fashion, date seed extract, an additional natural capping agent, has been employed in the synthesis of capped zinc oxide nanoparticles (ZnO NPs). These NPs have exhibited antimicrobial and UV protective properties on cotton fabric [85]. For NPs to be effectively stabilized on the surface of cellulose-based fabric, pre-activation or pre-treatment is necessary. In order to create a long-lasting functional textile, it is essential to examine the bio-adhesive substances or techniques which are safe for human skin. The antibacterial efficacy of ZnO NPs deposited simultaneously using sonochemical deposition with an enzymatic cross-linked phenolic network of gallic acid is very good even after significant number of washing cycles in hot water [86].

11.3 Copper-Based Agents

Synthesis of Cu-based nanoparticles (NPs) and CuO NPs coated fabrics have been accomplished by using a variety of techniques, including metal chelation, chemical reduction, and ultrasound irradiation [7]. The antimicrobial efficacy of textile materials coated with functionalized CuO NPs against both Gram-positive and negative bacteria can be ascribed by three primary mechanisms: copper ion release, contact of CuO NPs and bacteria, and generation of reactive oxygen groups [87, 88]. The selenium brooms have been synthesized with almond skin extract exhibited time-dependent alterations, transitioning from nanoparticles in 15 s to selenium brooms in 45 min. Fabric coated with cotton and featuring such structures demonstrated efficacy against Bacillus subtilis [88]. The resistance of a functional ecofriendly nanohybrid material formed from nano-silica (derived from rice husk) and oligo chitosan (derived from crab shells) to *Phytophthora infestans* fungal attack indicates that advance research in the application of green agriculture [89]. The achievement of consistent size, desired morphology, and prevention of agglomeration has consistently posed difficulties in research pertaining to nanoparticles. Generally, capping with ligands is employed to overcome such obstacles. A number of natural capping agents are used to coat the fabric effectively [10].

11.4 Titanium Dioxide

Several techniques exist for the fabrication of fabric coated with TiO_2, including the pad dry cure method and sono synthesis which are operated at low temperatures [7, 90]. Functional cotton fabric is coated with amino-capped TiO_2 nanoparticles via two-step sol–gel process. The hydrothermal method of fabrication has confirmed the efficacy of the fabric in inhibiting the growth of *S. aureus* and *E. coli* bacteria [10], Zhang et al. (2019).

11.5 Other Metal-Based NPs

Other metal-based antibacterial compounds, including those based on iron, cerium oxide, gold, nickel, silicon dioxide, and platinum have also been incorporated into the textiles. Iron-based nanoparticles (Fe_3O_4) have been synthesized and applied on a blended cotton-polyester fabric using ultrasonic mediation. The resulting nanoparticles have shown antibacterial, photocatalytic, and magnetic properties. DC magnetron sputtering has been employed to examine the antibacterial properties of Platinum (Pt) NPs on polyester fabric treated with various natural and synthetic pigments. Without antibacterial agent derivatives, platinum itself has not exhibited any antibacterial activity [7, 91].

12 Antimicrobial Based on Natural Compounds

Numerous natural active agents, apart from curcumin, have been extracted and incorporated into the formulation of antimicrobial fabrics. Additionally, fabrics are treated with honey, plant extracts, and essential oils to prevent the spread of wound infections. Sustainable antibacterial textile finishing can be made with cyclodextrins. The central cavity of these cyclic oligosaccharides is lipophilic, while the outer surface is hydrophilic. The textile industry is witnessing a rise in the utilization of cyclodextrin and its derivatives, including α-cyclodextrin, β-cyclodextrin, and γ-cyclodextrin (Singh and Sahu 2019). Lignin confers microbial resistance to botanical specimens. Generally, the dark-colored phenolic compound gets removed during the cellulose fiber manufacturing process. Fabric coated with lignin which is extracted from sugarcane bagasse demonstrated antibacterial efficacy against *Staphylococcus epidermidis*. The concentration of lignin coating has been determined by its minimum inhibitory concentration (MIC), which indicated that within six hours of contact, the bacterial growth could be impeded to prevent further propagation [15]. Chitosan is frequently used in the fabrication and processing of antimicrobial textiles owing to its antimicrobial and anti-odor characteristics. Chitosan is a cationic polysaccharide produced when chitin is deacetylated in an alkaline environment. Antibacterial activity has been exhibited by the application of chitosan hydrogel onto cellulosic fabric against *Listeria monocytogenes*, *S. aureus*, and *E. coli* [92]. On the contrary, a report has indicated that the application of chitosan in a combination of silica sols and chitosan to cotton and polyester fabrics could not have a substantial impact on their antimicrobial properties [93]. In contrast, films composed of silver nanoparticles and consumable chitosan/pectin derived from natural biodegradable polymers have exhibited efficacious antibacterial properties when examined against *E. coli*. Antibacterial films have the potential to increase the expiration life of products and can be utilized for packaging purposes [94].

13 Conclusions

In healthcare services, the detection of antimicrobial substances embedded in textiles is crucial for attributing efficacy to the reduction of microbial burden in these textiles and/or the incidence of healthcare-associated infections, according to the findings of numerous studies. An assortment of substances has been utilized for various purposes, including inanimate healthcare environments, patients during their hospitalization, health professionals, titanium nanoparticles, titanium nanoparticles doped with silver, quaternary ammonium, chlorhexidine, zinc oxide, alcohols and iso-thiazolone derivatives, and chitosan and dimethylol dimethyl hydantoin combined. While adverse effects such as pruritus, erythema, and pruritus have not been noticed in textiles treated with substances including copper, chlorhexidine, chitosan, dimethylol, and dimethyl hydantoin, zinc oxide, silver, and quaternary ammonium have proven to be detrimental. Antimicrobial textiles containing metal nanoparticles and synthetic chemicals are effective, but they also appear to be a hazard to the environment due to the paucity of information regarding the precise effects of chemicals that seep into the environment. Fabric treated with a nano-emulsion containing Tetrahydroxy Curcumin demonstrates antibacterial characteristics due to the presence of β cyclodextrin and polycarboxylic acid, which effectively impede the growth of microorganisms. Renewable and environmentally benign green natural biopolymers, including but not limited to chitosan, sericin, and alginate, which are derived from plants or animals, have the capacity to serve as indispensable resources for sustainable multifunctional textiles of the future. Thus, more research will be devoted to environmentally benign synthesis methods, and the use of green agents with potential long-term antibacterial activity will be promoted in an effort to improve the environment and public health.

References

1. Mitchell A, Spencer M, Edmiston C (2015) Role of healthcare apparel and other healthcare textiles in the transmission of pathogens: a review of the literature. J Hosp Infect 90:285–292
2. Sanders D, Grunden A, Dunn RR (2021) A review of clothing microbiology: the history of clothing and the role of microbes in textiles. Biol Lett 17:rsbl.2020.0700
3. Ohl M, Schweizer M, Graham M, Heilmann K, Boyken L, Diekema D (2012) Hospital privacy curtains are frequently and rapidly contaminated with potentially pathogenic bacteria. Am J Infect Control 40:904–906
4. Goyal S, Khot SC, Ramachandran V, Shah KP, Musher DM (2019) Bacterial contamination of medical providers' white coats and surgical scrubs: a systematic review. Am J Infect Control 47:994–1001
5. Mishra SK, Maharjan S, Yadav SK, Sah NP, Sharma S, Parajuli K et al (2020) Bacteria on medical professionals' white coats in a university hospital. Can J Infect Dis Med Microbiol. 2020:5957284
6. Owen L, Laird K (2020) The role of textiles as fomites in the healthcare environment: a review of the infection control risk. PeerJ 8:e9790. https://doi.org/10.7717/peerj.9790

7. Tan LY, Sin LT, Bee ST, Ratnam CT, Woo KK, Tee TT, Rahmat AR (2019) A review of antimicrobial fabric containing nanostructures metal-based compound. J Vinyl Add Tech 25(S1):E3–E27
8. Li J, He J, Huang Y (2017) Int J Biol Macromol 94:466
9. Paterson P (2009) Neem the wonder tree, its pesticide and medicinal application. University of Maryland, p 27
10. Gulati R, Sharma S, Sharma RK (2022) Antimicrobial textile: recent developments and functional perspective. Polym Bull 79(8):5747–5771
11. Bonaldi RR (2018) Functional finishes for high-performance apparel. In: High-performance apparel. Elsevier, London, pp 129–156
12. Simoncic B, Tomsic B (2010) Structures of novel antimicrobial agents for textiles—a review. Text Res J 80:1721–1737
13. Bueno J (2015) Antimicrobial models in nanotechnology. In: Nanotechnology in diagnosis, treatment and prophylaxis of infectious diseases. Elsevier, London, pp 19–38
14. Khatami M, Sharifi I, Nobre MAL et al (2018) Waste-grass-mediated green synthesis of silver nanoparticles and evaluation of their anticancer, antifungal and antibacterial activity. Green Chem Lett Rev 11:125–134
15. Sunthornvarabhas J, Liengprayoon S, Lerksamran T et al (2019) Utilization of lignin extracts from sugarcane bagasse as bio-based antimicrobial fabrics. Sugar Tech 21:355–363. https://doi.org/10.1007/s12355-018-0683-2
16. Cuce M, Kilinc M, Kilinc N et al (2020) Color, fastness and antimicrobial properties of wool fabrics dyed with *Helichrysum arenarium* subsp. Aucheri Extract J Nat Fibers 2020:1
17. Bancalari E, Martelli F, Bernini V et al (2020) Bacteriostatic or bactericidal? Impedometric measurements to test the antimicrobial activity of Arthrospiraplatensis extract. Food Control 118:107380
18. Igal K, Vázquez P (2020) Antimicrobial fabrics impregnated with Ag particles included in silica matrices. In: Waste in textile and leather sectors. IntechOpen
19. Nadi A, Boukhriss A, Bentis A et al (2018) Evolution in the surface modification of textiles: a review. Text Prog 50:67–108
20. Sharma JL, Dhayal V, Sharma RK (2021) White-rot fungus mediated green synthesis of zinc oxide nanoparticles and their impregnation on cellulose to develop environmental friendly antimicrobial fibers. 3 Biotech 11:269
21. Tzanov T (2017) Pre-commercial lines for production of surface nanostructured antimicrobial and anti-biofilm textiles, medical devices and water treatment membranes
22. Openshaw JJ, Morris WM, Lowry GV, Nazmi A (2016) Reduction in bacterial contamination of hospital textiles by a novel silver-based laundry treatment. Am J Infect Control 44:1705–1708
23. Staneva D, Vasileva-Tonkova E, Yordanova S et al (2020) Spectral characterization, antimicrobial and antibiofilm activity of poly(propylene imine) metallodendrimers in solution and applied onto cotton fabric. Int J Polym Anal Charact 25:374–384
24. Wang L, Natan M, Zheng W et al (2020) Small molecule-decorated gold nanoparticles for prepar-ingantibiofilm fabrics. Nanoscale Adv 2:2293–2302
25. Vasileva-Tonkova E, Staneva D, Medel S et al (2019) Antimicrobial, antibiofilm and cytotoxicity activity of a new acridine hyperbranched polymer in solution and on cotton fabric. FibersPolym 20:19–24. https://doi.org/10.1007/s12221-019-8687-4
26. Ren T, Dormitorio TV, Qiao M et al (2018) N-halamine incorporated antimicrobial nonwoven fabrics for use against avian influenza virus. Vet Microbiol 218:78–83
27. Seino S, Imoto Y, Kosaka T et al (2016) Antiviral activity of silver nanoparticles immobilized onto textile fabrics synthesized by radiochemical process. MRS Adv 1:705–710
28. Jeremiah SS, Miyakawa K, Morita T et al (2020) Potent antiviral effect of silver nanoparticles on SARS-CoV-2. Biochem Biophys Res Commun 533:195–200
29. Montagut AM, Granados A, Lazurko C et al (2019) Triazine mediated covalent antibiotic grafting on cotton fabrics as a modular approach for developing antimicrobial barriers. Cellulose 26:7495–7505

30. Gu J, Yuan L, Zhang Z et al (2018) Non-leaching bactericidal cotton fabrics with well-preserved physical properties, no skin irritation and no toxicity. Cellulose 25:5415–5426
31. Saif MJ, Zia KM, Rehman F et al (2015) An eco-friendly, permanent and non-leaching antimicrobial coating on cotton fabrics. J Text Inst 106:907–911. https://doi.org/10.1080/00405000.2014.952137
32. Li Z, Chen J, Cao W et al (2018) Permanent antimicrobial cotton fabrics obtained by surface treatment with modified guanidine. Carbohydr Polym 180:192–199
33. Prusty AK, Das T, Nayak A, Das NB (2010) Colourimetric analysis and antimicrobial study of natural dyes and dyed silk. J Clean Prod 18:1750–1756
34. El-Khatib EM, Ali NF, El-Mohamedy RSR (2020) Influence of Neem oil pretreatment on the dyeing and antimicrobial properties of wool and silk fibers with some natural dyes. Arab J Chem 13:1094–1104
35. Ganesan P, Karthik T (2017) Analysis of colour strength, colour fastness and antimicrobial properties of silk fabric dyed with natural dye from red prickly pear fruit. J Text Inst 108:1173–1179
36. Abd El Aty AA, El-Bassyouni GT, Abdel-Zaher NA, Guirguis OW (2018) Experimental study on antimicrobial activity of silk fabric treated with natural dye extract from neem (Azadirachtaindica) leaves. FibersPolym 19:1880–1886
37. Emam HE, Darwesh OM, Abdelhameed RM (2018) In-growth metal organic framework/synthetic hybrids as antimicrobial fabrics and its toxicity. Colloids Surf B Biointerfaces 165:219–228
38. El-Ola SMA (2008) Recent developments in finishing of synthetic fibers for medical applications. Des Monomers Polymr 11:483–533
39. Bhattacharjee S, Joshi R, Chughtai AA, Macintyre CR (2019) Graphene modified multifunctional personal protective clothing. Adv Mater Interfaces 6:1900622
40. YıldızVaran N, Eryuruk SH (2018) The effects of quat-silane antimicrobials on the physical and mechanical properties of cotton and cotton/elastane fabrics used for clothing. IOP Conf Ser Mater Sci Eng 460:012004
41. Swar S, Máková V, Horáková J et al (2020) A comparative study between chemically modified and copper nanoparticle immobilized Nylon 6 films to explore their efficiency in fighting against two types of pathogenic bacteria. EurPolym J 122:109392
42. Buşilă M, Muşat V, Textor T, Mahltig B (2015) Synthesis and characterization of antimicrobial textile finishing based on Ag:ZnO nanoparticles/chitosan biocomposites. RSC Adv 5:21562–21571
43. Schneider G, Vieira LG, Carvalho HEFD, Sousa ÁFLD, Watanabe E, Andrade DD, Silveira RCDCP (2023) Textiles impregnated with antimicrobial substances in healthcare services: systematic review. Front Public Health 11:1130829
44. Burden M, Keniston A, Frank MG, Brown CA, Zoucha J, Cervantes L et al (2013) Bacterial contamination of healthcare workers' uniforms: a randomized controlled trial of antimicrobial scrubs. J Hosp Med 8:380–385
45. Marcus EL, Yosef H, Borkow G, Caine Y, Sasson A, Moses AE (2017) Reduction of health care-associated infection indicators by copper oxide-impregnated textiles: crossover, double-blind controlled study in chronic ventilator-dependent patients. Am J Infect Control 45:401–403
46. Schneider G, Bim FL, Sousa AFL, Watanabe E, Andrade D, Fronteira I (2021) The use of antimicrobial-impregnated fabrics in health services: an integrative review. Rev Lat Am Enfermagem 29:e3416
47. Madden GR, Heon BE, Sifri CD (2018) Effect of copper-impregnated linens on multidrug-resistant organism acquisition and *Clostridium difficile* infection at a long-term acute-care hospital. Infect Control Hosp Epidemiol 39:1384–1386
48. Condò C, Messi P, Anacarso I, Sabia C, Iseppi R, Bondi M et al (2015) Antimicrobial activity of silver doped fabrics for the production of hospital uniforms. New Microbiol 38:551–558
49. Tufanaru C, Munn Z, Aromataris E, Campbell J, Hopp L (2020) Chapter 3: Systematic reviews of effectiveness. In: Aromataris E, Munn Z (eds) JBI manual for evidence synthesis JBI. The Joanna Briggs Institute, Adelaide

50. Wilson G, Jackson V, Boyken L, Puig-Asensio M, Marra AR, Perencevich E et al (2020) A randomized control trial evaluating efficacy of antimicrobial impregnated hospital privacy curtains in an intensive care setting. Am J Infect Control 48:862–868
51. Luk S, Chow VCY, Yu KCH, Hsu EK, Tsang NC, Chuang VWM et al (2019) Effectiveness of antimicrobial hospital curtains on reducing bacterial contamination—amulticenter study. Infect Control Hosp Epidemiol 40:164–170
52. Kotsanas D, Wijesooriya WRPLI, Sloane T, Stuart RL, Gillespie EE (2014) The silver lining of disposable sporicidal privacy curtains in an intensive care unit. Am J Infect Control 42:366–370
53. Zhao Y, Xu Z, Lin T (2016) Barrier textiles for protection against microbes. In: Sun G (ed) Antimicrobial textiles. Woodhead Publishing Series in Textiles, London, pp 225–245
54. Razic SE, Glogar MI, Peran J et al (2020) Plasma pre-treatment and digital ink jet technology: a tool for improvement of antimicrobial properties and colour performance of cellulose knitwear. Mater Today Proc 31:S247–S257
55. Ribeiro AI, Senturk D, Silva KK et al (2019) Antimicrobial efficacy of low concentration PVP-silver nanoparticles deposited on DBD plasma-treated polyamide 6,6 fabric. Coatings 9:581
56. Hanczvikkel A, Víg A, Tóth A (2019) Survival capability of healthcare-associated, multidrug-resistant bacteria on untreated and on antimicrobial textiles. J Ind Text 48:1113–1135
57. Irfan M, Perero S, Miola M, Maina G, Ferri A, Ferraris M et al (2017) Antimicrobial functionalization of cotton fabric with silver nanoclusters/silica composite coating via RF co-sputtering technique. Cellulose 24:2331–2345
58. Pietsch F, O'Neill AJ, Ivask A, Jenssen H, Inkinen J, Kahru A et al (2020) Selection of resistance by antimicrobial coatings in the healthcare setting. J Hosp Infect 106:115–125
59. Pal C, Asiani K, Arya S, Rensing C, Stekel DJ, Larsson DGJ et al (2017) Metal resistance and its association with antibiotic resistance. In: Poole RK (ed) Advances in microbial physiology microbiology of metal ions. Academic Press, New York, pp 261–313
60. Bengalli R, Colantuoni A, Perelshtein I, Gedanken A, Collini M, Mantecca P et al (2021) In vitro skin toxicity of CuO and ZnO nanoparticles: application in the safety assessment of antimicrobial coated textiles. Nano Impact 21:100282
61. Zhou C-E, Kan C, Yuen CM et al (2016) Plasma treatment applied in the pad-dry-cure process for making rechargeable antimicrobial cotton fabric that inhibits *S. aureus*. Text Res J 86:2202–2215
62. Ibrahim NA, Eid BM, Abdellatif FHH (2018) Advanced materials and technologies for antimicrobial finishing of cellulosic textiles. In: Yusuf M (ed) Handbook of renewable materials for coloration and finishing. Wiley, London, pp 303–356
63. Mogensen JE, Jørgensen P-E, Thomsen TR (2016) A microbiological evaluation of SiO_2-coated textiles in hospital interiors: the effect of passive coatings on the cleaning potential of interior textiles. J Ind Text 46:361–371
64. Murphy F, Tchetchik A, Furxhi I (2020) Reduction of health care-associated infections (HAIs) with antimicrobial inorganic nanoparticles incorporated in medical textiles: an economic assessment. Nanomaterials 10:999
65. Stiefel P, Schneider J, Amberg C, Maniura-Weber K, Ren Q (2016) A simple and rapid method for optical visualization and quantification of bacteria on textiles. Sci Rep 6:39635
66. Epstein L (2014) Fifty years since silent spring. Annu Rev Phytopathol 52:377–402
67. Morris H, Murray R (2020) Medical textiles. Text Prog 52:1–127
68. Massella D, Ferri A et al (2019) Bio-functional textiles: combining pharmaceutical nanocarriers with fibrous materials for innovative dermatological therapies. Pharmaceutics 11:403. https://doi.org/10.3390/pharmaceutics11080403
69. Atanasova D, Staneva D, Grabchev I (2021) Textile materials modified with stimuli-responsive drug carrier for skin topical and transdermal delivery. Mater (Basel) 14:930
70. Elnaggar M, Emam H, Fathalla M et al (2021) Chemical synthesis of silver nanoparticles in its solid state: highly efficient antimicrobial cotton fabrics for wound healing properties. Egypt J Chem 2021:1

71. Hedayati N, Montazer M, Mahmoudirad M, Toliyat T (2020) Ketoconazole and ketoconazole/β-cyclodextrin performance on cotton wound dressing as fungal skin treatment. CarbohydrPolym 240:116267
72. Jagadeesan K (2018) Antimicrobial textiles for long duration space flight. University of Rhode Island
73. Ardusso M, López ADF, Buzzi NS et al (2020) COVID-19 pandemic repercussions on plastic and COVID-19 pandemic repercussions on plastic and antiviral polymeric textile causing pollution on beaches and coasts of South America. Sci Total Environ 2020:1
74. Limpiteeprakan P, Babel S (2016) Leaching potential of silver from nanosilver-treated textile products. Environ Monit Assess 188:156
75. Taweehiransuwan N (2020) The Innovation of antimicrobial apparel affecting to consumer buying behaviour in Bangkok Metropolitan
76. Taghdiri M (2009) Antibacterial, anti-absorbent and antimicrobial breast supports, bras and related apparel
77. Klepp IG, Buck M, Laitala K, Kjeldsberg M (2016) What's the problem? Odor-control and the smell of sweat in sportswear. Fash Pract 8:296–317
78. Balakumaran MD, Ramachandran R, Jagadeeswari S, Kalaichelvan PT (2016) In vitro biological properties and characterization of nanosilver coated cotton fabrics—an application for antimicrobial textile finishing. Int Biodeterior Biodegradation 107:48–55
79. Polygiene (2021) Virus-fighting denim from diesel. https://polygiene.com/virus-fighting-denim-from-diesel/. Accessed 29 May 2021
80. Fernandez C (2020) Brands claim their clothes kill viruses. Will consumers buy it? In: Bus. Fash. https://www.businessoffashion.com/articles/retail/antiviral-fashion-coronavirus-diesel. Accessed 7 Jan 2020
81. Liva A (2020) Care made fashionable. In: Liva. https://www.livafluidfashion.com/livaantimicrobial/. Accessed 7 Jan 2020
82. Shahid-ul-Islam S-I, Butola BS, Mohammad F (2016) Silver nanomaterials as future colorants and potential antimicrobial agents for natural and synthetic textile materials. RSC Adv 6:44232–44247. https://doi.org/10.1039/C6RA05799C
83. Prasad V, Arputharaj A, Bharimalla AK, Patil PG, Vigneshwaran N (2016) Durable multifunctional finishing of cotton fabrics by in situ synthesis of nano-ZnO. Appl Surf Sci 390:936–940
84. Pandimurugan R, Thambidurai S (2017) UV protection and antibacterial properties of seaweed capped ZnO nanoparticles coated cotton fabrics. Int J BiolMacromol 105:788–795
85. El-Naggar ME, Shaarawy S, Hebeish AA (2018) Multifunctional properties of cotton fabrics coated with in situ synthesis of zinc oxide nanoparticles capped with date seed extract. Carbohydr Polym 181:307–316
86. Salat M, Petkova P, Hoyo J et al (2018) Durable antimicrobial cotton textiles coated sonochemically with ZnO nanoparticles embedded in an in-situ enzymatically generated bioadhesive. Carbohydr Polym 189:198–203
87. Román LE, Gomez ED, Solís JL, Gómez MM (2020) Antibacterial cotton fabric functionalized with copper oxide nanoparticles. Molecules 25:5802
88. Sadalage PS et al (2020) Sustainable approach to almond skin mediated synthesis of tunable selenium microstructures for coating cotton fabric to impart specific antibacterial activity. J Colloid Interface Sci 569:346–357
89. Kim BM, Van Minh N, Choi HY, Kim WG (2019) Coralmycin derivatives with potent anti-gram negative activity produced by the myxobacteria *Corallococ cuscoralloides* M23. Molecules 24:1390
90. Meftahi A, Yadegari F, Alibakhshi S, Vatandoost S, Mohammadi H (2015) Finishing of nylon/cotton fabric with ZnO/TiO2 nanocomposite. Proc Mater Sci 11:600–604
91. Rajakumar G, Gomathi T, Abdul Rahuman A, Thiruvengadam M, Mydhili G, Kim SH, Chung IM (2016) Biosynthesis and biomedical applications of gold nanoparticles using *Eclipta prostrata* leaf extract. Appl Sci 6(8):222

92. Benltoufa S, Miled W, Trad M et al (2020) Chitosan hydrogel-coated cellulosic fabric for medical end-use: antibacterial properties, basic mechanical and comfort properties. Carbohydr Polym 227:115352
93. Cheng VCC, Chen JHK, Leung SSM, So SYC, Wong SC, Wong SCY et al (2017) Seasonal outbreak of bacillus bacteremia associated with contaminated linen in Hong Kong. Clin Infect Dis 64:S91–S97
94. Akalin GO, OztunaTaner O, Taner T (2021) The preparation, characterization and antibacterial properties of chitosan/pectin silver nanoparticle films. Polym Bull. https://doi.org/10.1007/s00289-021-03667-0

Health and Hygiene Care Textiles

C. Rajshkumar and D. Raja

1 Introduction

Healthcare and hygiene textiles play a crucial role in promoting wellness and preventing the spread of infections in various healthcare and related environments. These textiles encompass a wide range of products designed to meet stringent standards for cleanliness, durability, and comfort. From hospital linens and surgical gowns to personal protective equipment (PPE) and patient apparel, these textiles contribute significantly to maintaining a safe and hygienic healthcare environment. The primary goals of health and hygiene care textiles are to provide a barrier against contaminants, facilitate infection control, and ensure the comfort and well-being of both healthcare professionals and patients. These textiles are often treated with antimicrobial agents or designed with specific features to enhance their resistance to bacteria, viruses, and other pathogens. In addition to their protective functions, health and hygiene care textiles are engineered to withstand frequent washing and sterilization processes without compromising their integrity. This durability is essential to maintain a high level of cleanliness in healthcare settings, where stringent hygiene standards are paramount [1].

As healthcare practices continue to evolve, the demand for innovative health and hygiene care textiles has grown. Advancements in material science and technology have led to the development of textiles with enhanced properties, such as moisture-wicking capabilities, breathability, and improved ergonomics. These features contribute to the overall comfort and functionality of the textiles, ensuring that they not only meet strict hygiene standards but also enhance the overall experience for both healthcare providers and patients. Their role extends beyond mere

C. Rajshkumar (✉) · D. Raja
Department of Fashion Technology, Sona College of Technology, Salem, India
e-mail: crajez@gmail.com

D. Raja
e-mail: crajez@gmail.com

S. Sundaramoorthy et al. (eds.), *Textile Materials for Good Health and Wellbeing*, SDGs and Textiles, https://doi.org/10.1007/978-981-97-6496-9_3

functionality, encompassing comfort, durability, and adherence to stringent hygiene standards to promote the well-being of individuals in healthcare settings [2].

1.1 Historical Perspectives

Healthcare and hygiene textiles have played a crucial role in promoting cleanliness and infection prevention across various civilizations. In ancient times, civilizations like the Egyptians, Greeks, and Romans recognized the significance of textiles such as linen and wool for clothing and bandages. These materials, prized for their natural properties like breathability and absorbency, formed the foundation of early hygiene practices, including the establishment of public baths. During the middle ages, hygiene practices fluctuated, but the use of wool and linen persisted, albeit with varying emphasis on personal cleanliness across different regions. However, the Renaissance period marked resurgence in attention to personal hygiene, coinciding with a renewed interest in fabrics like cotton and an improved understanding of infectious diseases. Clean clothing and bedding became increasingly valued for their contribution to health and well-being [3].

The eighteenth and nineteenth centuries witnessed significant advancements in textile manufacturing during the Industrial Revolution, making fabrics more accessible. Cotton emerged as a popular choice due to its affordability and comfort, further reinforcing the link between cleanliness and disease prevention. By the twentieth century, innovations in healthcare textiles led to the development of synthetic fibers and disposable medical textiles, revolutionizing infection control in healthcare settings. In the twenty-first century, ongoing advancements in textile technology continue to drive progress, with the emergence of antimicrobial textiles and smart textiles with sensing capabilities poised to further enhance healthcare and hygiene practices [1].

1.2 Global Health Challenges

In 2019, the world confronted significant global health challenges, foremost among them being the COVID-19 pandemic, caused by the novel coronavirus SARS-CoV-2. This unprecedented crisis reverberated across public health, economies, and societies globally, underscoring the critical importance of personal protective equipment (PPE) crafted from specialized healthcare textiles, such as masks and gowns, in stemming the spread of infectious diseases. Personal protective equipment, PPE, serves as a vital defense mechanism for frontline healthcare workers and the general populace against exposure to harmful pathogens like viruses and bacteria. Masks act as a shield against respiratory droplets containing infectious agents, mitigating airborne transmission risks, while gowns provide an additional layer of protection, minimizing direct contact with contaminated surfaces or bodily fluids. Amidst the pandemic, the surge in demand for PPE underscored the indispensable role of specialized healthcare textiles in epidemic preparedness and response efforts, prompting intensified production and research endeavors to enhance the efficacy and comfort of these essential products [4].

1.3 Applications of Healthcare and Hygiene Textiles

Healthcare and hygiene textiles find diverse applications across various sectors, playing a crucial role in maintaining cleanliness, preventing infections, and promoting overall well-being. Hospitals and healthcare facilities prioritize patient comfort and hygiene through specialized apparel such as gowns, robes, and pajamas, along with bed linens like sheets, pillowcases, and mattress covers engineered for durability and infection control. In surgical settings, sterile textiles including gowns, drapes, and wraps are crucial for maintaining aseptic conditions. Personal protective equipment (PPE) plays a critical role in healthcare settings, with items like surgical masks, N95 respirators, gloves, and isolation gowns designed to protect against airborne contaminants and prevent the transmission of infections among healthcare workers and patients [5, 6].

Long-term care facilities require specialized textiles to manage various needs, including incontinence products such as adult diapers and bed pads, as well as sanitary napkins for menstrual hygiene. Comfortable apparel tailored for elderly care ensures residents' well-being in assisted living or long-term care settings. In-home healthcare, antimicrobial bedding and washable under-pads provide added protection for individuals with compromised immune systems, while sports and fitness enthusiasts benefit from moisture-wicking fabrics and antimicrobial sportswear designed to enhance comfort and hygiene during physical activities [7, 8].

The hospitality and travel industries prioritize guest safety and cleanliness through the use of antimicrobial towels and bedding in accommodations, as well as disposable airline headrest covers to minimize the risk of transmission during travel. In the food industry, hygienic practices are upheld with the use of food handling gloves and protective aprons and smocks to prevent contamination during food preparation and handling. Military and emergency services rely on specialized textiles such as chemical and biological protective clothing, as well as field dressings in medical kits, to ensure personnel are adequately protected during hazardous situations. Childcare and educational settings maintain hygiene standards with nap mats and bedding for young children, along with antimicrobial school uniforms to promote cleanliness among students. Consumer products also offer antimicrobial solutions for everyday use, including bedding and clothing with antibacterial properties, promoting a clean and hygienic environment at home [9].

These applications underscore the versatility and importance of health and hygiene care textiles across a wide range of industries and settings, contributing to the maintenance of a clean and safe environment.

1.4 Importance of Healthcare and Hygiene Textile Products

Healthcare and hygiene textile products are indispensable across diverse sectors, from healthcare facilities to everyday life, owing to their pivotal role in infection prevention and personal well-being. In healthcare settings, these textiles serve as crucial barriers against pathogens, with specialized items like surgical gowns and masks playing a vital role in safeguarding patients and medical professionals from infections. The durability ensures rigorous sterilization processes can be effectively maintained, upholding stringent hygiene standards. Beyond healthcare, health and hygiene care textiles find widespread application in industries such as hospitality and food service, where they contribute to maintaining cleanliness and preventing cross-contamination. From antimicrobial linens in hotels to disposable gloves in food processing units, these textiles play a pivotal role in upholding hygiene standards, and enhancing safety for both customers and staff. Moreover, in everyday life, products like diapers and sanitary napkins utilize advanced textile technology to provide comfort and absorbency, significantly impacting the quality of life for individuals of all ages [10].

Additionally, the evolution of health and hygiene care textiles includes focus on sustainability, with the development of eco-friendly options such as organic materials and recycled fabrics. This aligns with increasing awareness of environmental sustainability and promotes a healthier, safer society. Overall, health and hygiene care textiles remain essential components in preserving health, preventing infections, and promoting hygiene across various environments, ultimately contributing to a higher quality of life for individuals worldwide [4].

1.5 Demand for Health and Hygiene Care Textiles

The global demand for health and hygiene products and practices has experienced a significant surge due to several intertwined factors. Firstly, heightened public health concerns, particularly in light of recent pandemics, have underscored the importance of personal hygiene in preventing the spread of infectious diseases. Individuals and communities are now more aware of the critical role, the cleanliness plays in safeguarding public health. Stringent healthcare standards and regulations have mandated the use of high-quality health and hygiene products, especially in healthcare settings. Compliance with these standards is crucial to ensure the safety of patients and healthcare professionals alike, driving the demand for specialized textiles and personal protective equipment. The increasing emphasis on occupational safety across various industries has led to a growing demand for protective gear and health and hygiene textiles for workers. This includes items such as gloves, masks, and specialized clothing designed to shield individuals from workplace hazards and contaminants. The rise in global travel and tourism has heightened concerns about disease transmission, particularly in hospitality and travel sectors where cleanliness and safety are paramount. This has driven a significant demand for health and hygiene products to meet regulatory requirements and ensure customer satisfaction [11].

Societal shifts such as an aging population and growing environmental consciousness have further propelled the demand for health and hygiene solutions. There is a need for products tailored to the needs of elderly individuals and a preference for sustainable and eco-friendly options. These factors, along with advancements in technology and increased consumer awareness, are expected to sustain and even accelerate the demand for health and hygiene products in the foreseeable future.

1.6 Characteristic of Materials for Health Care and Hygiene Textile Products

Health and hygiene care products utilized in the medical field demand meticulous adherence to stringent standards to guarantee safety, efficacy, and regulatory compliance. Paramount among the characteristics of materials employed in these products is biocompatibility, ensuring they do not elicit adverse reactions when in contact with living tissues, thereby averting allergic responses or irritation. Additionally, maintaining sterility is imperative to prevent infections and sustain aseptic conditions, safeguarding users' health and promoting positive outcomes. The materials in health and hygiene care textile products must exhibit non-toxic properties, releasing no harmful substances that could be absorbed by the body and potentially lead to health issues. Chemical stability is equally critical to uphold the product's integrity during storage and use, ensuring it remains effective without undesirable alterations over time. Resistance to corrosion and erosion becomes paramount for products in contact with bodily fluids, guaranteeing durability and preventing contamination [6].

Characteristics like hypoallergenic properties, ease of cleaning and disinfection, and compliance with regulations contribute significantly to ensuring user safety and product efficacy. Transparency and traceability in sourcing and manufacturing processes, along with compatibility with other materials, further bolster the quality and safety of medical products. Ultimately, considerations of durability, user-friendly design, and comfort enhance compliance and contribute to the overall effectiveness and positive health outcomes of health and hygiene care textile products [12].

2 Classification of Health and Hygiene Care Textile Products

Health and hygiene care textile products encompass a wide range of items designed to promote cleanliness, infection control, and overall well-being in various requirements. These products can be classified into several categories based on their intended use and applications [4].

Personal Hygiene Materials and Products:

- **Feminine Hygiene Products:** Products such as sanitary pads, tampons, and menstrual cups are essential for managing menstruation, maintaining comfort, and preventing leakage, odor, and infections.
- **Disposable Wipes**: Disposable wipes are convenient personal hygiene products for cleaning hands, surfaces, and even for personal care, such as makeup removal or refreshing during travel.

Personal Protective Equipment (PPE):

- **Surgical Gowns:** Sterile gowns worn by healthcare professionals during surgical procedures to prevent the transmission of pathogens.
- **Isolation Gowns:** Non-sterile gowns used in healthcare settings to protect against contamination during patient care and contact with potentially infectious materials.
- **Face Masks and Respirators:** Protective facial coverings designed to reduce the risk of inhaling or spreading airborne contaminants.
- **Gloves:** Disposable gloves made from various materials, including latex, nitrile, or vinyl, to protect hands from contamination.

Incontinence Products:

- **Adult Diapers:** Absorbent products designed for individuals with incontinence issues.
- **Incontinence Pads and Under-pads:** Disposable or washable pads placed on beds or furniture to protect against incontinence-related leaks.

Patient Apparel:

- **Hospital Gowns:** Garments worn by patients in healthcare facilities, designed for easy access during medical examinations or procedures.
- **Patient Robes and Pajamas:** Comfortable clothing for patients to wear during their hospital stay, promoting a sense of dignity and well-being.

Antimicrobial Textiles:

- **Antimicrobial Fabrics:** Textiles treated with agents to inhibit the growth of bacteria, viruses, and fungi.
- **Antimicrobial Clothing:** Everyday clothing is designed with antimicrobial properties to reduce the growth of odor-causing bacteria.

Bedding and Linens:

- **Hospital Bed Sheets and Pillowcases:** Textiles designed for hospital beds, often treated for durability and infection control.
- **Mattress Protectors:** Covers for mattresses to prevent contamination and protect against spills and stains.
- **Blankets and Bedspreads:** Comfortable and durable textiles for patient use in healthcare settings.

Specialized Textiles for Healthcare:

- **Surgical Drapes:** Sterile covers are used to create a barrier between the surgical field and the surrounding environment.
- **Disposable Textiles:** Single-use products like disposable bed sheets, pillows, and towels are used in healthcare settings for infection control.

Sports and Fitness Textiles:

- **Moisture-wicking Fabrics:** Textiles designed to manage sweat and moisture during physical activities.
- **Compression Apparel:** Clothing designed to provide support and enhance circulation during sports and fitness activities.

Childcare and Elderly Care Textiles:

- **Nap Mats and Bedding:** Hygienic textiles for young children in daycare settings.
- **Elderly Care Apparel:** Clothing designed for comfort and ease of use in elderly care environments.

Environmental and Eco-Friendly Textiles:

- **Sustainable Bedding:** Textiles made from organic or recycled materials, promoting environmental sustainability.
- **Eco-Friendly Towels and Linens:** Textiles with minimal environmental impact in terms of production and disposal.

Travel and Hospitality Textiles:

- **Travel Pillows and Blankets:** Compact and hygienic textiles designed for use during travel.
- **Hotel Linens:** Clean and durable textiles used in hospitality settings to ensure guest comfort and safety.

These classifications illustrate the diverse range of health and hygiene care textile products, each serving specific purposes in different industries and circumstances. The development of innovative textiles continues to contribute to improved hygiene practices and overall well-being.

3 Feminine Hygiene Care Products

3.1 Sanitary Napkin

Menstrual sanitary napkins have undergone a remarkable evolution over the years, becoming an essential aspect of women's health and hygiene. From their humble beginnings to the sophisticated products available today, sanitary napkins play a vital role in providing comfort, confidence, and cleanliness during menstruation.

This chapter explores the history, importance, and innovations surrounding menstrual sanitary napkins, offering valuable insights for women seeking optimal menstrual hygiene management.

3.2 Importance of Proper Menstrual Hygiene

Proper menstrual hygiene is of paramount importance for the overall health and well-being of individuals assigned female at birth. Menstruation is a natural and integral part of the reproductive cycle, and maintaining good menstrual hygiene practices is crucial for several reasons. Firstly, adequate menstrual hygiene helps prevent infections and reduces the risk of reproductive health issues. Regular changing of sanitary products, such as pads or tampons, prevents the buildup of bacteria and pathogens, minimizing the chances of urinary tract infections or other complications. Secondly, practicing proper menstrual hygiene contributes to physical comfort and mental well-being, fostering confidence and a positive attitude toward menstruation. Thirdly, it promotes social inclusivity and reduces stigma, as good hygiene practices enable individuals to engage in daily activities without fear of leakage or discomfort. Bharadwaj et al. [13] Furthermore, proper disposal of menstrual products is essential for environmental sustainability. Educating individuals about menstrual hygiene not only empowers them to make informed choices but also contributes to breaking societal taboos surrounding menstruation, fostering a more inclusive and supportive community. Ultimately, the importance of proper menstrual hygiene extends beyond individual health, influencing broader social and environmental considerations [14].

3.3 Anatomy of a Sanitary Napkin–Components, Layers, and Materials

Sanitary napkins, also known as menstrual pads, are essential personal hygiene products used by women during menstruation. These pads are designed with multiple layers, materials, and components to ensure comfort, absorbency, and protection. Users often choose products based on personal preferences, comfort, and specific needs, with eco-friendly options becoming increasingly popular due to their use of sustainable materials like organic cotton or bamboo fibers [15]. In modern sanitary napkins, the layers typically include an absorbent core, which captures menstrual flow, and a moisture-wicking layer to keep the surface dry. The bottom barrier layer is preventing leaks and ensures that menstrual fluid is contained within the pad. While the specific composition, materials, and layers may vary among different brands and types, below mentioned (Table 1) are some common components and layers that are configured and used in sanitary napkins manufacturing.

Table 1 Sanitary napkin—components, layers and materials

Layer	Description	Materials used	Function
Top sheet	The top sheet is the layer that comes in direct contact with the skin. It is designed to be soft, comfortable, and wick away moisture to keep the skin dry	Nonwoven fabrics are typically made from polypropylene or a combination of synthetic fibers	Provides a soft and comfortable surface against the skin. It is designed to quickly absorb and transfer moisture away from the body
Absorbent core	The absorbent core is the central part of the sanitary napkin responsible for absorbing menstrual fluid	Wood Pulp: Sourced from softwood trees, it provides absorbency	Absorbs menstrual fluid and retains it to keep the surface dry
		Superabsorbent Polymers (SAP): These are hydrophilic polymers capable of absorbing and retaining large amounts of liquid	
Back sheet	The back sheet is the outer layer of the sanitary napkin that comes in contact with clothing. It is typically made of a waterproof or water-resistant material to prevent leaks and ensure that menstrual fluid is contained within the pad	Typically made of a polyethylene film or similar waterproof material	Prevents leaks by acting as a barrier between the absorbed fluid and clothing
Adhesive strip	An adhesive strip is used to attach the sanitary napkin to the underwear, keeping it securely in place during use	Pressure-sensitive adhesive	Attaches the sanitary napkin to underwear, securing it in place during use
Wings	Some sanitary napkins come with wings, which are flaps on the sides that fold over the edges of the underwear to provide extra protection against leaks and help keep the pad in place	Similar to the top sheet and back sheet materials	Adds extra protection by wrapping around the sides of the underwear to prevent leaks
Perfume or fragrance (optional)	Some sanitary napkins may include a perfume or fragrance to mask odors and provide a sense of freshness. However, fragrance-free options are also available for those with sensitivity to scents	Typically a blend of fragrance compounds	Masks odors and provides a pleasant scent

(continued)

Table 1 (continued)

Layer	Description	Materials used	Function
Breathable layers (optional)	Some pads may incorporate breathable layers to enhance air circulation and reduce the risk of irritation	Micro-perforated films or breathable materials like microporous films	Enhances air circulation, reducing the risk of skin irritation
Wrap or packaging	The sanitary napkin is typically individually wrapped for hygiene and convenience. The packaging may also include information about the product and usage instructions	Typically made of paper or plastic	Provides individual wrapping for hygiene and convenience, often with information about the product

Eco-friendly options are gaining traction, employing biodegradable materials to reduce environmental impact. Manufacturers are also catering to users with sensitivities by offering fragrance-free and dye-free options. Overall, the menstrual hygiene product market is evolving to meet diverse needs while prioritizing sustainability and user comfort.

3.4 Absorption Technologies

Sanitary napkins have evolved significantly over the years, incorporating advanced absorption technologies to enhance comfort, performance, and overall user experience. Traditionally, these pads utilized a layer of wood pulp renowned for its absorbent properties, complemented by superabsorbent polymers (SAP) integrated into the absorbent core. SAPs form a gel-like substance upon contact with fluid, significantly boosting absorbency. Synthetic fibers like polyethylene or polypropylene are also common in modern sanitary napkins, strengthening the core structure and enhancing absorbency. Additionally, the integration of air-laid paper, a porous nonwoven fabric, further augments absorbency by efficiently retaining liquid and preventing pooling [16].

In response to growing environmental concerns, sanitary napkins increasingly incorporate natural fibers such as cotton and bamboo to manage menstrual flow effectively while ensuring comfort and reliability. Cotton, known for its absorbency, breathability, and biodegradability, offers a sustainable option for menstrual hygiene products. Bamboo fibers, with their rapid growth and minimal environmental impact, further contribute to sustainability efforts in the industry. By prioritizing the use of natural fibers, manufacturers aim to reduce the environmental footprint of disposable sanitary products while providing effective menstrual fluid management [17].

The continuous innovation in absorption technologies, combining natural and synthetic fibers along with advanced materials like superabsorbent polymers, reflects the industry's commitment to enhancing user comfort and addressing the environmental concerns. Top sheets designed for quick-drying and breathability contribute to maintain a dry surface and reduce skin irritation. As consumers increasingly prioritize the factors like absorbency, comfort, and sustainability, manufacturers strive to offer a diverse range of sanitary napkin options tailored to meet varied needs and preferences [18].

3.5 *Parameters Involved in Selection of Sanitary Napkins*

Selecting the right sanitary napkin is essential for comfort, hygiene, and overall well-being. Factors like absorbency level, material composition, and fit play crucial roles in choosing the most suitable product. It's important to match absorbency with menstrual flow to prevent leaks and ensure adequate protection throughout the day or night. Opting for products made from organic cotton or other breathable materials can reduce the risk of skin irritation, particularly for those with sensitivities. Comfort features like contoured shapes and flexible materials enhance satisfaction and prevent leaks, while environmentally friendly options such as biodegradable materials or reusable products contribute to sustainability efforts [19].

Health considerations should guide product selection, especially for those with specific concerns like sensitive skin. Choosing hypoallergenic products designed for sensitive skin can mitigate discomfort, and reading reviews from other users provides valuable insights into product performance. It's essential to maintain proper hygiene practices, follow usage instructions, and change the product regularly to prevent discomfort or irritation. Consulting with a healthcare professional can offer personalized advice and address any specific health concerns or questions regarding product selection [20].

3.6 *Different Types of Sanitary Napkins*

There are various types of sanitary napkins designed to cater to different needs and preferences during menstruation. The choice of a specific type depends on factors such as menstrual flow, lifestyle, comfort, and environmental considerations. Sanitary napkins come in various types to suit individual needs and preferences. Regular pads offer moderate protection for everyday use, while ultra-thin pads provide discreet comfort for light to moderate flow. Maxi pads offer extra absorbency for heavy flow, while overnight pads ensure protection during extended wear. Pantyliners offer minimal coverage for light flow or spotting. Some pads feature wings for added security, while organic or reusable options prioritize sustainability and comfort. Menstrual cups offer a reusable alternative, while specialized products may include features like quick-dry topsheets or odor control. Choosing the right pad involves considering absorbency, comfort, materials, and lifestyle preferences for a positive menstrual experience [21].

3.7 Innovations in Feminine Hygiene Products

In recent years, the field of feminine hygiene has witnessed notable innovations driven by technological advancements, sustainability efforts, and a focus on user comfort and convenience. One significant innovation is the introduction of menstrual cups, reusable bell-shaped devices made of medical-grade silicone, rubber, or latex. These cups offer a sustainable alternative to traditional pads and tampons, providing environmentally friendly, cost-effective, and long-wearing solutions for menstrual care. Similarly, period underwear has emerged as a popular option, featuring built-in absorbent layers for leak-proof protection and washable, reusable design, contributing to sustainability efforts while offering comfort and convenience. The industry has seen a rise in biodegradable and organic sanitary products, addressing environmental concerns and catering to individuals with sensitivities. Smart menstrual products, equipped with sensors to monitor saturation levels and provide real-time alerts, offer enhanced convenience and menstrual cycle tracking. Reusable cloth pads, crafted from soft, breathable fabrics, provide customizable and sustainable alternatives to disposable pads, emphasizing eco-friendliness and cost-effectiveness [22].

Ongoing research and development have led to the integration of advanced absorption technologies in disposable pads, such as quick-dry topsheets and superabsorbent polymers, improving comfort and leakage prevention. Wearable devices designed for period pain relief, utilizing heat therapy or transcutaneous electrical nerve stimulation (TENS), offer drug-free solutions for managing menstrual discomfort discreetly and portably. Additionally, customizable subscription services provide tailored delivery of feminine hygiene products, including eco-friendly options, enhancing convenience, and meeting individual needs. These innovations underscore a growing awareness of diverse user needs, environmental considerations, and advancements in materials and technology within the feminine hygiene industry. As the industry continues to evolve, further innovations are expected to focus on enhancing the overall user experience while addressing sustainability concerns and promoting individual well-being [23].

3.8 Biodegradable and Eco-friendly Options

The shift toward environmentally conscious sanitary napkins reflects growing consumer concerns over environmental impact and waste reduction, driving manufacturers to develop sustainable alternatives. These products utilize materials like organic cotton and bamboo fibers, minimizing plastic usage and promoting biodegradability to reduce landfill burden. Certifications such as GOTS and OEKO-TEX Standard 100 reassure consumers of sustainability, while brands prioritize eco-friendly packaging and manufacturing practices to minimize their carbon footprint and promote transparency. Environmentally conscious brands educate consumers about the environmental impact of traditional products and highlight the benefits of choosing eco-friendly options. They innovate with materials and absorption technologies to create effective yet environmentally responsible sanitary napkins. With increasing consumer awareness and demand, the market for these products is expanding, supported by corporate social responsibility initiatives and government regulations promoting sustainable practices in the menstrual hygiene industry [24].

3.9 Sanitary Napkin Usage and Disposal

Proper usage and disposal of sanitary napkins are essential for personal hygiene and environmental sustainability. Regular changing, matching absorbency levels, and secure placement in underwear help prevent bacterial growth and leaks. It's vital to wrap used napkins and dispose of them in designated bins, avoiding flushing to prevent plumbing issues and pollution. Staying hydrated and switching to hypoallergenic options if needed contribute to comfort and health. Educating others about eco-friendly choices and following waste disposal regulations further promote responsible practices, ensuring a cleaner and more sustainable environment [25].

4 Disposable Wipes

Disposable wipes have become integral to contemporary hygiene and cleaning practices, offering unmatched convenience and adaptability in various environments. From personal care to household cleaning, these pre-moistened, single-use products cater to a myriad of needs. Whether it's gentle cleansing with baby wipes or effective sanitization with disinfectant wipes, they address diverse tasks efficiently. Despite environmental considerations, their popularity persists due to ease of use and effectiveness. Made from a range of materials like nonwoven fabrics, cotton, and synthetic fibers, manufacturers carefully select materials to balance performance and environmental impact. Disposable wipes have become indispensable tools in modern life, providing convenient solutions for daily tasks across different settings [26].

5 Surgical Gowns

Surgical gowns are indispensable protective gear for healthcare professionals, offering a crucial barrier against potential contamination during surgical procedures. Crafted from materials like polyester, polyethylene, and polypropylene, these gowns strike a balance between comfort and protection, covering the front of the body and arms with long sleeves featuring elastic or knit cuffs. Designed with tie or hook-and-loop closures at the back, they allow for easy donning and removals without compromising sterility, ensuring aseptic conditions are maintained. Sterility is paramount in surgical environments, with surgical gowns packaged to preserve aseptic conditions. These gowns are fluid-resistant, safeguarding healthcare workers from exposure to blood and bodily fluids during procedures. By adhering to infection control protocols, surgical gowns contribute significantly to patient safety by preventing cross-contamination between different patients, providing vital protection for both the surgical team and patients alike.

Isolation gowns are essential protective garments used in healthcare and laboratory settings to mitigate the transmission of infections and contaminants. They form a critical part of personal protective equipment (PPE), shielding healthcare workers from exposure to blood, bodily fluids, and infectious materials. Constructed from materials like polyethylene, polypropylene, or polyester, these gowns cover the front and back of the body, including the arms, with elastic cuffs ensuring a secure fit and preventing skin exposure. Isolation gowns are available as either disposable or reusable, with the latter designed for laundering and sterilization to facilitate multiple uses. Categorized by levels of fluid resistance and protection, these gowns play a pivotal role in infection prevention, ensuring safety for both healthcare professionals and patients. Manufacturers utilize a variety of raw materials to produce isolation gowns, each selected to provide the necessary protective barrier against fluids and contaminants. Materials such as polyethylene, polypropylene, polyester, SMS, CPE, Tyvek, and laminated films offer distinct properties including water resistance, durability, breathability, and chemical resistance. Adhering to industry standards and regulations, manufacturers carefully select materials to ensure compliance with performance and safety requirements. Factors influencing material choice include intended use, required protection level, and whether the gown is for single or multiple uses [27].

Isolation gowns find application across a spectrum of settings, including healthcare facilities, laboratories, dental practices, emergency response situations, industrial environments, and clean room settings. Healthcare workers wear them during patient care, surgeries, and medical procedures, while laboratory technicians use them to shield against hazardous materials. Dental professionals employ isolation gowns to prevent contact with blood and saliva during dental procedures, and emergency responders utilize them to protect against infectious diseases. Additionally, isolation gowns are utilized in industrial settings where exposure to hazardous materials is possible and in clean rooms to maintain sterile conditions. Serving as a physical barrier, isolation gowns reduce the risk of contamination and infection transmission, tailored to specific work environments and potential exposure hazards [28].

6 Face Masks and Respirators

Face masks and respirators serve distinct purposes in protecting individuals from airborne contaminants, including respiratory droplets and other particles. Face masks, such as cloth masks and surgical masks, are commonly used in public settings to control the spread of respiratory droplets from asymptomatic or mildly symptomatic individuals. They are also standard in healthcare environments, where surgical and procedure masks are worn by healthcare professionals to prevent the transmission of respiratory droplets to patients during medical procedures and to protect the wearer from larger particles. Additionally, face masks offer personal protection in everyday activities like shopping or commuting, as well as protection from environmental factors like dust and pollen for individuals with respiratory sensitivities or allergies.

In contrast, respirators, particularly N95 respirators, are crucial for healthcare workers and individuals in industrial settings exposed to airborne infectious agents, dust, and other contaminants. N95 respirators provide a higher level of filtration efficiency, filtering out a minimum of 95% of airborne particles. They are essential during pandemics or outbreaks of infectious diseases, where individuals may be at a higher risk of exposure. Furthermore, respirators are used in various occupational settings, such as construction and manufacturing, to protect against inhaling hazardous substances like chemical fumes or biological agents [29].

The choice between face masks and respirators depends on factors such as the level of protection required, the nature of the airborne hazards present, and the specific application. While face masks offer source control to protect others from the wearer's respiratory droplets, respirators provide personal protection against inhaling smaller airborne particles. Both play critical roles in infection control and protection, and their selection is often guided by regulations and standards to ensure adequate protection. Additionally, manufacturers use a range of raw materials in the production of face masks and respirators, including polypropylene, polyethylene, polyester, and metal strips, to meet performance and safety requirements. These materials vary based on the type of mask or respirator, intended use, and regulatory guidelines, ensuring the effectiveness and safety of the final products [30].

7 Adult Diapers

Adult diapers, also known as adult briefs, are vital for managing urinary or fecal incontinence in adults facing challenges with bladder or bowel control. Comprising multiple absorbent layers like fluff pulp and superabsorbent polymers, these diapers effectively contain urine and feces, ensuring the wearer's comfort and dryness. Available in disposable and reusable options, they feature adjustable tabs or elastic waistbands for a secure fit, with some types offering odor-control technology. Proper usage involves wearing them snugly and changing them regularly to maintain hygiene and prevent skin irritation. Caregivers should handle changes with sensitivity, prioritizing the individual's dignity and privacy.

Constructed with materials like fluff pulp, superabsorbent polymers, and moisture-wicking fabrics, adult diapers provide optimal absorbency and leakage protection. They feature elastic leg gathers, waterproof backing, and closure options like adhesive tapes for a secure fit. Manufacturers may use cloth-like materials for enhanced comfort, ensuring a practical solution for managing incontinence while promoting comfort and well-being. Choosing the right type and size of diaper is crucial for enhancing confidence and allowing users to engage in daily activities without concern for accidents or leakage [31].

8 Incontinence Pads and Under-Pads

Incontinence pads and under-pads are essential for managing urinary or fecal incontinence, offering protection against leaks and accidents. Incontinence pads, worn inside regular underwear, cater to light to moderate urinary incontinence with various sizes and absorbency levels. Underpads, larger rectangular pads, safeguard mattresses and furniture from leaks, available in disposable and reusable options. Crafted from materials like fluff pulp and superabsorbent polymers, these products prioritize comfort and confidence, enabling individuals to carry out daily activities without any discomfort. In maternal care, underpads serve critical roles during pregnancy, postpartum recovery, and childbirth, offering protection against postpartum bleeding and maintaining cleanliness. They also assist in managing urinary or fecal incontinence during pregnancy or postpartum, ensuring hygiene and comfort. Proper selection and usage of these products are essential for effective incontinence management, promoting comfort, dignity, and well-being for individuals in need, with healthcare professionals providing guidance for tailored care [32].

9 Antimicrobial Textiles

Antimicrobial textiles play a vital role across diverse industries, from healthcare to everyday apparel, by impeding the growth of various microorganisms like bacteria and fungi. Achieved through methods such as chemical treatments, natural antimicrobial agents, and nanotechnology, these textiles maintain cleanliness and reduce odors in different settings. Chemical treatments embed agents like silver nanoparticles or zinc oxide into fabrics, while natural substances like essential oils and chitosan offer alternative antimicrobial properties. Nanoparticles, particularly silver nanoparticles, release ions that hinder microbial growth on textile surfaces, contributing to enhanced hygiene and durability [3].

The benefits of antimicrobial textiles extend to various domains, including sportswear, healthcare, and household textiles. They effectively control odor-causing bacteria in sportswear, prevent microbial contamination in healthcare settings, and promote prolonged freshness in household items like bedding and towels. However, it's crucial to remember that while antimicrobial textiles offer valuable benefits, they're not a substitute for proper hygiene practices and regular cleaning. Concerns about environmental and health impacts have prompted the exploration of safer alternatives, emphasizing the importance of coupling antimicrobial textile usage with sustainable and health-conscious practices for long-term benefits [33].

10 Specialized Textiles for Healthcare

Specialized textiles for healthcare are meticulously engineered to meet the exacting demands of medical environments, patients, and healthcare professionals. These textiles encompass a diverse array of fabrics and materials, each tailored with specific properties such as comfort, durability, safety, and hygiene. For instance, antimicrobial fabrics undergo treatments to inhibit bacterial growth, crucial for infection control in hospitals. Additionally, fluid-resistant textiles act as barriers against bodily fluids, ensuring cleanliness and minimizing contamination. Fire-retardant fabrics comply with safety regulations, providing protection against fire-related hazards, while moisture-wicking textiles manage perspiration, particularly beneficial for patients with mobility issues. These advanced materials contribute to enhanced patient care, safety, and overall healthcare outcomes by meeting the unique requirements of medical settings.

Surgical drapes play a pivotal role in maintaining sterility and safety in operating rooms, serving multiple functions to ensure the success of surgical procedures. They establish a physical barrier between the surgical site and non-sterile areas, preventing microbial contamination and reducing the risk of surgical site infections. Furthermore, surgical drapes manage fluids generated during surgery, maintain visibility and access for the surgical team, and prioritize patient comfort and safety. Crafted from specialized fabrics and materials, surgical drapes adhere to stringent standards for sterility, fluid resistance, durability, and patient comfort, making their proper selection, positioning, and use essential for successful surgical outcomes [34].

Disposable textiles offer convenient, hygienic solutions across various industries and applications where single-use functionality, cleanliness, and cost-effectiveness are paramount. These textiles find extensive use in healthcare facilities, food service establishments, household cleaning, hospitality, travel, and industrial environments. Their versatility contributes to maintaining hygiene standards, preventing cross-contamination, and promoting safety by providing temporary or short-term functionality in diverse settings. From hospitals to restaurants, households, and manufacturing facilities, disposable textiles play a pivotal role in facilitating convenience, hygiene, and efficiency, ensuring compliance with stringent hygiene standards and regulations [1].

References

1. Zahir A, Mahmood U, Nazir A, Hussain T, Abid S (2022) Biomaterials for medical and healthcare products. In: Medical textiles from natural resources. Woodhead Publishing, London, pp 43–86
2. Ajmeri MCJ, Mr JRA (2006) Application of nonwovens in healthcare and hygiene sector. In: Medical textiles and biomaterials for healthcare. Woodhead Publishing, London, pp 80–89
3. Jeong SH, Hwang YH, Yi SC (2005) Antibacterial properties of padded PP/PE nonwovens incorporating nano-sized silver colloids. J Mater Sci 40:5413–5418
4. El-Ghazali S, Khatri M, Kobayashi S, Kim IS (2022) An overview of medical textile materials. Med Text Nat Resources 2022:3–42
5. Chellamani KP, VigneshBalaji RS, Veerasubramanian D (2013) Medical textiles: the spunlace process and its application possibilities for hygiene textiles. J Acad Indus Res 1(12):735–739
6. Rajendran S, Anand SC, Rigby AJ (2016) Textiles for healthcare and medical applications. In: Handbook of technical textiles. Woodhead Publishing, London, pp 135–168
7. Guru R, Kumar A, Kumar R (2022) Healthcare and hygiene products application in medical textile. In: Next-generation textiles. IntechOpen
8. Montazer M, Rangchi F, Siavoshi F (2010) Preparation of protective disposable hygiene fabrics for medical applications. In: Medical and healthcare textiles. Woodhead Publishing, London, pp 164–170
9. Wang WY, Chiou JC, Yip J, Yung KF, Kan CW (2020) Development of durable antibacterial textile fabrics for potential application in healthcare environment. Coatings 10(6):520
10. Oche MO, Umar AS, Gana GJ, Ango JT (2012) Menstrual health: the unmet needs of adolescent girls' in Sokoto, Nigeria. Sci Res Essays 7(3):410–418
11. Gultie TK (2014) Practice of menstrual hygiene and associated factors among female Mehalmeda high school students in Amhara Regional State, Ethiopia. Sci J Public Health 2(3):189–195
12. Ajmeri JR, Ajmeri CJ (2010) Nonwoven personal hygiene materials and products. In: Applications of nonwovens in technical textiles. Woodhead Publishing, London, pp 85–102
13. Bharadwaj S, Patkar A (2004) Menstrual hygiene and management in developing countries: taking stock. Junct Soc 3(6):1–20
14. Saranya J, Prakash C, Kubera SKS (2019) Menstrual hygiene practices among adolescent girls. J Innov Technol Explor Eng 8(12):364–366
15. Ajmeri JR, Ajmeri CJ (2011) Nonwoven materials and technologies for medical applications. In: Handbook of medical textiles. Woodhead Publishing, London, pp 106–131
16. Bachra Y, Damiri F, Berrada M, Tuteja J, Sand A (2023) Introduction of superabsorbent polymers. In: Properties and applications of superabsorbent polymers: smart applications with smart polymers. Springer, Singapore, pp 1–18
17. Rajesh Kumar C et al (2022) Study on moisture behavior properties of milkweed and milkweed/cotton blended sanitary napkins. J Nat Fibers 19(8):2849–2860
18. Mounika K, Ilango KB, Baranitharan AK, Dhivyatharshini S, Gowtham K, Swetha V, Vijaykumar S (2023) Biodegradable sanitary napkins an innovative approach towards menstrual hygiene using bamboo fiber and lemon grass. Int J Pharmaceut Sci 1(10):1–1
19. Raina D, Balodi G (2014) Menstrual hygiene: knowledge, practice and restrictions amongst girls of Dehradun, Uttarakhand, India. Glob J Interdiscip Soc Sci 3(4):156–162
20. Ali TS, Rizvi SN (2010) Menstrual knowledge and practices of female adolescents in urban Karachi, Pakistan. J Adoles 33(4):531–541
21. Sharma S, Nagar A, Singh L, Singh S (2018) Low cost sanitary napkins for health and hygiene of remote/slum feminine population of India. In: Humanizing work and work environment (HWWE 2016): English, p 231
22. Selvasudha N, Sweety JP, Dhanalekshmi UM, Devi NSD (2021) Smart antimicrobial textiles for healthcare professionals and individuals. In: Antimicrobial textiles from natural resources. Woodhead Publishing, London, pp 455–484

23. Sanjanwala D, Londhe V, Trivedi R, Bonde S, Sawarkar S, Kale V, Patravale V (2023) Polysaccharide-based hydrogels for medical devices, implants and tissue engineering: a review. Int J Biol Macromol 2023:128488
24. Zhang Y, Yao J (2016) High-value utilization of natural cellulose: cellulose-based biocomposite materials. Handb Sustain Polym Process Appl 99:1
25. Kaur R, Kaur K, Kaur R (2018) Menstrual hygiene, management, and waste disposal: practices and challenges faced by girls/women of developing countries. J Environ Public Health 2018:1
26. Briain OÓ, Mendes ARM, McCarron S, Healy MG, Morrison L (2020) The role of wet wipes and sanitary towels as a source of white microplastic fibres in the marine environment. Water Res 182:116021
27. Thadepalli S (2022) Review of multifarious applications of polymers in medical and health care textiles. Mater Today Proc 55:330–336
28. Gogoi N, Bhuyan S (2020) Medical textiles: it, s present and prospects. Pharma Innov J 9(9):1
29. Ajmeri JR, Ajmeri CJ (2016) Developments in the use of nonwovens for disposable hygiene products. In: Advances in technical nonwovens. Woodhead Publishing, London, pp 473–496
30. Wiesemann F, Adam R (2011) Absorbent products for personal health care and hygiene. In: Handbook of medical textiles. Woodhead Publishing, London, pp 316–335
31. Bitencourt GR, Santana RF (2021) Evaluation scale for the use of adult diapers and absorbent products: methodological study/Escala de avaliacao do uso de fraldas e absorventes: estudometodologicoEscala de evaluacion del uso de panales y absorbentes: estudiometodologico. Online Braz J Nurs 20(1):1
32. Bostan LE et al (2019) The influence of incontinence pads moisture at the loaded skin interface. J Tissue Viab 28(3):125–132
33. Gulati R, Sharma S, Sharma RK (2022) Antimicrobial textile: recent developments and functional perspective. Polym Bull 79(8):5747–5771
34. Getu A, Sahu O (2014) Technical fabric as health care material. Biomed Sci Eng 2(2):35–39

Masks and Respirators for Non-medical Applications (Masks and Respirators Against Dust, Chemicals and Others)

T. Gobikannan, B. A. Patil, and S. J. Pawar

Abstract The chapter describes the role of masks and respirators in preventing respiratory infections, mainly focusing on non-medical applications. Egyptians first used masks during mummification as a precaution against inhalation of harmful substances. During the Bubonic Plague and Industrial Revolution, masks were used as a precautionary method. The masks can be classified based on their use, design, and features. Masks can be used during surgical procedures or while interacting with patients to prevent exchanging fluids and particulate materials. Cloth and disposable masks are reusable, readily available, low cost, and often made with nonwoven polypropylene material. The masks evolve based on the classifications and recommendations made occasionally. Chemical, biological, and radiological agents also pose health risks and can be avoided by gas masks. Respirators like N95, R99, and P100 are highly effective respirators. Further, the respirators that filter out air contaminants and use filters or canisters are also discussed. Respirators offer clean air via remote sources (compressed air tanks) like airline respirators or have their oxygen supply, which is to be carried by the user. The respirators are used where the surrounding air is dangerous to life or if the oxygen level is too low. Half-face and full-face respirators are variations of respirators that provide protection. Fit, type, comfort, type of hazard, maintenance, and replacement are factors that help in mask selection. Moreover, breathability, straps, and weight should be checked before prolonged use. Various physical and mechanical principles, like electrostatic filtration, diffusion, inertia, etc., filter out the air contaminants. The masks are assessed by bacterial filtration efficiency, particle filtration efficiency, differential pressure or breathing resistance, flow resistance, fluid or splash resistance, flame resistance,

T. Gobikannan (✉) · B. A. Patil
Handloom and Textile Technology, Indian Institute of Handloom Technology, Foulk's Compound, Thillai Nagar, Salem, Tamil Nadu 636001, India
e-mail: gobikannant@gmail.com

B. A. Patil
Department of Mechanical Engineering, Government College of Engineering and Research, Avasari Khurd. Dist. Pune. Maharashtra, India 412405

S. J. Pawar
Department of Applied Mechanics, Motilal Nehru National Institute of Technology Allahabad, Prayagraj, UP, India 211004

S. Sundaramoorthy et al. (eds.), *Textile Materials for Good Health and Wellbeing*, SDGs and Textiles, https://doi.org/10.1007/978-981-97-6496-9_4

fit well and sit comfortably, and reusability. Nonwoven fabrics like spun bond and melt-blown fabric are used to produce masks for medical applications and respirators. The cartridges and canisters used for removing gas and chemical vapours use activated carbon for purification. In continuation, mask standardization is also discussed. The International Organization for Standardization, International Electrotechnical Commission, and American Society for Testing and Materials are the standards used in the production of masks. In the current scenario, the organization must consider a life cycle perspective for better environmental management. Hence, the life cycle assessment of masks has been given, referring to various phases such as the Goal and Scope, Life Cycle Inventory, Life Cycle Impact Assessment, and Interpretation.

1 Introduction

The current chapter introduces masks and respirators for non-medical applications and provides a brief overview of different types of masks and respirators used for filtration against dust particles, chemicals, and other similar objects. The simplest difference between medical and non-medical masks is based on the intended use of the masks, whereas non-medical masks are for the general public. Non-medical masks are also used in non-healthcare-related environments. Additionally, they should be worn in case of low risk of exposure and outside of healthcare facilities. On contrary, respirators are designed to defend user by filtering the air and fitting closely on the face to filter out particles, along with dust particles, chemicals, and other similar objects. Both masks and respirators can ensure different kinds of protection depending on the type of mask and respirator. Moreover, how they are used is also very significant. On one side, loosely woven cloth products provide least protection, however, layered finely nonwoven products ensure more protection. Medical masks are considered as 'Personal Protective Equipment' that are mainly used in medical environments, on contrary, non-medical masks are for the general public or use in non-healthcare related environments. The chapter starts with the historical background of masks and respirators which is followed by the classification of masks and respirators. In this chapter masks and respirators are discussed, in reference to the Ancient Civilization, Industrial Revolution, and Post-World War II, including the Modern Era. As aforementioned, masks are categorized into medical and non-medical masks based on their usage. Non-medical masks include the cloth and disposable single-use masks, dust masks, face shields, valved masks, and specialized masks (industrial respirators). Similarly, the classification of respirators mainly includes air-purifying respirators (APRs), atmosphere-supplying respirators, and other respirators. Specially, masks and respirators against dust and chemicals are discussed in separate sections. Considering the sole purpose of masks or respirators to separate or filter out particles, the mechanism of filtration must be understood. Filtration is the separation of one material from another with the purpose to improve the purity of the filtered material.

So, in the next section common principles of filtration have also been demonstrated by explaining the various filtration mechanisms, viz. electrostatic filtration, diffusion, inertial impaction, interception, and gravitational. Primary purpose of the filter medium is to maximize the collision and retention of the suspended particles in the fluid (air/gas or liquid) stream which can be effectively achieved by fibrous structure of the filter media while minimizing the energy loss of the stream of the fluid. The characteristics of mask are an essential aspect of both medical and non-medical face masks. The characteristics mainly include efficiencies and resistances. This chapter talks about the filtration efficiency, bacterial filtration efficiency, particle filtration efficiency, differential pressure or the breathing resistance, flow resistance, fluid resistance (splash resistance), and flame resistance. It provides an overview about the factors responsible for the effectiveness of filtration process and also discusses about fitting factors of a mask like its sit and fit, along with reusability. An attempt is made to introduce the mask capability index in the chapter that helps the designers, manufacturers, and end users for the selection and assessment of appropriate masks. The design of the mask is further explained in brief which is mainly focuses on various kinds of masks and respirators. The standard is an acceptable level of quality of any material, product, process, and service. This chapter also gives the reader an overview of the International Standards involved in the making and designing of a mask, however, a detailed explanation is categorically avoided considering the scope of this chapter. The standards are developed by organizations that are familiar with the subject under consideration and its technical aspects in the form of technical standard document that are accepted and followed globally. These standards are used worldwide for deliberation, attention, and usage related to face masks. Each product, service, material, or in general activity affects environment. So the life cycle assessment of the such activities is considered noteworthy for the measurement of the environmental impact of different activities by highlighting the most harmful one so as to offer the data and recommendations essential to reduce or offset these impacts. It is a most crucial tool for eco-design and eco-planning for making more environmentally efficient products and processes. The life cycle assessment of masks and respirators is compiled by referring to a few published case studies relating to medical and non-medical face masks because of widespread awareness of the research community for sustainability and environment friendliness of product, process, or service. ASTM F2100, ASTM F2299, ASTM F1671/F1671M-22 are some of the international standards that deal with the material for mask, test for measuring the efficiency of materials to penetration and the test method for the resistance offered by material. Moreover, ISO 14040 is the main standard for LCA studies. Finally, a few concluding remarks are made that highlights general comments and observations.

2 Historical Background of Masks and Respirators

Masks and respirators have a long and diverse history, serving various purposes across different cultures and time periods. A brief overview of their historical background is given below. This overview includes the usage of mask and similar products in ancient civilization which has been followed similar products during the era of 'Industrial Revolution' and 'Modern Era' that has been counted post World War II.

2.1 Ancient Civilizations

In ancient Egypt (circa 3000 BCE), priests and embalmers used cloth masks to protect themselves from inhaling harmful substances during the mummification process. Chinese physicians used masks made of greased paper to filter out dust and other particles during the warring states period at circa 500 BCE [1]. The Black Death, a devastating pandemic, led to the use of masks with beak-like extensions filled with sweet-smelling substances in the fourteenth Century [2]. This was an early form of the respirator designed to protect against miasma, or 'bad air' believed to carry disease. The plague doctor's outfit, with its distinctive beak mask, became a common sight during outbreaks of Bubonic Plague in Europe during the seventeenth century. During this time people used masks containing aromatic substances to counteract the foul odours. But it offered limited protection against the plague outbreaks [1].

2.2 Industrial Revolution and Early Respirators

With the onset of the Industrial Revolution, workers faced increasing exposure to harmful airborne particles. Early respirators such as the 'fireman's respirator' used a wet sponge as a filter. Its usage was started by a worker in the nineteenth Century [3]. The development of the 'smoke helmet' by Garrett Morgan, an African American inventor, marked an important milestone in the growing industry [4]. It was an early version of a gas mask that provided protection against smoke and harmful gases, shown in Fig. 1. In continuation of the industrial revolution, twentieth century faced the outbreaks of Spanish Flu in 1918. The outbreak of the Spanish Flu (influenza pandemic) prompted widespread use of masks. However, many of these were simple cloth masks with limited effectiveness against viruses. It was mandatory to wear it for police force and medical workers during the influenza pandemics in USA [3]. A photographic image of Red Cross workers using a reusable face mask is shown in Fig. 2. The Dust Bowl and World War II accelerated advancements in respiratory protection. The medical mask was replaced by disposable paper during 1930s. As a further advancement, high-filtration masks and gas masks were developed and widely used by military personnel and civilians during 1930–40 [3].

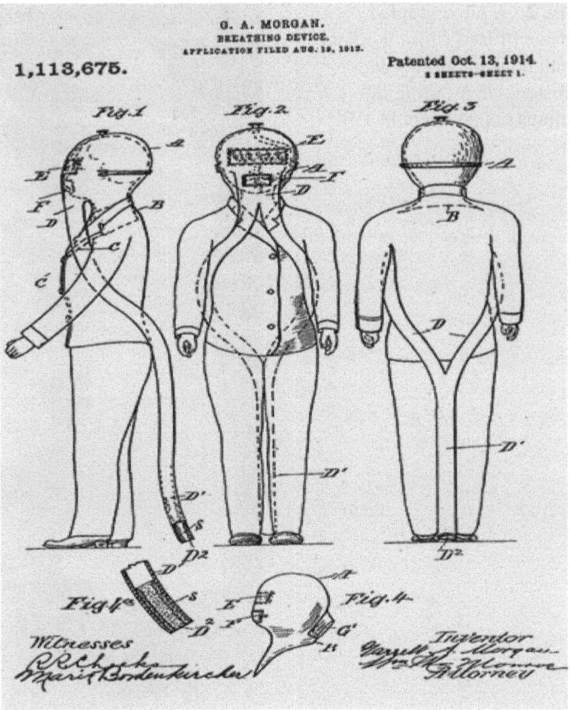

Fig. 1 A photographic image of first gas mask invented by Garrett A. Morgan [4]

2.3 *Post-world War II and Modern Era*

After World War II, the development of disposable masks made from synthetic materials like polypropylene revolutionized the mask production during 1950s–1960s. The 'N95' respirator was introduced, which had high-level filtration capabilities [3]. The N95 respirator mask became particularly popular as it was deemed effective in filtering out airborne particles. However, widespread everyday use of masks as a fashion statement or as a common preventive measure, as seen more recently, had not yet become a cultural norm in many parts of the world at that time. In 2019, face masks were used in construction work, or by individuals in regions affected by air pollution. However, the latter part of 2019 marked the emergence of a novel coronavirus, now known as SARS-CoV-2. They continue to evolve with the ongoing advancements in materials, design, and filtration technologies [6].

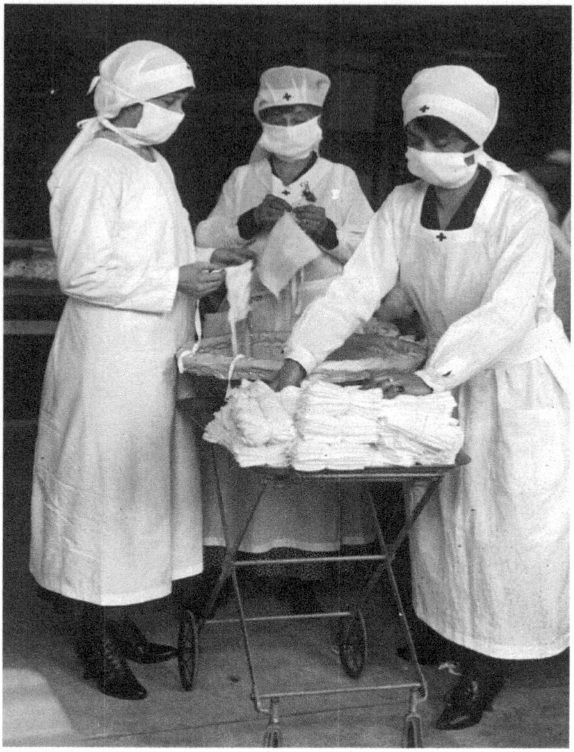

Fig. 2 A photographic image of Red Cross workers with reusable masks at Boston, USA during the influenza pandemic in 1919 [5]

3 Classification of Masks (Medical and Non-medical)

Masks can be classified into various categories based on their intended use, design, and features. The two primary categories are medical and non-medical masks (Fig. 3), each serving different purposes. A breakdown of these classifications is given below; however, information about medical mask is given in brief since it's out of the scope of this chapter.

3.1 Medical Masks

Medical masks are specifically designed for use in healthcare settings to protect both healthcare professionals and patients from transmission of infectious diseases [7–9]. They are subjected to strict regulatory standards and are typically classified into several subtypes.

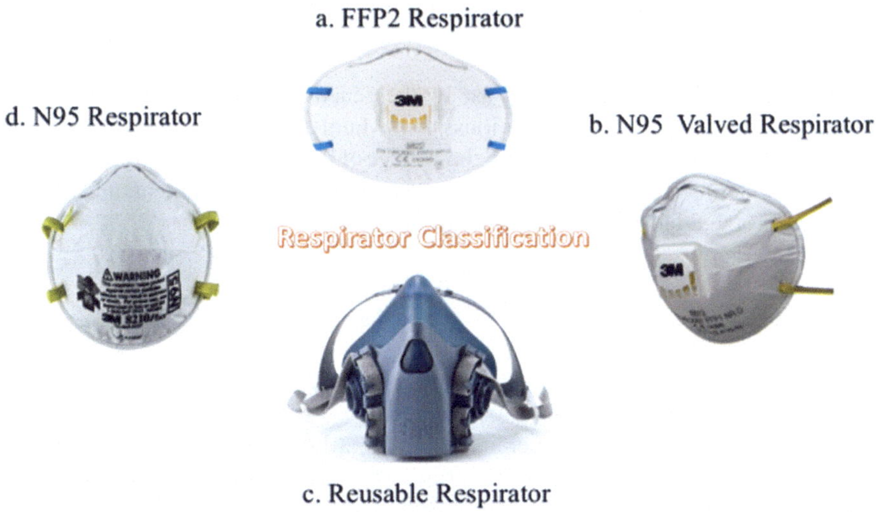

Fig. 3 A photographic image of different types of respirator [13,14,15,16]

3.1.1 Respirators

The respirators are categorized into N95, N99, N100, R95, R99, R100, P95, P99, P100, etc. by different countries. The respirators are highly efficient masks that are intended to filter approximately 95% of airborne particles, including small respiratory droplets and infectious agents. Additionally, they are tight-fitting and provide a higher level of protection and are essential in situations where airborne transmission is a significant concern. These respirators provide varying levels of filtration efficiency, with N100 and P100 offering the highest protection (filter out almost 99.97% airborne particles). They are often used in healthcare settings and other environments with high levels of airborne contaminants [10].

3.1.2 Nasal Mask

People with sleep apnea may have problems with continuous breathing in their day-to-day life. In such situations, the nasal mask offers a better option for continuous positive airway pressure (CPAP) therapy. In this way, the supply of nonstop air flow to the nose with the support of the mask is ensured [11, 12].

3.2 *Non-medical Masks*

Non-medical masks are intended for use by the general public or in non-healthcare settings including industrial applications [17]. Mainly, they are aimed to decrease the spread of respiratory droplets from the wearer of the mask to others or to prevent the wearer inhaling smokes and dust from the surrounding environment, and can be further categorized based on their materials and design.

3.2.1 Cloth and Disposable Single-Use Masks

These masks are typically made from fabric materials like cotton (most common), polyester, or an arrangement of various materials [18]. They are reusable and can be homemade or purchased commercially. Cloth masks can have multiple layers for improved filtration and may come in various styles, like pleated, fitted, and neck gaiters. The cloth mask usage has increased significantly during the COVID-19 pandemic due to the shortage of supply of medical face masks in the global market. It may not be as efficient as the surgical or N95 masks for reducing the spread of aerosols. However, it occupied the market due to its easy availability and low cost [19, 20]. The non-medical disposable masks are often made from nonwoven polypropylene material. They are designed for single use and come in various styles, including 3-ply masks and ear-loop masks. The cloth and disposable single-use masks are frequently used in public settings to reduce the spread of respiratory droplets by blocking them.

3.2.2 Dust Masks and Face Shields

Dust, pollens, animal dander, and moulds can be filtered out by dust masks. They are more flexible and may not provide any protection against any virus or chemicals (*Department of Environmental Health and Safety (EH&S) at UC Santa Cruz, Disposable N95 for Health Care Workers* [21]). While not technically masks, face shields [an item of personal protective equipment (PPE)] are transparent plastic shields which are covering the face of wearer from his forehead to below the chin. The face shields are often used in combination with masks for added protection and are commonly used in healthcare and some non-medical settings [22].

3.2.3 Valved Masks and Specialized Masks (Industrial Respirator)

In case of some of the non-medical masks, an exhalation valves are provided which make breathing convenient and easier, these are called as valved masks. Nevertheless, these masks may not effectively prevent the wearer from spreading respiratory droplets to others and are not recommended in situations where source control is essential [17].

In addition to the above categories, there are specialized masks designed for specific purposes, such as the industrial respirators for workplace safety, gas masks for chemical protection, and high-filtration masks for certain occupational settings [23, 24]. It is important to note that the classifications and recommendations for mask usage may evolve based on public health guidelines, scientific research, and the specific context of the pandemic or health crisis. Therefore, it's essential to follow the local health authorities' guidelines regarding the use of masks in day-to-day life and industrial purpose, especially during outbreaks of infectious diseases.

4 Classification of Respirators

Respirators are generally referred as PPE. They are normally designed and developed to protect the wearer (user) from inhaling harmful matters that may comprise air borne particulate matter which may be a mixture of many chemical species, may be a complex mixture of solids and aerosols composed of small droplets of liquid, dry solid fragments, and solid cores with liquid coatings, gases, and vapours. They can be broadly classified into two main categories: air-purifying respirators and atmosphere-supplying respirators [25–27].

4.1 Air-Purifying Respirators (APRs)

Air-purifying respirators have filters, cartridges, or canisters. They filter out contaminants from the air before the wearer breathes it in. They do not supply additional oxygen and rely on the surrounding atmosphere.

4.1.1 Particulate Respirators

These respirators (N95, N99, N100, P95, P99, and P100) filter out solid and liquid aerosols, but are not effective against gases or vapours. Mainly, N95 respirators are close-fitting (tight-fitting) face-piece respirators. With proper fitting, they filter out nearly 95% of suspended airborne particles. On the contrary, N99 respirators offer higher level of protection as compared to N95 respirators i.e. it filters out 99% of particles. N100 respirators filter out 99.97% of particles, thus, they deliver the utmost degree of respiratory safeguard among the disposable respirators. Typically, P95, P99, and P100 respirators are oil-proof respirators that provide the same or higher level of protection as compared to N95, N99, and N100 respirators; they also provide protection against oil-based particles. However, R95, R99, and R100 respirators are

Fig. 4 A Photographic image of gas mask respirator [28]

somewhat oil-resistant respirators and provide protection against both the oil and non-oil-based particles.

4.1.2 Other Air-Purifying Respirators

Figure 4 shows the photographic image of a gas mask respirator. They can be further classified as gas and vapour respirator and combination respirator. The gas and vapour respirators use specific cartridges or canisters filled with sorbents to filter out specific gases or vapours. However, the combination respirator combines particulate filters with chemical cartridges or canisters, offering protection against both the particles and specific gases or vapours.

4.2 Atmosphere-Supplying Respirators

Atmosphere-supplying respirators offer a clean mean of breathing air to the user which is independent of the ambient atmosphere (work area). They can be further divided into two subcategories: airline respirators and self-contained breathing apparatus (SCBA). Normally, airline respirators deliver a clean air from remote sources (e.g. compressed air tanks, airline hoses, etc.) to the user. However, SCBAs have their own air supply carried by the user in a backpack or harness. They are used in circumstances wherein the ambient air is immediately dangerous to life or health (IDLH), or when the oxygen levels are too low (hypoxemic) due to a lack of healthy red blood cells. These respirators normally guard the user from many types of airborne contaminants viz. particles, gases, and vapours.

4.3 Other Respirators

Other respirators include escape respirators, powered air-purifying respirators (PAPRs), supplied air respirators (SAR), specialized respirators, etc. The disposable and reusable respirators are also occupying the market depending on the industrial requirements. Escape respirators are designed for emergency situations. Normally, PAPRs use a fan to draw air through filters before delivering it to the user. However, SCBAs provide a continuous supply of breathable air from a compressed air tank worn on the back. They are used in situations where the air is too contaminated to be filtered. SAR delivers clean air from a stationary source, often using a hose. They are commonly used in situations where a portable respirator may not be suitable. Some respirators are designed for specific purposes or industries, such as those used in healthcare settings (e.g. N95 respirators for protecting against infectious diseases), or for industrial applications where specific hazards need to be mitigated. Respirators can also be classified based on their intended lifespan. Disposable respirators are designed for single use and are discarded after a single use and reusable respirators can be cleaned, maintained, and reused multiple times. It is noteworthy to mention that the selection of a respirator is very vital for its proper use. It must be based on a thorough assessment of the workplace environment, the type of hazard present, and the individual's specific job task. Fit testing and training are also crucial components of respirator use to ensure proper protection.

5 Masks and Respirators Against Dust and Chemicals

Masks and respirators are crucial PPE when working in environments with dust and chemicals. They help protect the respiratory system from inhaling harmful particles or fumes. Here are some types of masks and respirators commonly used for this purpose [29–31] which are explained further.

5.1 Dust Masks

Dust mask may be used to provide the protection against various dust particles present in the working atmosphere. It could be realized by the proper use of N95, N99, and N100. They filter out the particles down to 0.3 microns in size with the range of 95–99.97% of airborne particles. The dust masks are mainly produced by the amalgamation of spun bond and melt-blown fabric, where tight and fully closed fabric structure is used in the form of different layers. These masks are usually suitable in construction, metal working, etc.

Fig. 5 A photographic image of gas mask [32]

5.2 *Half-Face and Full-Face Respirators*

Half-face respirators mainly shield the nose and mouth of user. They have replaceable cartridges which are responsible to filter out specific particles or gases. They cover the nose and mouth, and are equipped with cartridges or filters to purify the air. They are often used in environments where there may be a mix of harmful particles and gases. Full-face respirators provide protection to eye, nose, and mouth, and come with a wide range of filter options for different types of hazards. These respirators guard the whole face of the user, thus provide protection to the eye in addition to the respiratory protection. They are used in situations where in the eyes need to be shielded from hazardous substances.

5.3 *Gas Masks*

Figure 5 shows the photographic image of gas mask. These masks are specifically designed to protect against chemical, biological, radiological, and nuclear agents. They use special filters to neutralize or block these hazardous materials. These use cartridges to filter out specific gases and vapours. They are commonly used in industrial settings where there is a risk of exposure to chemical agents.

5.4 *Powered Air-Purifying Respirators*

PAPRs have a fan that actively filters air through a cartridge before delivering it to the wearer, providing a higher level of protection. They can be used with a variety of filters, including particulate filters, gas and vapour cartridges, or a combination of both. These respirators use a blower which is engaged to pass contaminated air via a highly efficient filter and it then delivers purified air to the face piece of user. They are often used in environments with very high levels of contamination.

5.5 Escape Respirators

Escape respirators are compact, self-contained units used for emergency evacuation from hazardous environments, where the user needs to quickly escape from an environment with contaminated air. They are typically small, portable units that provide a short-term supply of clean air.

The selection of mask and respirator depends on many factors that include fit, type of hazard, filter type, comfort & wearability, maintenance & replacement, regulations & standards, training & education, etc. It is crucial that the mask must forms a tight seal around the face to be effective. Fit testing may be required for some workplaces. Different masks are designed for different types of hazards. Some are better for particulate matter, while others are designed for chemical vapours. It needs to choose filters appropriately for the specific hazards faced by the worker. There are different cartridges and filters for dust, chemical vapours, gases, and biological hazards. Prolonged use can be uncomfortable, so factors like straps, weight, and breathability should be considered. It needs to follow manufacturer guidelines for maintenance, cleaning, and filter replacement. It must be ensured that the masks or respirators comply with the relevant industry and its safety standards. Proper training on how to use, fit, and maintains masks and respirators is also an important crucial factor to select the mask and respirator.

6 Principles of the Filtration

Filtration mechanism (Fig. 6) in masks, particularly in the context of respiratory masks like N95 respirators or surgical masks, are essential for providing protection against airborne particles, including bacteria, viruses, dust, and other pollutants. These masks utilize various physical and mechanical principles to filter out contaminants from the air. The key filtration mechanisms [33, 34, 35] in masks include electrostatic filtration, diffusion, inertial impaction, interception, and gravitational wherever the size of particles is smaller than the pore sizes.

6.1 Electrostatic Filtration

Some masks, particularly N95 respirators, employ an electrostatic charge to enhance filtration. The mask material is treated or manufactured with electrostatically charged

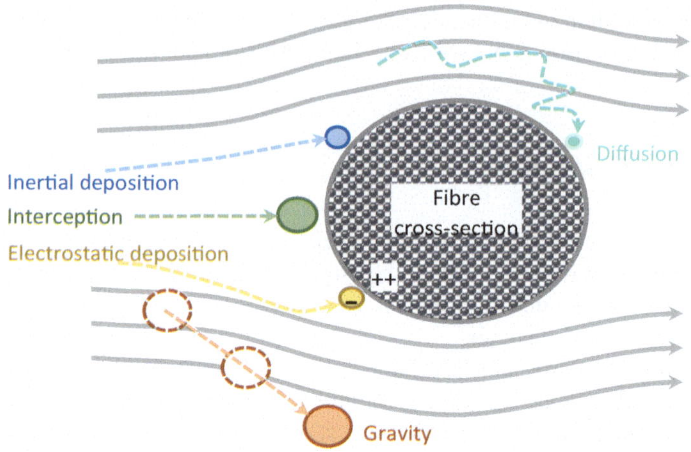

Fig. 6 A schematic diagram of various filtration mechanisms of the face mask [33]

fibres. This charge attracts and captures particles, even those smaller than the individual pores in the material. This mechanism increases filtration efficiency.

6.2 Diffusion

For very small particles, diffusion plays a role in their removal. Diffusion is the random movement of particles in the air, normally under a concentration gradient. While the pores in the mask may be larger than these tiny particles, the particles move erratically and are more likely to collide with mask fibres and become trapped due to Brownian motion.

6.3 Inertial Impaction

This mechanism is more relevant for large-sized particles. When particulate matter is too large and heavy to follow the airflow around mask fibres, they tend to move in a straight path and collide with the fibres, effectively being trapped. This mechanism is not as significant for respiratory masks but plays a role in some industrial masks designed for larger particles.

6.4 Interception and Gravitational

Interception mechanisms might involve fibres or surfaces designed to capture particles as they pass through the mask material. Some particles are captured when they come in contact with the fibres due to their size or chemical properties. This mechanism is particularly relevant for particles of intermediate size. This interception enhances the efficiency of the filtration process, ensuring that particular particles or substances are effectively captured or intercepted as the fluid or gas passes through the filter material. The concept of gravitational filtration mechanism in masks refers to how gravity assists in filtering out particles or droplets from the air as they pass through the mask material. Gravity, though not the primary force at play in most mask filtration systems, can aid in the removal of larger particles or droplets from the air. When someone wears a mask, gravity pulls the larger particles downwards, away from the breathing zone, reducing the chances of these particles entering the respiratory system.

Proper mask selection and usage are crucial for maximizing the filtration effectiveness and protecting against airborne and other gaseous contaminants. The filtration efficiency of the commonly used industrial face mask may depend upon the porosity of filtration media and particle size of the filtrant [36]. The effectiveness of a filter depends on its porosity and permeability. Porosity refers to the percentage of open space or voids in the filter medium, while permeability is a measure of how easily fluid can flow through the filter. A filter with higher porosity and permeability allows for better flow but may capture fewer particles. Filtration is designed to capture particles above a certain size while allowing smaller particles and the fluid to pass through. The size of particles retained by the filter is typically specified in micrometres (microns) or mesh size. The efficiency of a filter in retaining particles of a specific size is a critical consideration. The choice of filtration method and equipment depends on the specific application, such as liquid or gas filtration, water treatment, air purification, or chemical processing. Factors like temperature, pressure, and the nature of the particles being filtered must also be considered.

7 Characteristics of Masks

In the following section, the characteristics of masks are given for both medical and non-medical face masks viz. filtration efficiency, bacterial filtration efficiency (not discussed in this chapter), particle filtration efficiency, differential pressure or breathing resistance, flow resistance, fluid or splash resistance, and flame resistance. Additionally, two factors viz. fit well and sit comfortably and reusability are discussed. In extension, a mask capability index has been introduced which aids the designers, manufacturers, and end users for the assessment of appropriate masks for designated purpose.

7.1 Filtration Efficiency

Filtration efficiency is usually referred to the percentage of designated-sized particles that can be retained or stopped or filtered by a filter medium. High filtration efficiency (lesser penetration efficiency) along with a low flow resistance is the qualities of a better filter material. High GSM and high weight correlate with better filtration, but lower breathability can be caused due to the heavier weight.

7.2 Particle Filtration Efficiency (PFE)

The particle filtration efficiency determines the up-or downstream efficiency (competence) of filter materials and devices to retain sub-micron level particles of interest. The CWA 17553:2020 classified PFE (for particles size 3 ± 0.5 μm), into two categories: Category 1, FE is ≥ 90% and Category 2, FE is ≥ 70% [37]. Filtration efficiency requires a particle size of 3 μm as per the AFNOR SPEC S76-001. Considering the downstream particle concentrations of the mask (C_{down}) and the upstream particle concentrations of the mask (C_{up}), the PFE (in percentage) is calculated as [37].

$$\text{PFE} = 1 - \left[\frac{C_{\text{down}}}{C_{\text{up}}}\right] \times 100 \tag{1}$$

7.3 Differential Pressure or Breathing Resistance

The breathing resistance is utilized to estimate and quantify the flow resistance offered by the facemask. It is also an amount of airflow impeded by the facemask [38]. Technically, differential pressure is an amount of pressure where the reading and reference values are variable. Generally, a constant airflow is necessarily needed to be preserved for investigating the differential pressure of the sample during testing. The higher airflow resistance specifies lesser breathability of the face mask wearer [11]. As per EN 14683 standard, differential pressure analysis measures the pressure drop across a surgical mask [11]. The factors like atmospheric conditions, temperature, humidity, airflow, etc. can cause this pressure drop during measurements. Typically, lower differential pressure resistance increases breathability and comfort of user [11, 39]. An indicator of comfort during breathing throughout a material is called as breathability. The pressure drop (pressure differential) across the two opposite sides of a mask material gives a measure of its breathability. Air flows through the mask material should be maintained similar to that of normally available during breathing. Moreover, the tests should be performed at a specified face velocity perpendicular to the plane of the tested material or the tests can be performed at a specific flow rate across a surface area of a specified face mask material [38]. In short, a face mask must filter out particles on one side and permit a person to breathe effortlessly for the effectiveness of face mask.

7.4 Flow Resistance

The flow resistance and breathability are closely related. It is associated with effectively limiting and preventing the spread of virus during human activities like breathing, coughing, sneezing, speaking, singing, etc. It results in reduced danger of contamination in the surrounding area of an infected person [40]. The flow resistance of filter materials is used to assess and gauge breathability. Low flow resistance indicates high breathability and vice versa. Normally, the flow resistance is estimated by air flow measurement across a section of the material under a given pressure difference.

7.5 Fluid or Splash Resistance

Fluid resistance is a key feature of masks. It represents the mass resistance to penetration by the synthetic blood under pressure (mm of Hg). As per ASTM F2100, it is classified into three levels viz. ASTM level 1: 80 mm Hg and Bacterial Filtration Efficiency (BFE) $\geq$ 95%, ASTM level 2: 120 mm Hg and BFE $\geq$ 98%, and ASTM level 3: 160 mm Hg and BFE $\geq$ 98%. Fluid or splash resistance inspects the resistance offered by the face mask material against the high-velocity fluid streams at various velocity ranges [11]. The highest flash resistance confirms that during surgical operations, the surgical face mask protects the operator in a better way against flashes of potentially contaminated fluid [41].

7.6 Flame Resistance

The time needed for the spreading the fire to a specific length of sample is termed as a flame resistance. It is a parameter used to assess the mask behaviour to promote the ignition in the surrounding area. Normally, the flammability test detects such behaviour. The fire test for surgical fabric is used to assess the oxygen requirement to transmit the flame while ignition is triggered by an electro surgery unit or laser unit. The amount of oxygen require is directly proportional to the flame resistance [41].

7.7 Fit Well and Sit Comfortably

The close facial fit [42] makes sure that the respirator fits appropriately on user's face [43] by evaluating the seal between the face and respirator by identifying gaps or leaks across a respirator and the face. The close-fitting of respirators enables good self-protection against droplet infection than simple smooth mouth and nose covers [40]. Typically, quantitative fit test (QNFT) or qualitative fit test (QLFT) is necessary to safeguard the proper make, model, size, and style of respirator. Typically, a QLFT comprises the passage of test agents like gas, vapour, aerosol, etc. into the surrounding head of a person wearing the respirator. If during the test, the test agent is not detected, the respirator is correct, else it is wrong. The purpose of QNFT is to sense the quantity of air (test agent) leakage into a respirator [43].

7.8 Reusability

In case of non-medical masks, the primary objective is not only to design a single-use mask, but also, to take care of the environment, sustainability, and cost aspect so as to promote circular economy. After assessing the material's circularity indicator, washable masks and 3D-printed masks are found to be justifiable on the ground of sustainability. Additionally, they also reduce the environmental impact [44]. In general, face masks with higher GSM fabric exhibit higher reusability [38].

7.9 Assessment of Mask Capability Index

The mask capability index is derived from the assessment factors and features of the available masks in the market. This index is proposed by authors as a measure of the capability of a mask to provide valuable information within certain specification limits in reference to three stakeholders of mask (designer, manufacturer, and end user). It is also the quantitative index for the end use of mask. Total six assessment factors are decided from the existing literature, and looking into the human safety and environment (one can increase assessment factors as per need). The assessment factors are as follows: reusable, ease of disposal, protection against gases, vapours, and particles, eye protection, useful in immediately dangerous to life and health (IDLH), and oxygen deficient environment. The designer, manufacturer, and end user can determine the mask capability index as illustrated in the Table 1 as a case, wherein the weight is assigned for each assessment factor. If the mask under consideration satisfies the assessment factor, then the score is equal to the weight assigned and vice versa. This index will assist the designer and manufacturer to arrive at an optimal design of the mask as per the specific requirements. In addition, this index can also assist end user to evaluate desired mask for his intended end use.

Table 1 Mask capability index

S. no.	Assessment factor	Weight	Yes/No	Score
1	Reusable	2	Yes	2
2	Ease of disposal	1	No	0
3	Protection against gases, vapours, and particles	2	Yes	2
4	Eye protection	1	Yes	1
5	Useful in IDLH	2	No	0
6	Oxygen deficient environment	2	Yes	2
	Total weight (maximum possible score)	10		
Mask capability index				7

8 Mask Design Parameters

Non-medical application masks used by general public are commonly manufactured from woven and knitted fabric. But, they may not provide good protection against infections. Respirator used against chemicals and dust mainly contains an inlet covering. This inlet covering acts as a barrier against the contaminant atmosphere. Inlet coverings are mainly classified into two categories such as tight and loose fit coverings. Tight-fit coverings are produced from silicone, neoprene, and moulded rubber, with woven elastic or rubber head straps attached to them. Helmets, hoods, suits, and blouses are usually called as loose fit coverings. The inlet covering selection is based on the exposure atmosphere and need of covering the various parts of body [45].

Particulate filtering mask and respirators mainly use three categories of particulate filter medium viz. (1) Machine-made flat disk of random laid nonwoven fibre material, (2) Flat disk of compressed natural wool or synthetic fibre felt, and (3) Loosely packed filter container. The machine made flat disk of random laid nonwoven fibre material are the most commonly used filter materials. It can achieve high-filtration efficiency as compared to other medium. The flat disk of compressed natural wool or synthetic fibre felt, or a blend, impregnated with resin or mechanically needled is another one optional medium used for particulate filtration. Third one is loosely packed filter container that is made from non-felted glass fibre and resin impregnated wool fibre. It is not as common as the felted type. Interception, diffusion, and electrostatic capture mechanism or combination mechanism are used for removing the particles by these fabrics [45].

Respirators used for removing gas and chemical vapours mainly contain cartridges and canisters, designed by using activated carbon [46]. These respirators are called air-purifying respirators. Cartridge and canisters differ in volume of sorbent contained in them. Canisters contain a larger volume of sorbent as compared to the cartridge. Adsorption, absorption, and catalysis are the three basic removal mechanisms used in gas respirators. The PAPR respirators are designed with a blower to allow fresh air with cartridges and canisters. Other atmosphere respirators have a similar setup arrangement in addition to fresh air back up with the user. Continuous airflow is supplied with the additional arrangement of hoses. A combination of all respirators is used for special chemical and dust exposures.

9 International Standards

As per the Oxford Dictionary, 'Standard' is a level of eminence, particularly thought of as acceptable by the people. It is also a document providing requirements, guidelines, specifications, characteristics, etc. that are used to guarantee that materials, processes, products, and services are tailor-made for their intended purpose. An International Standard is a technical standard that has been established by one or more International Standards Organizations and is available for worldwide deliberation and usage. Such prominent organizations include the International Organization for Standardization (ISO), the American Society for Testing and Materials (ASTM), the International Telecommunication Union (ITU), the International Electrotechnical Commission (IEC), etc. Without prejudice, a few International Standards are listed in the following in brief as a quick reference. ASTM International, formerly known as the American Society for Testing and Materials is an international standards organization that describes and sets standards for a multiplicity of materials, systems, products, services, etc. More than 12800 ASTM standards are operating throughout the World currently. In the following, a few most significant ASTM standards (for both medical and non-medical masks) are deliberated in the brief, however, other standards like EN and ISO are also available and can be referred to.

ASTM F2100-21: Standard Specifications for Performance of Materials used in Medical Face Masks. It explains the required materials and their testing requirements for the production of medical face masks. It does not address requirements for respirators. In continuation, ASTM F1494 defines all the terminology concerning to protective clothing. ASTM F1670/F1670M-17a includes the customary testing procedure and method for the determination of resistance offered to penetration by synthetic blood by the material used in protective clothing. Similarly, ASTM F1671/F1671M-22 comprises the typical measurement and test techniques for the resistance offered to penetration by blood-borne pathogens (microorganisms like viruses and/or bacteria that are carried in blood and can cause disease in people) by materials used in protective clothing. The ASTM F1862 specifies the method for determining the resistance offered by the medical face masks to penetration by synthetic blood (horizontal projection of fixed volume at a known velocity). Also, ASTM F2101 discusses the test process for evaluating the bacterial filtration efficiency of the mask materials employed for making medical face masks by using a biological aerosol consisting of *Staphylococcus Aureus*. ASTM F3050 decides the monitoring for the conformation assessment of personal protective clothing and related equipment. Various levels of masks are decided in ASTM F2100 based on certain well-defined parameters and risks involved [47]. Specifically, ASTM Level 1 is perfect for processes wherein there is a possibility of low risk of exposure to fluid (no sprays and no splashes at all). Whereas, ASTM Level 2 is perfect for techniques which involve a moderate risk of exposure to fluid (sprays and/or splashes can be produced). On the other hand, ASTM Level 3 is suitable for actions wherein there is a high risk of exposure to fluid (sprays and/or splashes will be definitely produced). The ASTM F2100 standard can also be useful for other personal protective equipment (PPE) like respirators. For various ASTM Levels, a few significant parameters are as follows. For Level 1 masks, the differential pressure in mm H_2O/cm^2 could be < 4.0 and the filtration by sub-micron particulate @ 0.1 μm could be $\geq$ 95%. Moreover, Level 1 masks are suitable for low-risk conditions that may involve procedures like visiting a doctor's office or travelling on public transportation, such that they can protect against respiratory and large particle droplets. The Level 1 masks may be made from lightweight and breathable materials so that they can assist in the prevention of the spread of infection. For Level 2 and Level 3 masks, the allowed differential pressure in mm H_2O/cm^2 could be < 5.0 and the filtration by sub-micron particulate @ 0.1 μm could be $\geq$ 98%, respectively. Whereas Level 2 masks are suitable for moderate risk conditions (dental procedures or when visiting a hospital) and Level 3 masks are appropriate for high-risk conditions (during surgery or when there is a high risk of exposure to body fluids or blood), such that they should provide protection against the respiratory and very small particle (0.1 micron) droplets. The Level 2 masks may be made from more substantial and thicker materials than ASTM Level 1 masks so that they can assist in the prevention of the spread of infection. However, Level 2 masks should be made from more substantial and thick materials than earlier ASTM Level 1 masks so that they should assist in the prevention of the spread of infection in high-risk situations or when recommended by a healthcare provider.

10 Life Cycle Assessment of Masks and Respirators

In accordance with ISO 14001, the life cycle is '*Consecutive and interlinked stages of a product (or service) system, from raw material acquisition or generation from natural resources to final disposal.*' A life cycle perspective by the organization comprises environmental consideration of an organization's activities (including products and services) such that the organization can control or influence better environmental management. The life cycle embraces various stages, starting from the acquisition of raw materials, design, production, transportation/delivery, use, end-of-life treatment, and final disposal. Thus, with a life cycle perspective, the organization can deliberate on particular stages in the life cycle over which the organization can exercise the most significant control or influence. Hence, the organization may deal with the utmost prospects to lessen the use of resources and minimize pollution or waste.

Life cycle assessment (LCA), or 'life cycle analysis' is a method to assess the environmental impact accompanying with the different phases of a commercial product, process, or service. It is most frequently the cradle-to-grave type investigation that mainly focuses on the environmental impact assessment. The LCA study encompasses a detailed and a systematic study of the materials and energies (two main factors) that are essential across the supply chain and value chain involving a product, process or service of entities involved. Thus, LCA computes the conforming environmental emissions due to product, process, or service by assessing the accumulative possible impact on the environment with the objective to record, document, and increase the complete environmental profile of the product, process or service. This document and recommendation can serve as an all-inclusive baseline to accurately compare carbon footprints during the product, process or service under consideration. Most important document regarding LCA could be sequence of ISO standards for the environmental management, in particular, ISO 14040 (principles and framework) and ISO 14044 (requirements and guidelines summary). The LCA assess the environmental influences related with all the stages of a product's life (raw material extraction, material processing, manufacturing (conversion of raw material to finished product), distribution of the finished product, use of product, maintenance and repair of the product, and final disposal or recycling of the product) so that the result helps decision-makers to make appropriate decision for the least environmental impact.

According to ISO 14040 and ISO 14044, an LCA is accomplished in four different but interdependent phases, viz. Goal and Scope, Life Cycle Inventory (LCI), Life Cycle Impact Assessment (LCIA), and Interpretation. The first phase includes objectives and parameters of the LCA along with boundaries. The second phase comprises requisite data (inputs and outputs for materials and energies) for the calculation. The third phase encompasses the evaluation of the significance of potential environmental impacts obtained based on the findings of the earlier phase. However, the fourth phase involves the combination of findings of the second and third phases of LCA for conclusions and recommendations in light of the defined goal and scope. The variants of LCA include Cradle-to-grave type (full LCA), Cradle-to-gate type (partial LCA), Cradle-to-cradle type (closed loop production), Gate-to-gate type (partial LCA), Well-to-wheel type (specific LCA), Economic input–output LCA, Ecologically based LCA, and Exergy-based LCA, etc.

LCA of cotton vs. surgical face mask is reported by [48] using Open-LCA 1.10.3 software. To design textile masks more sustainably so as to identify relevant ecological factors for design decisions, a simplified LCA was carried out on surgical masks and 2-layered cotton masks. Various sources (journal articles, scientific literatures, reliable online resources, policy and media reports, online survey with users, etc.) have been used for data collection pertaining with mask usage and disposal. Mainly, LCA is comprised of four steps viz. (a) ISO 14040 (Environmental management, LCA, Principles and Framework), (b) ISO 14041 (Environmental Management, LCA, Goal definition and inventory analysis), (c) ISO 14042 (Environmental Management, LCA, Life cycle impact assessment), and (d) ISO 14043 (Environmental management, LCA, Life cycle interpretation). The system boundaries for cotton mask include the production of cotton fabric and elastic band, its transportation and usage, and finally its incineration and landfill, whereas, system boundaries for surgical mask are same like cotton mask except production of nonwoven polypropylene spun bond and melt blown, production of aluminium nose wire, and elastic band. In the inventory analysis, data collection and calculations were done for single face mask of both types to quantify material and energy inputs and outputs of the facemask cases within the defined system boundaries. In continuation, documentation and quantification of appropriate material and energy flows (inputs and outputs) were done. In LCIA, various impact categories were considered, which include global warming potential, heat waste, climate change, human toxicity, acidification, ozone layer depletion, carbon monoxide, particulate matter, and cancer effect with the open burning and incineration as a major disposal method. By considering the most largely recognized environmental problems which include acidification, carbon monoxide, global warming potential, etc., the final results were derived. These results are having a close association with the transport, production, energy generation, and waste management. The conclusion indicates that the surgical mask causes higher impression on the environment than cotton mask. In an another study [49], the author has carried out life cycle environmental influences of disposable medical masks to recognize the life cycle (cradle to cradle) stages that effect the uppermost influence on the environment by using actual and comprehensive statistics (information) obtained from the medical mask producer in Turkey. The goal of the study was to evaluate the life cycle environmental influences of the throw-away medical face mask (single) to recognize the maximum environmental impact producing stage using GaBi v.10.5 (LCA software). Being a cradle-to-grave type LCA, the system boundaries starts from raw material extraction and processing and ends with the final disposal. The system boundaries also involve all other activities like manufacturing, packaging, distribution, and use. In addition, transport and waste management beside the associated supply chain were also included in the study. All boundaries were having few exclusions as well due to the lack of data. The inventory analysis was based on the primary data and secondary data for Type I and Type II medical face mask. The primary data is taken from the manufacturer, however, the secondary data taken from Ecoinvent database, 2021. The detailed LCA includes material, processing, transportation and disposal in the form of end of life (incineration/burning). In LCIA, various identified impact categories were abiotic depletion potential, abiotic depletion potential (fossil), acidification potential, global warming potential, ozone layer depletion potential, photochemical oxidant creation potential, and terrestrial eco-toxicity potential, freshwater aquatic eco-toxicity potential, marine aquatic eco-toxicity potential, human toxicity potential, and eutrophication potential. The findings from LCA show that raw material supply is the main for of the first 7 factors, incineration/burning of waste is the foremost contributor to the next 3 factors, and the transportation is only a noteworthy contributor for the last factor. The findings of the study were projected for the entire environmental impacts of the Worldwide usage of disposable medical face masks in 2020 for single mass and estimated that the total 52 billion disposable medical masks produced Worldwide consumed 22 TJ of energy with a global warming potential at around 1.1 Mt CO_2.

Lee et al. [50] have carried out LCA (cradle-to-grave type) of surgical face mask (single-use) and face mask with embedded filtration layer (reusable). Total of nine impact categories are considered along with the waste generated by each face mask in the context of Singapore. The 3-layered single-use surgical face mask (single person, 31 days, 31 face masks) and the embedded filtration layered reusable face mask (single person, single face mask, 30 washes) were considered in this study. The main goal of the LCA study was to notify the policymaker about the environmental influence of the use face masks under consideration. In inventory analysis, Ecoinvent 3.6 database was used to obtain the emission factors and the same is also used to estimate waste generation. The impact assessment involves nine pertinent midpoint impact classifications those includes climate change, fossil fuel depletion, metal depletion, water depletion, freshwater eco-toxicity, freshwater eutrophication, marine eco-toxicity, marine eutrophication, and human toxicity. Additionally, the generation of waste was furthermore calculated. Finally, it was found that the embedded filtration layer reusable face mask would produce fewer waste and have a reduced impact of at least 30% among the nine aforementioned impact categories except a few impact categories like freshwater eutrophication, marine eutrophication, water depletion, and human toxicity. Also, it was concluded that predominant emission occurs at the stage of raw material acquisition for most of the impact categories of both face masks with the exception of (1) the water depletion for embedded filtration layered reusable face mask and (2) the freshwater eutrophication of a single-use face mask. Additionally, with the inclusion of incineration, the usage of a single-use face mask might lead to a net positive freshwater eutrophication. In the waste generated, the usage stage in the LCA was acknowledged as the foremost contributor to the aggregate waste. Finally, the LCA study has concluded that the embedded filtration layered reusable face mask has a lesser emission for most of the nine impact categories. Chau et al. [51] conducted an LCA in United Kingdom for the environmental impacts of different mask options (medical grade (surgical mask and reusable cloth mask) and shielding device grade (respirator (FFP3), and rigid half masks) for healthcare sector. The LCA study was done using ISO standards viz. 14040:2006 and 14004:2006 along with the Product Environmental Footprint (PEF) guidelines [48] in GaBi Software and mixture of GaBi and Eco-Invent. The goal and scope definition includes an evaluation of the environmental influences connected with various face mask choices within the healthcare segment. The system boundaries comprise material sourcing and production. It also includes transportation, mask use and reuse (if applicable). Further, the waste processing at the end of life along with associated activities was also considered. The life cycle inventory includes material, manufacturing and packaging, transportation, decontaminated for reuse, waste management, allocation (recycling and incineration), etc. However, the life cycle impact assessment comprises the Environment Footprint (EF) 3.0 impact method. It is supplemented by the assemblage of impact classifications and indicators established by the EC-JRC (European Commission's Joint Research Centre). In conclusion, this study accomplishes that recycling of both medical and protective-device grade settings propose to produce lower amount of waste and leads to lesser environmental influences in maximum classifications than traditional single-use one.

Rodríguez et al. [44] attempted the LCA of five different types [3D-printed model (uses disposable FFP2 filters (8 h change cycle), reusable structure), surgical mask (single use, 8 h use only), respirator (FFP2) with exhalation valve (single use, 8 h use only), respirator FFP2 without an exhalation valve (single use, 8 h use only), and washable mask (reusable, 50 washes)] of masks considering the Italian scenario using SimaPro 9.0.0.49. The set objectives of the study were an eco-design actions guide for masks to decrease the undesirable influences of these masks on the environment. Total 16 indicators have been considered during LCIA with all materials and energy flows (inputs and outputs) that can be quantified related to the desired functional units. The detailed study concludes that the reusable masks have flawless advantages over the disposable counterparts. Maceno et al. [52] evaluated handmade reusable cotton fabric face masks (3 h and 30 reuses after wash) and single-use nonwoven fabric face masks (3 h without reuses) for LCA to support government public policies in the Brazilian context. However, Material Circularity Indicator (MCI) studies were also carried out. The purpose of the LCA (ISO 14040) was to recognize and enumerate the environmental influences of usage of two non-medical face masks with cradle-to-grave [extraction of raw materials to disposal (destined for landfills)] approach and Ecoinvent 3.6 database. For LCIA the software SimaPro v. 9.00.49 was used and the ReCiPe 2016 and IMPACT World + methods were chosen.

11 Concluding Remarks

It can be concluded from this chapter that masks and respirators play an important part in the prevention of infections spreading via droplets, suspended particles, microorganisms, gases etc. The type of mask, usage for it, and its fit should be checked before using a mask. However, not significant for the user, technically filtration efficiencies and certain resistances are very significant characteristics of the masks and respirators. Due to the environmental concerns, it is best to use a reusable mask if it fulfils the criteria. The increased pollution levels make the use of mask mandatory. Now with the increased awareness about the use of mask due to SARS, individuals have started using them in case of cold and coryza. The dust mask can be used in cases of dust allergy as well. Industrial respirators are mandatory for the workers exposed to the hazardous environment. The type of exposure changes the type of mask being used.

References

1. Lizzie Enfield (2022) A history of medical masks. Retrieved from https://wellcomecollection.org/articles/XwMmcBQAAGwR9GY8. Accessed Jan 2024
2. Byrne J (2006) Daily life during the Black Death. Greenwood Press, London.
3. Strasser BJ, Schlich T (2020) A history of the medical mask and the rise of throwaway culture. Lancet 396(10243):19–20. https://doi.org/10.1016/S0140-6736(20)31207-1
4. Morgan Garrett Augustus (1914) US patent (1,113,675). https://patents.google.com/patent/US1113675A/en

5. NARA US (2024) Records of the war department general and special staffs, 1860–1952. https://catalog.archives.gov/id/45499363
6. Feng S, Shen C, Xia N, Song W, Fan M, Cowling BJ (2020) Rational use of face masks in the COVID-19 pandemic. Lancet Respir Med 8(5):434–436. https://doi.org/10.1016/S2213-2600(20)30134-X
7. Gobikannan T, Pawar SJ, Kumar SKS, Chavhan MV, Navinbhai AH, Prakash C (2023) Importance of antiviral and antibacterial face mask used in pandemics: an overview. J Nat Fibers. https://doi.org/10.1080/15440478.2022.2160407
8. Kumar S, Karmacharya M, Joshi SR, Gulenko O, Park J, Kim GH, Cho YK (2021) Photoactive antiviral face mask with self-sterilization and reusability. Nano Lett 21(1):337–343. https://doi.org/10.1021/acs.nanolett.0c03725
9. Robertson P (2023) Comparison of mask ratings, standards, and filtration effectiveness (N95, KN95, FFP1, FFP2). Retrieved from https://smartairfilters.com/en/blog/comparison-mask-standards-rating-effectiveness/. Accessed Jan 2024
10. Salvi SS (2020) In this pandemic and panic of COVID-19 what should doctors know about masks and respirators? Retrieved from http://apiindia.org/wp-content/uploads/pdf/corona-virus/review-article-on-mask.pdf. Accessed Jan 2024
11. Deshpande S, Joosten S, Turton A, Edwards BA, Landry S, Mansfield DR, Hamilton GS (2016) Oronasal masks require a higher pressure than nasal and nasal pillow masks for the treatment of obstructive sleep apnea. J Clin Sleep Med 12(9):1263–1268
12. Patel SR, White DP, Malhotra A, Stanchina ML, Ayas NT (2003) Continuous positive airway pressure therapy for treating sleepiness in a diverse population with obstructive sleep apnea results of a meta-analysis. Arch Intern Med 163(5):565–571
13. 3M (2024) FFP2 respirator. Retrieved from https://www.3mindia.in/3M/en_IN/p/d/v000081497/. Accessed Mar 2024
14. 3M (2024) N95 respirator. Retrieved from https://www.3mindia.in/3M/en_IN/p/d/v000057604/. Accessed Mar 2024
15. 3M (2024) Reusable respirator. Retrieved from https://www.3mindia.in/3M/en_IN/p/d/v000057593/, Accessed Mar 2024
16. 3M (2024) Valved respirator. Retrieved from https://www.3mindia.in/3M/en_IN/p/d/v000061052/. Accessed Mar 2024
17. 3M (2023) Understanding valved filtering facepiece respirators in non-healthcare workplaces—United States. Technical Bulletin. Retrieved from https://multimedia.3m.com/mws/media/1994317O/understanding-valved-respirators-united-states-faq.pdf. Accessed Jan 2024
18. Konda A, Prakash A, Moss G, Schmoldt M, Grant G, Guha S (2020) Correction to aerosol filtration efficiency of common fabrics used in respiratory cloth masks. ACS Nano 14(5):10742–10743
19. Carias C, Rainisch G, Shankar M, Adhikari BB, Swerdlow DL, Bower WA, Pillai SK, Meltzer MI, Koonin LM (2015) Potential demand for respirators and surgical masks during a hypothetical influenza pandemic in the United States. Clin Infect Dis 60:S42–S51
20. Ramakrishnan D (2020) Covid-19 and face masks—to use or not to use! Indian J Commun Health 32:240–243
21. Department of Environmental Health and Safety (EH&S) at UC Santa Cruz, Disposable N95 for health care workers (2024). Retrieved from https://ehs.ucsc.edu/programs/safety-ih/documents/n95.pdf. Accessed Jan 2024
22. Mishra T (2021) What does face shield mean? https://www.safeopedia.com/definition/9058/face-shield
23. Adanur S, Jayswal A (2022) Filtration mechanisms and manufacturing methods of face masks: an overview. J Ind Text 51(3_Suppl):3683S-3717S
24. FDA (2024) N95 respirators, surgical masks, face masks, and barrier face coverings. Retrieved from https://www.fda.gov/medical-devices/personal-protective-equipment-infection-control/n95-respirators-and-surgical-masks. Accessed Jan 2024
25. CDC (2024) Types of respiratory protection. Retrieved from https://www.cdc.gov/niosh/npptl/topics/respirators/disp_part/respsourceTypes.html. Accessed Jan 2024

26. CEC (2024) Infection Prevention and Control Manual: COVID-19 and other acute respiratory infections. Retrieved from https://www.cec.health.nsw.gov.au/__data/assets/pdf_file/0018/644004/%0ACOVID-19-IPAC-manual.pdf%0A. Accessed Jan 2024
27. FDA (2021) Stakeholders for imported, non-NIOSH-approved FFRs made in China. Retrieved from www.fda.gov/media/136664/download. Accessed Jan 2024
28. 3M (2023) 3M™ ultimate FX full facepiece reusable respirator FF-400 series. Retrieved from https://www.3m.com/3M/en_US/p/d/v000057685/. Accessed Jan 2024
29. CDC (2023) The National Personal Protective Technology Laboratory (NPPTL) types of respiratory protection. Retrieved from https://www.cdc.gov/niosh/npptl/. Accessed Jan 2024
30. OSHA (2020) Respiratory protection. Retrieved from https://www.osha.gov/respiratory-protection. Accessed Jan 2024
31. OSHA (2020) Respiratory protection. Retrieved from https://www.osha.gov/laws-regs/regulations/standardnumber/1910/1910.134. Accessed Jan 2024
32. 3M (2024) Gas mask. Retrieved from https://www.3mindia.in/3M/en_IN/p/d/v000179827/. Accessed Mar 2024
33. Adanur S (1995) Wellington sears handbook of industrial textiles. In: Wellington sears handbook of industrial textiles. CRC Press, Florida. https://doi.org/10.1201/9780203733905
34. Maduna L, Patnaik A (2017) Textiles in air filtration. Text Prog 49(4):173–247. https://doi.org/10.1080/00405167.2018.1461921
35. 3M (2020) Respiratory protection for airborne exposures to biohazards. Technical Data Bulletin. Retrieved from https://multimedia.3m.com/mws/media/409903O/respiratory-protection-against-biohazards.pdf. Accessed Jan 2024
36. Wang CS, Otani Y (2013) Removal of nanoparticles from gas streams by fibrous filters: a review. Ind Eng Chem Res 52(1):5–17
37. Whyte HE, Montigaud Y, Audoux E, Verhoeven P, Prier A, Leclerc L, Sarry G, Laurent C, Coq LL (2022) Comparison of bacterial filtration efficiency vs. particle filtration efficiency to assess the performance of non-medical face masks. Sci Rep 12:1188
38. Kwong LH, Wilson R, Kumar S, Crider YS, Reyes Sanchez Y, Rempel D, Pillarisetti A (2021) Review of the breathability and filtration efficiency of common household materials for face masks. ACS Nano 15(4):5904–5924. https://doi.org/10.1021/acsnano.0c10146
39. Sathish T, Saravanan R, Sharma K, Zahmatkesh S, Muthukumar K, Panchal H (2023) A novel investigations on medical and non-medical mask performance with influence of marine waste microplastics (polypropylene). Mar Pollut Bull 192:115004. https://doi.org/10.1016/j.marpolbul.2023.115004
40. Hain R, Kahler CJ (2020) Fundamental protective mechanisms of face masks against droplet infections. J Aerosol Sci 118:105617
41. Chellamani KP, Veerasubramanian D, Vignesh Balaji RS (2013) Surgical face masks: manufacturing methods and classification. J Acad Ind Res 2(6):320–324
42. Freeman C, Burch R, Black C, Strawderman L, Rickert J, Wilson J, Saucier D, Smith B (2022) Filtration efficiency and differential pressure of fabrics used in non-medical masks based on SARS COVID-19 particle size. Text Res J 92(17–18):3234–3242. https://doi.org/10.1177/00405175211046056
43. Schmauder M (2015) Respiratory protection Occupational Safety and Health Administration (OSHA) construction. Arbeitsmedizin Sozialmedizin Umweltmedizin 50:33
44. Rodríguez B, Formentini G, Favi C (2021) Engineering design process of face masks based on circularity and life cycle assessment in the constraint of the COVID-19 pandemic. Sustainabilty 13:4948
45. OSHA (2024) Chapter 2—Types of respirators. Occupational Safety and Health Administration. Retrieved from https://www.osha.gov/otm/section-8-ppe/chapter-2#:~:text=Respiratorsareoftwogeneral,halfmask%2Candfullfacepiece. Accessed Jan 2024
46. M96 Cartridge_ADG (2024) Retrieved from https://www.adg.com/products/filters/m96-filter. Accessed Jan 2024
47. Kansas Department of Health and Environment (2020) Understanding ASTM levels for facemasks. Retrieved from https://www.coronavirus.kdheks.gov/DocumentCenter/View/1416/Understanding-ASTM-Levels-for-Facemasks-PDF---7-30-20. Accessed Jan 2024

48. Nimmy CJ, Ravi R, U AK, Venu D (2021) Life cycle assessment of cotton vs surgical face mask using open-LCA software. Int Res J Eng Technol 8(6):3196–3201
49. Atılgan Türkmen B (2022) Life cycle environmental impacts of disposable medical masks. Environ Sci Pollut Res 29:25496–25506. https://doi.org/10.1007/s11356-021-17430-5
50. Lee AWL, Neo ERK, Khoo ZY, Yeo Z, Tan YS, Chng S, Yan W, Lok BK, Low JSC (2021) Life cycle assessment of single-use surgical and embedded filtration layer (EFL) reusable face mask. Resour Conserv Recycl 170:105580. https://doi.org/10.1016/j.resconrec.2021.105580
51. Chau C, Paulillo A, Ho J, Bowen R, La Porta A, Lettieri P (2022) The environmental impacts of different mask options for healthcare settings in the UK. Sustain Prod tion 33:271–282. https://doi.org/10.1016/j.spc.2022.07.005
52. Maceno MMC, João S, Voltolini DR, Zattar IC (2023) Life cycle assessment and circularity evaluation of the non-medical masks in the Covid-19 pandemic: a Brazilian case. Environ Dev Sustain 25(8):8055–8082. https://doi.org/10.1007/s10668-022-02388-2

Masks and Respirators for Medical Applications

S. Subramanian and S. Vaidheeswaran

Abstract Aerosol is a fine suspension made of liquid or solid particles suspended in air with a size range usually below 100 μm. Airborne particles with a size less than 7 μm are considered hazardous as they are insoluble in body fluids. They reach the respiratory organ, lungs. Larger particles are of less concern as they usually settle out in the upper respiratory tract, from where normal lung clearance mechanisms remove them. Personal protection equipment against aerosols, such as respirators and masks are aimed at arresting particles of size smaller than 7 μm. Pathogenic microbial agents are viruses, bacteria, protozoa and fungi. This chapter describes the mechanism of filtration by fibrous media, the material and construction of face masks and N95 respirators, and tests to be carried out for respirators.

1 Introduction

Aerosol is a fine suspension made of liquid or solid particles suspended in air with a size range usually below 100 μm. Airborne particles have many different forms, such as dust, fume, mist, smoke, smog, fog, or pathogens agglomerated with liquid particles. Airborne particles with a size less than 7 μm are considered hazardous as they are insoluble in body fluids. They reach the respiratory organ, lungs. Larger particles are of less concern as they usually settle out in the upper respiratory tract, from where normal lung clearance mechanisms remove them. Personal protection equipment against aerosols, such as respirators and masks are aimed at arresting particles of size smaller than 7 μm. Pathogenic microbial agents are: (i) viruses (20–300 nm; rod, spherical shape; cause common cold, influenza, SARS, HIV, SARS-CoV-2 and its variants, (ii) bacteria (0.5–5 microns, few bacteria in mm; spherical, rod, spiral, single, chain, cluster shape; cause typhoid, diphtheria, tuberculosis), (iii)

S. Subramanian (✉)
Department of Textile Technology, A.C.Tech., Anna University, Chennai, India
e-mail: ssubbu.anna@gmail.com

S. Vaidheeswaran
Ordnance Clothing Factory, Ministry of Defence, Chennai, India

S. Sundaramoorthy et al. (eds.), *Textile Materials for Good Health and Wellbeing*, SDGs and Textiles, https://doi.org/10.1007/978-981-97-6496-9_5

protozoa (1 microns to several mm; cause amoebiasis, malaria) and (iv) fungi (2–10 microns in diameter and cm in length; cylindrical, thread-like structure; cause contamination of food, skin diseases). This chapter describes the mechanism of filtration by fibrous media, the material and construction of face masks and N95 respirators, and tests to be carried out for respirators.

2 Filtration by Fibrous Media

In the fibrous filter, the aerosol particles attach to a fiber due to a number of different mechanisms while they navigate through the filter media. The most common mechanisms are gravitational settling, inertial impaction, interception, diffusion, and electrostatic deposition [1–3].

Aerosol particle flow around the fibers inside the filter media and the particle capture by the fibers depends on particle size, electric charge, and face velocity. In the case of inertial impaction, the aerosol particle, along with the air, approaches the filter media. The particles follow a straight path, impact fiber, and are captured when they cannot follow a sudden change in the direction of air stream lines [4, 5]. In interception, the particle is captured when the distance between the center of the particle and the fiber is less than the particle radius. Particles with irregular diameters have a greater chance of being captured than those with a circular diameter, and the diameter of the filter fiber is an important factor in particle capture. The lower the density of particles, the better the particle capture will be. Brownian motion causes the particles to deviate from the air streamline, increasing the probability of contact with a filter fiber. If the time of a particle's close contact with fiber is longer, the chance of capture is higher. Low face velocity increases particle capture. Electrostatic attraction is an effective filtration mechanism that causes the attraction of very small aerosol particles by the fibers when they have opposite charges [6]. One side of the electrostatic filter has a positive charge, while the other side is negatively charged. As a result, the filter media attracts the charged particles. The stronger the charge, lower the face velocity, and smaller the particle size, the greater the chance for capture. In gravitational forces of attraction, particles above 5 μm are effectively trapped [4]. Lower the face velocity, better will be the capture. Particles between 1 and 5 μm have a high degree of inertial impaction and interception, and smaller particles are trapped by Brownian motion. The 0.3 μm particles are neither trapped by Brownian motion nor by impaction or interception. Hence, 0.3 μm particle size is referred to as the Most Penetrating Particle Size (MPPS). Hence, the filters are tested with aerosols of this MPPS to evaluate filtration efficiency [7–10]. The novel plasma textile filters offer improved filtration for submicron particles of diameters between 30 and 300 nm [11].

3 Masks and Respirators

3.1 Face Masks and Respirators

Facemasks are the devices worn on the face, mainly using ear bands to cover the nose and mouth area. The face masks are available in two types: non-reusable and reusable. In the case of reusable mask, it can be washed, sterilized (optiional), and reused. The difference between these two is mainly the material and structure of the material used.

The non-reusable masks are available with two-layered or three-layered fabrics based on the filtration efficiency required and the applications. Examples of face masks are pollution control masks, surgical masks, etc. Facemasks prevent the spread of droplets by the person wearing them. They are not designed to protect against breathing in small pathogens. The face masks are not tested for fit. There is also the possibility of breathing air through non-filter area.

The advantages of a reusable mask are: (i) less cost per usage, (ii) less contamination and pollution to the environment as it is reusable; (iii) confidence to the user, as the mask can be washed and used after purchasing from the retailer without the fear of possible contamination during production, packing, and selling, as found in the case of use and throw masks and (iv) comfortable in breathing; wearer can wear continuously without removing or touching the mask frequently as done in the case of N95 respirators.

Respirators are of two types: air-purifying respirators and atmosphere-supplying respirators. In the case of air-purifying respirator, the air is breathed from atmosphere and it uses filters, cartridges, or canisters to remove aerosols and contaminants from the air. This respirator is designed to protect the wearer from breathing in very small particles, which might contain pathogens. The respirators fit tightly to the face so that most of the air is inhaled through the filter material, unlike face masks, where breathing through non-filter area is higher. An example of this type of respirator is the N95 used for medical applications. In the N95 respirator, N stands for 'Not resistant to Oil', and it is used for the filtration of aerosols; 95 stands for filtration efficiency (https://www.cdc.gov/niosh/docs/96-101/default.html). The atmosphere-supplying respirator provides clean breathing air from a separate source. These respirators protect the wearer from many types of airborne contaminants and are used in oxygen-deficient atmospheres.

3.2 Materials for Masks and Respirators

The materials used for the manufacture of wash-and-use type masks are cotton and viscose rayon, whereas the materials used for non-reusable masks and respirators are mainly polypropylene. The functional requirements of fibers used in the respirators are: fibers that can be spun into the diameter of micrometers and are chargeable to impart electrostatic attraction.

3.3 Structure of Fabric Used for Masks and Respirators

Different structures are used in textile product development: woven, knitted fabrics, nonwovens, and electrospun mats. Woven and knitted fabrics are mainly used for apparel and other medical textile applications like wound dressings, bandages, hospital gowns, bed linens, caps, gloves, etc.

Woven and knitted fabric are used for the development of reusable masks, making them convenient to wash and reuse, which is practically impossible for nonwoven masks available in the market. Since the nonwoven masks are of the single-use type and are made up of non-biodegradable materials, they create environmental issues, and the cost of usage is also high. Cloth masks have relatively low filtration efficiency and limited protective effects and are generally recommended for protection against community transmission [12]. But getting higher filtration efficiency is practically difficult.

Nonwoven fabrics are described as a random fibrous web formed by means of air-laid, wet-laid, or mechanical means and bonded by mechanical, thermal, or chemical means. The advantages of nonwoven filter media are higher filtration efficiency, lower cost, and lower pressure drop. Nonwoven fabrics are extensively used for masks and respirators because of their multilayered fiber structure. The spun-bonded nonwovens used for masks and respirators have weights in the range of 25–40 g per square metre (GSM) and thickness of about 0.1–0.3 mm. The melt-blown nonwovens have the areal density of about 30–50 GSM and thickness of 0.4–0.6 mm. The filtration efficiency of melt-blown nonwoven is higher than that of spun-bonded nonwoven, but the breathability is lower.

A material designed for higher filtration efficiency reduces its breathability, and vice versa. The number of layers and their thickness are optimized to have the required filtration efficiency with sufficient breathability.

4 Construction of Masks and Respirators

4.1 Masks

The reusable mask is made from woven or knitted fabric of different layers to optimize filtration efficiency and breathability. They are stitched at the sides, and ear loops are attached at the sides, as shown in Fig. 1a. The non-reusable masks are made from spun-bonded nonwovens of two to three layers, with or without melt-blown

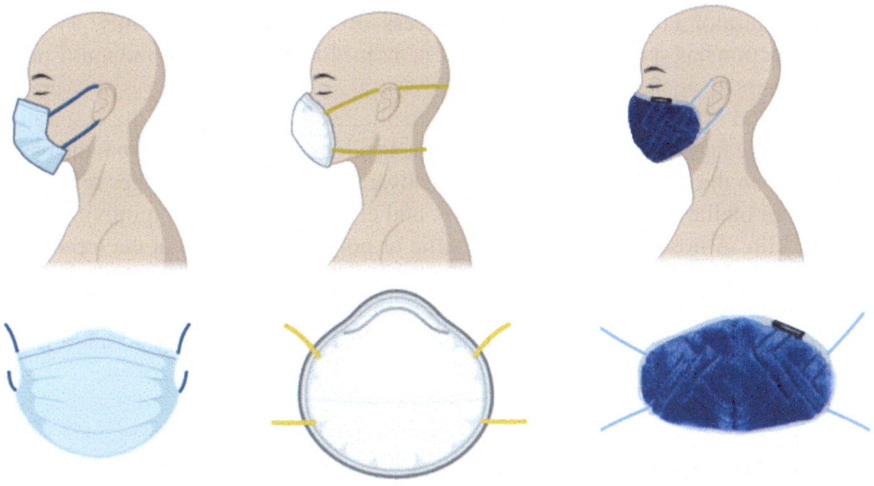

(a) (b) (c)

Fig. 1 **a** Non-reusable mask **b** N95 respirator, **c** reusable mask.

nonwoven. The respirators are made from a combination of spun-bonded and melt-blow nonwovens of different numbers of layers in different sequences. The number of layers of spun-bonded nonwoven and melt-blown nonwoven is decided based on the requirements of filtration efficiency, breathability, and application. The selection of face masks by the user should be based on application, size, and material.

The structure of surgical mask described as SMS, consists of spun-bonded (the outermost layer), melt-blown (the middle layer), and spun-bonded (the inner layer) layers, wherein the melt-blown layer is the main filter layer, and the spun-bonded layers are the filters and waterproof layers [13]. Face masks are available in different combinations of fabrics, such as SSS and SS, without melt-blown nonwoven. The layers of fabrics are bonded or molded together. The medical face mask shall not disintegrate, split, or tear during intended use. In the selection of the filter and layer materials, attention shall be paid to cleanliness. These masks may or may not have nose clips, which are used to fit the mask to the nose closely and prevent a gap between the nose and chin. It may have a face shield (to protect the wearer against splashes and droplets) with or without anti-fog function [14].

4.2 *Respirators*

Respirators are made in different forms by different manufacturers. The main components of a respirator are:

1. The respirator consists of different layers of nonwovens and films. Typical construction has spun-bonded nonwoven, melt-blown nonwoven, and multi-pore film layers.
2. Head band or ear band: The headband facilitates the gripping of the respirator against the face by wearing the band on the head. In the case of ear band, the band is worn on the ear. They have their own advantages and disadvantages related to pressure on the ear, disturbance of hair, and ease of wearing and removing.
3. Nose clip: A nose clip is a small metallic or polymer strip kept at the top of the respirator. It is pressed against the nose to fit tight and prevent gap between the nose and chin so that air is not breathed in through the non-filter area.

   Figure 1 (a) shows the non-reusable surgical mask with ear band and Fig. 1 (b) N95 respirator with headband.

5 Tests for Masks and Respirators

NIOSH 42 CFR 84 prescribes the tests needed for breathing system filters and should include a rating system such as N95, N99, or N100.

5.1 Particulate Filtration Efficiency

Charge-neutralized particles at the most penetrating particle size (0.3 μm) have become the standard particle for respirator testing and certification, according to NIOSH 42CFR Part 84. Sodium chloride solution is aerosolized and evaporated to the MPPS of 0.3 μm. Then they pass through a charge neutralizer to neutralize the charge of the aerosols. The number of sodium chloride particles both downstream and upstream of the test filter is calculated, and filtration efficiency can be calculated from the below equation.

$$\text{Filtration efficiency} = \frac{(\text{No. of particles entering filter} - \text{No. of particles exiting the filter}) \times 100}{\text{No. of particles entering filter}}$$

The particles higher or lower than the MPPS (0.3 μm) are trapped more effectively, so the filtration checked at the MPPS gives the minimum efficiency of the filter. Most of the filters show minimum filtering efficiency at this particle size. Hence, respirators with N95 filter media are tested for particles with a mass median aerodynamic diameter of 0.3 microns.

5.2 *The Bacterial and Viral Filtration Efficiency Tests*

The tests for the respirators include: (i) BFE (bacterial filtration efficiency): efficiency to filter out bacteria when challenged with a bacteria-containing aerosol. ASTM (F2101-07, F2107-19) specifies testing with a droplet size of 3.0 microns containing *S. aureus* (average size 0.6–0.8 microns). Non-pathogenic microorganisms similar to M tuberculosis (0.6 × 1.3 μm) or similar to viral pathogens (0.03 μm) are cultured and diluted in a solution to a known concentration. Then the diluted solution is aerosolized and subjected to filtration efficiency. The penetration of microbes is evaluated by calculating colonies formed in the collection medium. The size of test droplets is generally around 3.0 μm. So they are easy to filter by inertial impaction [10].

5.3 *Breathability Test*

Differential pressure is a measure of the air flow resistance of the mask or respirator, and it is an objective measure of breathability. This is measured by (i) the difference in pressure of air before and after the mask or (ii) the air flow at a particular pressure difference across the filter media. This can be measured by air/gas permeability tester.

Bacterial filtration efficiency, breathability, splash resistance (ISO 22609:2004), microbial cleanliness (EN ISO 11737-1:2018), and biocompatibility tests are prescribed for testing the performance requirements of surgical masks (European Standards [14]).

There is a challenge in optimization between Filtration Efficiency (FE) and breathability. When the facemask is aimed at a higher FE, the breathability reduces. A facemask or respirator with a higher FE and low breathability makes the wearer uncomfortable while breathing. The wearer may remove the mask from the nose area to breathe comfortably, which defeats the very purpose of wearing a facemask. The facemask with low FE is not preferable for protection.

5.4 *Recent Research on Development*

Corona (Covid 19) is a SARS-CoV-2 (2019) + RNA virus with a profile of spikes on the sphere. They cause the common cold, to SARS. They have an average diameter of 125 nm (85 + 20 envelope + spikes). As per the research at this stage, there is never a naked virus floating in the air or being released by people, and the virus attaches to water droplets or aerosols (i.e., really small droplets) that are generated by breathing, talking, coughing, etc. These consist of water, mucus, protein, and other biological material and are all larger than 1 micron. It is also revealed that droplets of size greater than 5 microns are the primary sources of infection; however, droplets of size less than 1 micron stay as aerosols in the air for 3–8 h. Though N95 respiratore is suitable for better protection, there was a scarcity of respirators during pandemic conditions, and the entire public in common places could not wear this respirator because of discomfort due to breathability.

The European Commission Directorate-General for Health and Food Safety: Health Systems, Medical Products, and Innovation Medical Devices [17] have issued guidance on regulatory requirements for medical face masks. During COVID-19, many types of masks were produced, starting from simple woven fabrics to nonwovens of different layers and coatings. A comparison of the air permeability and filtration efficiency of different masks and respirators is given in Table 1.

It can be noted from the table that the filtration efficiency at 0.3 mcirons is above 95% in the N95 respirator only; however, the breathability is low. In the case of 3 microns, a higher FE could be achievable with a reusable mask as well. The reusable mask with nonwoven loses its FE after washing due to damage caused to the nonwoven structure. Rubio-Romero et al. [15] stated that washing with soapy water also affects the electrostatic charges.

Wang et al. [16] suggested the direction for the development of masks for the future, i.e., reducing pollution to the environment, addressing the problem of tightness and fitting masks to different face shapes, ensuring thermal comfort when wearing masks, and developing materials with higher filtration efficiency and breathability.

Table 1 Air permeability and filtration efficiency of face masks and respirators

Mask type	Air permeability[a]	Filtration efficiency[b]	
	$cm^3/s/cm^2$	0.3 microns	3 microns
Reusable mask—three layered tribo charged warp pile fabric	13.3	41.05	97.65
Reusable mask—three layered, cotton-melt-blown nonwoven-cotton	4.39	36.50	98.37
N95 respirator	6.56	97.17	99.50
Non-reusable three ply surgical mask—SMS[c]	> 23	13.64	27.46
Reusable cotton 2 layers	> 23	6.05	39.42
Reusable cotton two layer + (Spun bond + 2 Melt Blown + Spun bond)	10.8	43.63	98.61

[a]Measured at pressure of 1 cm water; higher air permeability indicates easier breathing
[b]Measured at normal environment; higher FE indicates higher prevention of aerosols
[c]Spun-bonded, melt-blown, and spun-bonded nonwoven layers

6 Conclusions

In this chapter, the requirements for the filtration of aerosols and the filtration mechanisms of fibrous media are discussed. The difference between face masks and respirators, materials used, construction details and tests to be carried out are presented.

References

1. Hinds WC (1982) Aerosol technology: properties, behavior and measurement of airborne particles. Wiley, New York, pp 164–186
2. Lee KW, Liu BYH (1982) Theoretical study of aerosol filtration by fibrous filters. Aerosol Sci Technol 1(2):147–161. https://doi.org/10.1080/02786828208958584
3. Lee KW, Liu BYH (1980) On minimum efficiency and the most penetrating particle size for fibrous filters. J Air Pollut Cont Assoc 30(4):377–381
4. Maduna L, Patnaik A (2017) Textiles in air filtration. Text Prog 49:173–247. https://doi.org/10.1080/00405167.2018.1461921
5. Purchas DB, Sutherland K (2002) Handbook of filter media, 2nd edn. Elsevier, Oxford, pp 1–34
6. Adanur S, Jayswal A (2022) 2024 Filtration mechanisms and manufacturing methods of face masks: an overview. J Ind Text 51(Suppl. S3):3683S-3717S. https://doi.org/10.1177/1528083720980169
7. Brown RC (1993) Air filtration: an integrated approach to the theory and applications of fibrous filters. Pergamon Press, London
8. Ikkruvinaska (2005) Textiles for respiratory protection. In: Scott RA (ed) Textiles for protection. Woodhead Publishing Limited, London, pp 473–378
9. Kothari VK, Das A, Singh S (2007) Filtration behavior of woven and nonwoven fabrics. Indian J Fiber Text Res 32:214–220
10. Thiessen RJ (2006) Filtration of respired gases: theoreitical aspects. Respir Care Clin N Am 12:183–201
11. Jasper WJ, Rasipuram SC (2017) Plasma textiles as fibrous filter media. In: Brown PJ, Cox CL (eds) Fibrous filter media. Woodhead Publishing, London, pp 191–210
12. Zhao M, Liao L, Xiao W, Yu X, Wang H, Wang Q, Lin YL, Kilinc-Balci FS, Price A, Chu L (2020) Household materials selection for homemade cloth face coverings and their filtration efficiency enhancement with triboelectric charging. Nano Lett 20:5544–5552. https://doi.org/10.1021/acs.nanolett.0c02211
13. Yao B, Wang Y, Ye X, Zhang F, Peng Y (2019) Impact of structural features on dynamic breathing resistance of healthcare face mask. Sci Total Environ 689:743–753. https://doi.org/10.1016/j.scitotenv.2019.06.463
14. Belgian Standard (2019) EN 14683:2019; AC:2019 NBN Medical face masks—requirements and test methods
15. Rubio-Romero JC, Pardo-Ferreira MDC, Torrecilla-García JA, Calero-Castro S (2020) Disposable masks: Disinfection and sterilization for reuse, and non-certified manufacturing, in the face of shortages during the COVID-19 pandemic. Saf Sci 129:104830. https://doi.org/10.1016/j.ssci.2020.104830
16. Wang AB, Zhang X, Gao LJ, Zhang T, Xu HJ, Bi YJ (2023) A review of filtration performance of protective masks. Int J Environ Res Public Health. https://doi.org/10.3390/ijerph20032346
17. European Commission Directorate-General for Health and Food Safety: Health systems, medical products and innovation medical devices

Textile Materials for Air-Purifiers and Cleanroom

N. K. Palaniswamy, Suresh Kumar, and N. Gobi

Abstract This chapter sets the context for the development of textile materials for air purification and cleanroom applications. The first part discusses the effect of various air pollutants on the human body and the basic operation of air purifiers. The second part summarizes the cleanroom technology including a basic introduction, classification, and working of various cleanroom components. The third part reviews the design and development of textile materials for air purification and cleanroom applications. Factors affecting the performance of an air filter, designing of air filters for maximum performance, filters used in cleanroom technology, and the latest research findings are discussed.

1 Introduction

People's health, comfort, and productivity at the workplace are highly affected by indoor air quality, which is concerned primarily with the presence of pollutants detrimental to human health. Major air pollutants are suspended particulate matter, microorganisms, and toxic gasses including volatile organic compounds. These pollutants should be filtered adequately because inhaling of polluted air can cause discomfort, irritation, and headache and long-term exposure can affect the working of vital internal organs such as the lungs and heart [1–3]. Indoor air quality and a pleasant working environment are ensured by using air purifiers and HVAC systems (Heating Ventilation, and Air conditioning) in workplaces such as offices, shops, and residential buildings. A controlled environment is created by ensuring the adequate filtration of the incoming air and quick removal of the air-suspended particles and byproducts generated during the manufacturing process.

N. K. Palaniswamy (✉) · S. Kumar
Department of Textile Technology, Dr. B. R. Ambedkar National Institute of Technology, Jalandhar, India
e-mail: nkpalaniswamy@nitj.ac.in

N. Gobi
Department of Textile Technology, A.C.Tech., Anna University, Chennai, India

S. Sundaramoorthy et al. (eds.), *Textile Materials for Good Health and Wellbeing*, SDGs and Textiles, https://doi.org/10.1007/978-981-97-6496-9_6

In the case of biological laboratories, cleanroom technology ensures sterilization for the prevention of diseases caused by microorganisms [4, 5].

Textile materials are integral parts of air purifiers and cleanrooms due to their unique structural features. Textile fibers arranged in a specific structure are used as filtration media in most modern-day filters [6]. Advanced textile materials like micro-fibers and nanofibers and functional material-embedded fiber structures give very high filtration efficiency for particulate matter and microorganisms and are used in HEPA filters.

2 Air Purification

The concentration of the pollutants in the air determines its purity and safety for living organisms. Major air pollutants in atmospheric air include particulate matter (PM2.5 and PM10), ground-level ozone (O_3), Nitrogen dioxide (NO_2), Sulfur dioxide (SO_2) and carbon monoxide (CO). Their major sources are smoke produced by the combustion of normal and fossil fuels (Source of particulate matter, NO_2, SO_2), reaction of smoke and sunlight (O_3), and incomplete combustion of carbon containing fuels (CO) [7]. The Air Quality Index (AQI) is a standardized tool for communicating air pollution levels and their potential impact on the health of the public. AQI is based on the concentration of a few major pollutants present in the air and its value lies in the range of 0–500. The AQI value between 0 and 50 is considered good for health, and up to 100 is also acceptable with moderate health risks. AQI 100–150 is considered unhealthy for children and elder people and 150–200 for everyone. AQI range of 301–500 is classified as hazardous. Poor air quality affects health, comfort, and productivity of the people. Major health issues caused by poor air quality are respiratory problems (asthma, bronchitis, and allergies), cardiovascular problems (heart attack, high blood pressure, stroke, inflammation, and damage of blood vessels), and lung cancer due to long-term exposure to benzene and formaldehyde.

The comfort of the indoor environment mainly depends on the air quality. Deterioration in air quality can cause odors, stuffiness, and discomfort. Poor air quality can interfere with the temperature distribution within a space due to the greenhouse effect and hence can cause local places with high or low temperatures causing discomfort. Poor air quality can decrease human productivity by impairing cognitive functions like decision-making, and problem-solving. Health issues caused by poor air quality can promote issues like absenteeism, and decrease in productivity due to minor illness [8] Hence, it is very important to ensure a continuous supply of good quality air for maintaining people's health, comfort, and productivity regardless of where they live, work, or play indoors or outdoors. The best way to do so is sustainable development with controlled pollution and air purification by using suitable filtration media.

The process of air purification involves the removal of contaminants and polluting substances from the air to improve air quality. Air purification is very important in the case of indoor settings, where contaminants can accumulate and have a detrimental effect on human health. Air purification is a three-stage process in which

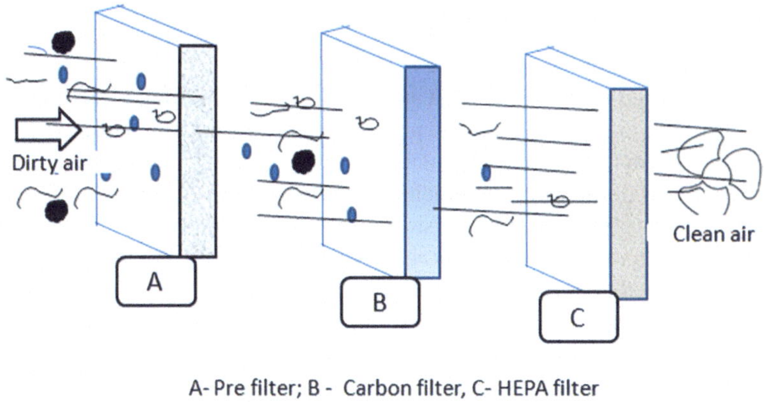

Fig. 1 Basic elements of an air purifier

coarse contaminants are filtered out in the initial pre-filtration stage. Volatile organic compounds and odors are absorbed in the intermediate stage and the remaining fine particles are filtered out in the final stage using a high-efficiency filter.

Important elements of an air purifier are shown in Fig. 1. Air purifiers can be designed according to specific requirements based on the size, type, and concentration of contaminants present in indoor settings [9].

3 Cleanroom

3.1 Introduction to Cleanroom

As per ISO standard 14644-1, a cleanroom represents a room in which the concentration of airborne particles is controlled and is constructed and used in such a way as to minimize the introduction, generation, and retention of airborne particles inside the room and in which other relevant parameters, e.g., pressure, temperature, and humidity are controlled as necessary. A clean room is an artificial environment, designed for maintaining clean and sterile conditions by controlling the pollutants, airborne particles, and contaminants to the minimum possible extent. This is done by supplying clean filtered air from ceiling vents to dilute the concentration of airborne particles present in the inside air of the cleanroom and then changing it by using exhaust suction at the bottom of the floor. Cleanrooms are used in industries like drug manufacturing, hardware fabricating, biotechnology, aviation, and medical services [4].

3.2 Importance of Cleanrooms

Cleanrooms are used by industries, that manufacture products having very high sensitivity toward foreign particles, and the presence of even very tiny particles or toxins can cause defects in the product. Regulations for this type of product manufacturing are very strict and compliance can be achieved only by using cleanrooms. The importance of the cleanroom can be understood from the examples of various industry types. In the food industry, food items are susceptible to the pollutants present in the outside air, and food contaminated by these pollutants can cause serious food-borne diseases or even death of the consumer, which is not acceptable at all and can result in the cancelation of the license of the manufacturing unit. Similarly in a biological research laboratory, research results may be affected badly by the presence of pollutants and harmful microorganisms like bacteria or viruses. In the pharmaceutical industry, the presence of harmful chemical pollutants or even particulate matter can change the composition of the medicine, which can cause some serious harm to the health of the patients. Similarly, in the semiconductor manufacturing industry, a small particle remaining in the circuit can make the entire circuit useless. In all the industries mentioned above, cleanrooms are used for maintaining the desired conditions which shows that cleanrooms are very important for precision manufacturing [10].

3.3 Various Elements and Operations in Cleanroom

Key elements of a cleanroom include a HEPA Filter, ceiling vent, HVAC system, and exhaust system. The basic cleanroom layout is shown in Fig. 2. While working in the cleanroom, outside air is sucked in through a HEPA filter by using a suction fan and pollutants are filtered. Filtered air is then directed to the clean room through a ceiling vent. By supplying air from the outside, the concentration of the particles (Airborne Particles per unit volume of air) decreases. A suction-based exhaust system is provided at the bottom of the cleanroom for the removal of polluted air with airborne particles. In this way, pollutants suspended in the air are removed from the cleanroom [4].

Cleanroom working is a complex process and fundamental operations required in the cleanroom are air quality management, temperature and airflow control, pressure differential maintenance. Air quality is managed by using advanced filtration and ventilation systems. HEPA (High-Efficiency Particulate Air Filter) and ULPA (Ultra Low Penetration Air filter) are examples of advanced filters. HVAC system (Heating, Ventilation, and Air Conditioning system) having precise temperature control is used to ensure the required relative humidity (RH) and temperature. Air flow control and regulation is an important operation as it affects the movement of particulate matter. Hence, the airflow pattern is designed according to the requirement of the cleanroom. To reduce particle deposition on the surfaces, laminar airflow is commonly applied

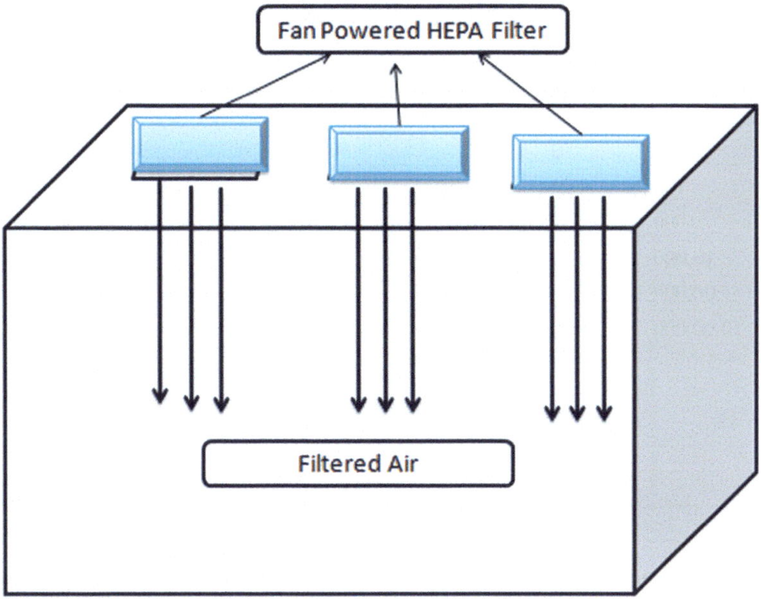

Fig. 2 Cleanroom layout

in which air flows through a unidirectional path. Pressure differentials (positive or negative air pressure levels) are maintained relative to adjacent spaces to prevent the entry of contaminants from outside space and block harmful substances inside the controlled environment [4, 11].

3.4 Classification of Clean Room

Cleanrooms can be classified based on various cleanliness standards, methods of ventilation, and application areas.

3.4.1 Classification Based on Various Standards.

The standardization process of cleanrooms started in the USA in 1963 with US Federal Standard F.S. 209 which was further upgraded to F.S. 209A (1966), F.S. 209B (1973), revised F.S. 209B (1976), F.S. 209C (1987), F.S. 209D (1988) and F.S. 209E (1992). Cleanrooms are mainly classified according to their cleanliness level, which is assessed by the maximum allowable number of airborne particles and other pollutants per unit volume of air. US Federal standard 209E was replaced by ISO 14644-1 in 1999. Cleanroom classification based on ISO 14644-1 and equivalent classes as per US Federal standard 209 E are in Table 1. Other major standards include

Table 1 Classification of cleanroom based on ISO 14644-1 and FED STD 209E

Class	Maximum particles/m^3						FED STD 209 equivalent
	≥ 0.1 μm	≥ 0.2 μm	≥ 0.3 μm	≥ 0.5 μm	≥ 1 μm	≥ 5 μm	
ISO 1	10	2					
ISO 2	100	24	10	4			
ISO 3	1000	237	102	35	8		Class 1
ISO 4	10,000	2370	1020	352	83		Class 10
ISO 5	1,00,000	23,700	10,200	3520	832	29	Class 100
ISO 6	10,00,000	2,37,000	1,02,000	35,200	8320	293	Class 1000
ISO 7				3,52,000	83,200	2930	Class 10,000
ISO 8				35,20,000	8,32,000	29,300	Class 100,000
ISO 9				3,52,00,000	83,20,000	2,93,000	Room air

the Japanese standard JIS B-9920, German Standard VDI 2083, British Standard: BS 5295, Australian Standard AS 1386, French Standard AFNOR X 44101, Dutch Standard: VCCN-RL-1, Russian Standard: GOST R 50766-95.

3.4.2 Classification Based on Method of Ventilation

Based on the type of ventilation, cleanrooms are classed as turbulently ventilated or unidirectional flow clean rooms. In turbulent-flow cleanrooms, fresh air is filtered by using a HEPA filter at the entrance and after that filtered air enters through vents in the walls or ceiling, travels randomly in all directions, and finally exits through exhausts on the other side. This causes the mixing of fresh air taken in with inside room air having particulate matter or pollutants and diluting of pollutants. These cleanrooms are suitable for processes that do not require extremely low contamination levels. Fig. 3 represents the airflow system within a turbulent-flow cleanroom. These cleanrooms are used in offices, shopping complexes, and a few industries [4].

Unidirectional flow (UDF) cleanrooms are advanced cleanrooms with a very high level of cleanliness and are also called laminar flow cleanrooms due to the laminar flow of air. The airflow system of a UDF cleanroom is shown in Fig. 4. The speed of the air inside UDF is high due to the directed air movement which causes higher air changes and hence higher air consumption. HEPA filters are installed at the entrance point i.e., at the ceiling in a localized manner for air filtration and a unidirectional flow of air blows away the contaminants at a faster rate. These cleanrooms are used in processes where ultra-clean environments are required. Semiconductor fabrication, medical device manufacturing, and aerospace industries are a few examples. Designing, fabrication, and operating costs of UDF cleanrooms are higher as compared to turbulent-flow cleanrooms [4].

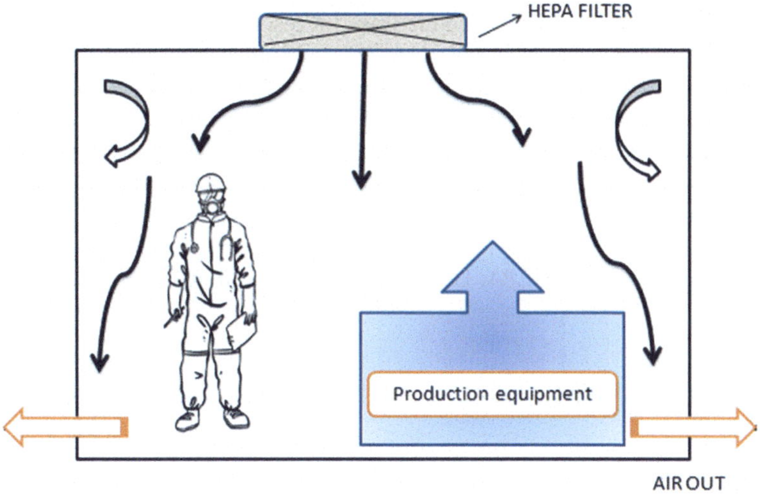

Fig. 3 Turbulent ventilated cleanroom

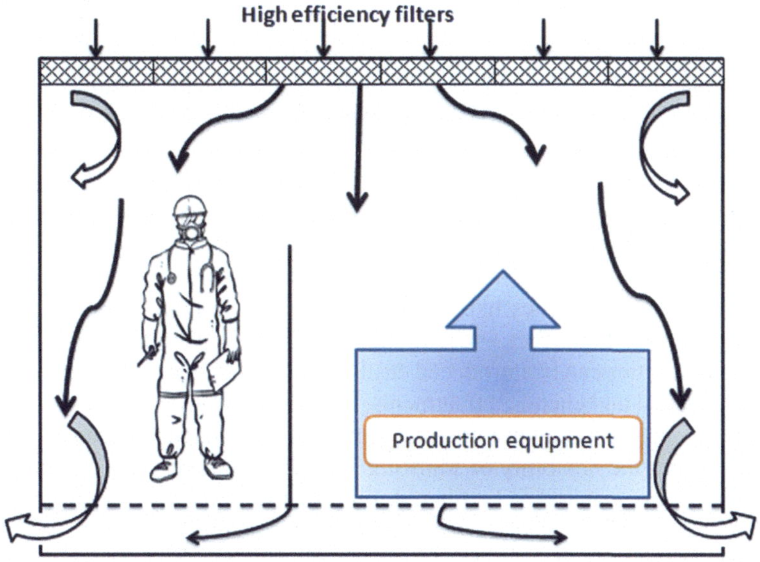

Fig. 4 Unidirectional flow cleanroom

3.4.3 Classification Based on the Area of Application

Based on the application area, cleanrooms can be classified into industrial cleanrooms and biological cleanrooms. The major objective of industrial cleanroom is to control the airborne contaminations in industrial manufacturing processes, whereas biological cleanroom are specialized facilities for research, development, and manufacturing of biopharmaceuticals, cell treatments, and other active chemicals. In industrial cleanroom, major focus is given on controlling particulate contaminations like dust, particles, and airborne microbes, whereas in biological cleanroom, along with controlling particulate contamination, maintaining sterility, and preventing microbial contamination are equally important. In industrial cleanrooms, turbulent and laminar flow systems are used based on the type of industry, and HEPA and ULPA filters are used for filtration. In biological cleanrooms, laminar flow systems are used and filtration is done by using HEPA, ULPA, and sterilizing filters [9, 10, 12, 13].

4 Role of Textile Material in Air Filtration and Cleanroom

4.1 General Overview

Clean room technology is used in specific industrial settings with no margin for pollutants beyond certain permissible limits. Textile materials are used for making clean room filters and clean room garments. Filters purify the external air before entering the cleanroom and special garments worn by workers prevent the transfer of the contaminations from the human body to the clean room environment. Filtration is a process of separating undesired components of a mixture/solution/suspension by using a suitable filtration media. Separation can be solid–solid, solid–liquid, solid–gas, liquid–liquid, liquid–gas, or gas–gas. Filter media has permeability to only one or two selected components and the remaining are separated. Fibrous textile materials are used as filtration media due to a unique combination of properties like large surface area, interstices, minimum required strength and elongation, and high elasticity. Addition properties like flame retardancy, high melting point, chemical resistance, antimicrobial effect, and hydrophilicity or hydrophobicity, adsorption and absorption properties can be introduced easily by using high-performance fibers and giving mechanical and chemical treatments. Few textile materials have electrostatic properties, so they can capture charged particles. The major advantage of using textile material is the flexibility of adjusting interstices or pore size according to the particle size of the material to be filtered. Fibrous materials can be used in various structural forms i.e., fibrous web, woven and knitted structures, nonwoven and fibrous membranes for making filter media. Fibrous web and nonwoven materials are very popular in the case of high-efficiency filters because they provide a large surface area with a combination of small interstice space and tortuous path. Fibrous membranes are made from very fine fibers (micro and nanofibers) having desired mechanical properties, and thermal and chemical resistance [14, 15].

Air purifiers and cleanroom technology have one common objective of maintaining good indoor air quality and gas filters are used in both. Filtration technology used in these two applications can be classified as capture type (mechanical capturing of pollutants) and reactive type (ionization or chemical reactions to remove pollutants) [2]. The interaction of particles with single fibers and the structural arrangement of fibers are the main factors responsible for the arresting of particles during filtration. However, the combination of some other factors like the mode of filtration, filter unit design, and operating parameters also affect particle arresting. In the case of cleanroom and air purifiers, the solid–gas filtration system is used, in which the interaction of particles with single fibers plays a major role in particle capturing. Fibers are cylindrical objects and particle capturing on single fibers is based on various mechanisms depending on the type and size of the particles as shown in Fig. 5. Particles below 0.3 μm are captured by diffusion and Brownian motion mechanisms. Relatively larger particles are captured by direct interception and inertial impaction depending on their momentum and inertia. The combined effect of high momentum and inertia of large-size particles deviates the path of the particles from the air streamline and causes the inertial impaction of particles on the fiber surface. In case of insufficient inertia, particles move along with the air streamline very close to the fiber and get intercepted. Gravitational settling takes place in the case of heaviest particles such as dust and no major role of fiber is there in this case. Electrostatic attraction is used for capturing electrostatically charged particles (0.1–1.0 μm) on oppositely charged fiber surfaces. The electrostatic charge may be present inherently on the particles based on their nature or can be introduced by using the electric field. Along with single-fiber interaction, the role of fiber assembly is equally important as it decides the size and shape of the interstices. So, it is the fiber assembly that decides the place of particle capturing i.e., at the surface (surface filtration) or inside the filtration media (depth filtration) as shown in Fig. 5. Particles captured at the surface form the cake (layers of captured particles) which can be cleaned after a particular period but in the case of depth filtration periodic cleaning is not possible and filters are replaced on clogging. Surface filtration is used mainly for air processing and de-dusting in the case of industrial filters. A highly porous structure with fine pore size and low thickness is used in surface filtration and filter cakes formed are cleaned for re-use of the filters.

Depth filtration is used for HVAC systems and cleanroom technology where very fine particles also need to be captured. Filters used have highly porous structures with low density and progressive inward direction pore size. Depth staining and depth retaining mechanisms are used for capturing particles in depth filtration [2, 14].

4.2 Classification of Air Filters

Air filters are classified based on so many factors like material used, filter design, mechanism of filtration, etc. However, international standards available are based on the final performance of the filter in terms of filtration efficiency, which is tested under standard pre-defined conditions. Classification is provided in European standard EN799:2012 [Coarse particle filters (G1–G4), Medium and fine particle filters

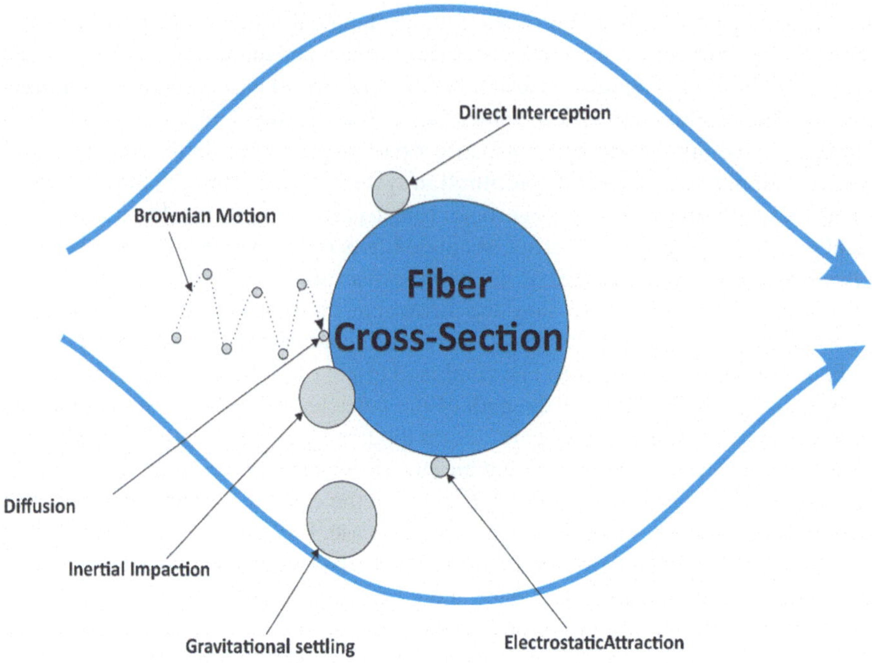

Fig. 5 Particle capture mechanism in case of single-fiber interaction

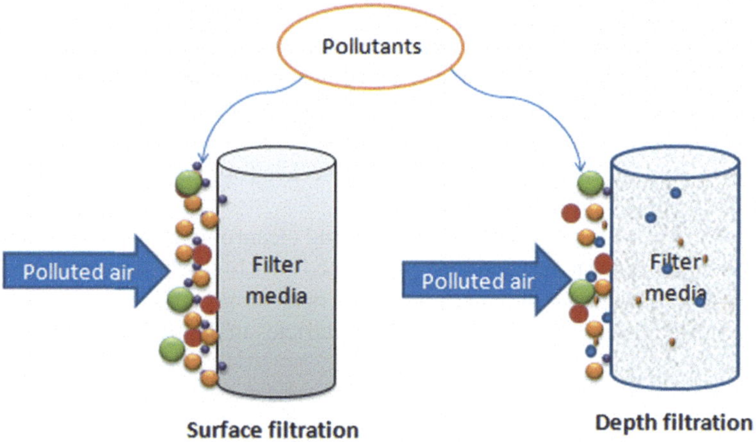

Fig. 6 Surface filtration and depth filtration

Table 2 Classification of filters (Coarse, medium, and fine) based on EN 799 standard

Filter class	Average arrestance (Am) of ASHRAE test dust (%)	Average efficiency (Em) of 0.4 μm particles (%)	Minimum efficiency of 0.4 μm particles (%)	Final test pressure drop (Pa)
G1	50 ≤ Am < 65			250
G2	65 ≤ Am < 80			250
G3	80 ≤ Am < 90			250
G4	90 ≤ Am			250
M5		40 ≤ Em < 60		450
M6		60 ≤ Em < 80		450
F7		80 ≤ Em < 90	35	450
F8		90 ≤ Em < 95	55	450
F9		95 ≤ Em	70	450

Table 3 Classification of filters (EPA, HEPA, ULPA) based on EN 1822 standard

Filter type	Filter class	Filtration efficiency
EPA (efficient particulate air filter)	E10	> 85
	E11	> 95
	E12	> 99.5
HEPA (high-efficiency particulate air filter)	H13	> 99.95
	H14	> 99.995
ULPA (ultra low penetration air filter)	U15	> 99.99955
	U16	> 99.99995
	U17	> 99.999995

(M1–M2 and F7–F9)] and EN 1822 (EPA/HEPA filters—E10–H14, ULPA filters-U15–U17) [14, 16]. Classifications of air filters based on standards EN 799 and EN 1822 are given in Table 2 and Table 3 respectively.

4.3 *Designing and Development of Cleanroom Filters*

There are some important parameters like flow rate, working pressure, etc., which must be considered by the designer during designing of the air filter. Pore sizing and internal layers of the filtration system are designed according to the required flow rate and suspended solid load. Material of the filtration system is selected according to the working temperature and velocity of the air passing through the filter. Velocity of air, decides the impact force exerted by the solid-suspended particles on filter media which can damage the internal structure of the filter. Pressure-drop should not be less than 20% of the total external static pressure. Thickness of the filtration system is decided according to the working pressure. In addition to the factors described above, the following significant guidelines may be used to design air filters efficiently [2, 16].

- The use of prefilters in combination with high-efficiency filters can increase the service life of the filters.
- The filter system should provide enough space for hassle-free servicing of the filters.
- In case of the external air intake, bird screens and weather louvers should be included.
- To identify when a filter needs to be serviced, a differential pressure drop gage should be installed throughout the system.
- The filtration system should be designed smartly by considering future needs along with present requirements. Though cost is a very important factor, priority should be given for the filter performance.

4.3.1 Construction of HEPA and ULPA Filters

HEPA and ULPA filters have folded or pleated filter media so that the maximum area of filter media can be offered to the air stream passing through the filter. Spacer construction is the most important part of the design as it affects the configuration of airflow inside the filter and the ratio of the effective filter area and gross filter area these two factors further affect the pressure drop of the HEPA or ULPA filters at a given filtration efficiency. Spacer construction techniques (filter media folding) used for HEPA and ULPA filters are the separator technique, mini-pleat, cold-forming, and thermal embossing. In the separator technique, corrugated aluminum foil is used as a separator between the layers of the filter media (superfine fiberglass). This technique has the disadvantage of slightly higher pressure drop as compared to others as folding decreases the effective filtering area. Secondly, sharp-edge separators need careful handling during installation as they can puncture the filter medium. In the mini-pleat technique, filter media (superfine fiberglass) is folded to avoid the use of metallic separators. Threads having adhesive are attached to the filter media before pleating, and after folding these threads get attached and the filter media remains separated. In this method, the metallic separators are eliminated but maximum pleat depth remains restricted up to 130 mm. These pleats are arranged in a V-shaped configuration to overcome the problem of low pleated depth. The pressure drop is lower in this case. V-configuration pleated filters are suitable for turbulent-flow conditions only and have the problem of double deflection of air stream in case of laminar flow conditions. In the cold-forming technique, wet un-bonded filter paper with convex dimples is folded into pleats and the dimples touch each other and keep the filter paper separated. After fixing the package is dried. This technique has two disadvantages, firstly the filter paper production and fixation process cannot be separated and rubbing of the dimples during actual operation can cause passing of the glass fibers into the cleanroom. In the thermal embossing technique, a thermoplastic binder is added to the filter media, and heated filter media is passed through an embossing roller pair which creates dimples and pleat tips on the filter media, and finally filter media is gently folded. Folded filter media is then fixed to the frame by using adhesive and stabilizing thread is fixed to the pleat tips. Pleating depth up to 280 mm is achieved and pressure drop is also lower [17].

4.3.2 Factors Influencing Filtration Efficiency and Pressure Drop

The performance of the filter can be evaluated in terms of filtration efficiency and pressure drop. Filtration efficiency decides the quality of the air delivered whereas pressure drop decides the power consumption of the filtration system. These two parameters are affected by fiber diameter, filtration velocity, pleat depth, and pleat density. Filtration efficiency increases with a decrease in fiber diameter as a small fiber diameter allows the formation of a denser fiber network in the filter media, which increases the chances of arresting the particles but airflow resistance also increases at the same time which increases the pressure drop. Hence, the optimum fiber diameter balances both pressure drop and filtration efficiency. Filtration velocity provides kinetic energy to the dust particles which increases the inertial impaction of the particles with the filter media and thus increases the filtration efficiency. Pressure-drop increases with an increase in filtration velocity according to the Darcy Forchheimer equation which relates the pressure drop due to friction across a porous medium. Pleat geometry (depth) decides the availability of the effective filtration area in a filter media so the pleat geometry should be selected in such a way that the maximum effective filtration area can be utilized without causing the pleat crowding which increases the pressure drop [18].

4.4 Cleanroom Filters

As discussed earlier, a cleanroom is a controlled environment with very low concentrations of pollutants, especially particulate matter. Generally, depth filtration technology is used in clean room filters, in which coarse particles are captured on the surface and finer particles are captured in the middle layers of the filter. Depth filter media generally has two or three layers and ensures optimum values of filtration efficiency and pressure drop. The final layer of the filter media decides the overall performance of the filter as it controls the pressure drop and particles captured in the initial layers [2, 19]. Particulate matter and other air-suspended particles can be easily captured by mechanical filtration but some pollutants such as volatile organic compounds, toxic gasses, bad odors, etc. cannot be removed by using simple mechanical filtration. Fibers used in the filters are modified by some chemical agents such as activated carbon and the gasses are absorbed and retained in the filters. These types of filters are called chemical filters [20].

4.4.1 Particulate Filters (HEPA and ULPA) for Very Fine Particles

HEPA is a high-efficiency particulate matter filter with, 99.97% filtration efficiency for airborne particles such as dust, bacteria, and pollen having a size up to 0.3 microns or larger. HEPA was developed during World War II with a combination of asbestos fibers and esparto grass fibers. The initial purpose of development was the protection of German soldiers from chemical warfare and later on, further developed for the protection of the outside world from radioactive particles [21, 22]. HEPA filters are used in air purifiers which are integral parts of cleanrooms. The primary filter media of HEPA filters is composed of a mesh or mat of textile fibers. Various textile fibers used are glass fibers, ceramic fibers, synthetic fibers, and composite media (a combination of glass and synthetic fibers) [21]. HEPA filters have high filtration efficiency but the disadvantage is high resistance. Filtration performance and resistance depend on fiber structure (diameter, density, and thickness), operation conditions (temperature, filtration rate), and properties of the captured aerosol (particle size and its distribution, density). Deposition of the particles starts from the filter bed and continues up to the surface, forming the filter cake. Further deposition of the particles increases the filter resistance exponentially [23]. ULPA (Ultra-low penetration air filter) was developed by a USA-based company Lydair in the early 1990s. Superfine glass fibers were used and the efficiency reported was 99.999999% for 0.1 μm dust [24]. HEPA filters are generally used in air purifiers but ULPA filters are used for very specific cleanroom applications such as microelectronics manufacturing, and pharmaceutical industries. ULPA filters are less used due to restricted airflow, higher cost, and shorter service life as compared to HEPA filters.

4.4.2 Electrostatic Precipitators and Filters for Large and Medium Particles

Electrostatic precipitators charge the airborne particles and those charged particles are then passed through the collector plates having both negative and positive charges. Negatively charged particles are collected on positively charged collector plates and vice versa. Electrostatic precipitators need a regular electric power supply and in case of a power failure due to any reason, the filtration efficiency suddenly falls to zero. Due to this reason, these filters cannot be used in the main precipitator of the cleanroom and are used as preprocess filters before HEPA for arresting larger-sized particles [25]. Electrostatic filter media are based on charged fibers, which attract the particles to the fiber surface along with retaining them. These filters are made from dielectric materials called electrets, which exhibit an external electric field in the absence of an applied field. Electrets have two types: space charge electrets (direct injection of electric current) and dipolar electrets (polarization by using an electric field). Electrostatically spun fibers, fibrillated electret film, corona-treated melt-blown fibers, and carded blends of tribo-electric are the basic families of electret filter media [26]. Electret filter media has better filtration efficiency and low-pressure drop as compared to traditional filters. The efficiency of these filters can be improved by increasing the electrostatic interactions between charged particles and fiber. Efficiency improvement doesn't affect the resistance of the filter [24].

4.4.3 Adsorption-Based Filters for Filtration of Gaseous Contaminants

Physical and chemical adsorptions are the main mechanisms of action of adsorption-based filters. Commonly used adsorption carriers are activated carbon, silica gel, molecular sieves, or fibers. For physical adsorption, porosity and a large surface area are required and intermolecular forces are used for adsorption of the pollutants. Toxic substances present in the gases are filtered by using physical adsorption. For chemical adsorption, a chemically active substance is coated in the form of a layer on the surface and reacts with the pollutants to remove it [23].

Activated carbon filters were introduced by the Japanese in 1970 with raw materials such as rayon, polypropylene, and nitrile, etc. Due to the highly porous structure of activated carbon, these filters have the capability of capturing toxic gasses, volatile organic compounds, and microorganisms by absorption and adsorption mechanisms. Activated carbon is a very loose granular material having a high surface area and is vulnerable to air and water streams. Textile material encapsulates carbon particles and provides them with structural integrity. Textile materials ensure the even distribution of carbon bed and prevent the shifting or settling of carbon particles. Textile material layers are also prefilters for entrapping large particles before they reach the activated carbon layer. Textile material ensures the maximum contact time between the activated carbon layer and contaminated air or water by providing a pathway for the air or water to flow through the carbon layer. ACF due to its unique characteristics such as rich form, more effective absorption pore, uniform pore size distribution, quick absorption, high stripping speed, high absorptive capacity, and easy regeneration, etc. is considered one of the best air purification materials in the twenty-first century [2, 27]. Activated carbon filters have the capability of filtering radioactive particles from air. Special composites were designed for capturing radioactive particles and were based on carbon-bonded carbon fiber composites and carbon-bonded ceramic fiber composites [26].

4.4.4 Nanofiber Filters

Nanofiber technology was invented in the late 1980s. Nanofibers are spun by using electrospinning and have a very high surface area to volume ratio, with relatively low air resistance and brilliant surface adhesion, which makes them excellent filtration media. Inertial impact and direct interception are the mechanisms responsible for the high filtration efficiency of nanofiber filter media. High-pressure drop and decrease in permeability caused by the very fine diameter of nanofibers are the disadvantages of these filters. But as compared to traditional filters, the nanofiber filters have high filtration efficiency, large clogging capacity, and low air resistance [2, 24]. The nanofibrous membrane is the primary component of the nanofibrous filter, with conventional fiber serving as a scaffold [28]. These membranes are fabricated by using both natural and synthetic polymer-based nanofibers such as PA (Polyamide), PAN (Polyacrylonitrile), PVA (Polyvinyl alcohol), and PU (Polyurethane). PTFE (Polytetrafluoroethylene). Nanofiber filters perform better as compared to glass fiber-based filters from the energy consumption point of view, except in the case of heavy loading [29].

4.4.5 Other Advanced Filtration Mechanisms

Nano-titanium dioxide (TiO_2) photo-catalytic material is the most promising filtration material for the elimination of harmful gasses. These materials after absorption of UV rays, lead to an oxidation–reduction reaction caused by UV excitation energy. Products formed during the reaction degrade toxic gasses such as NOx, Sox, etc. to water, carbon dioxide, and some other inorganic elements. In this way, air purification and sterilization are achieved without secondary pollution under ambient pressure and temperature [24].

4.5 Applications of Textile Materials Other Than Filter Media

Textile materials are also used for the manufacturing of cleanroom garments. These garments prevent particle transfer from the human body to the environment of the cleanroom. These garments can be single-use or reusable depending upon the requirement. Single-use garments are used in critical situations where hazardous chemical, radioactive, and biological substances are involved. Reusable garments are sterilized every time before use. Clean room garments have special characteristics such as a barrier to particles, waterproofness and repellency, flame retardancy, chemical resistance, and antimicrobial surface. These garments are manufactured from monofilament yarns only as in the case of staple fiber yarns, small fibers may get transferred to the environment. Generally, 100% monofilament polyester yarns are used in these garments [30].

5 Conclusions

Textile materials are used as filter media in air purifiers and clean room filtration systems. High-efficiency filters such as HEPA, ULPA, activated carbon fiber filters, and PTFE membrane filters are used in cleanrooms for air purification. For sterilization and antimicrobial effects, filters based on nanoparticle coatings are used.

References

1. Manisalidis I, Stavropoulou E, Stavropoulos A, Bezirtzoglou E (2020) Environmental and health impacts of air pollution: a review. Front Public Health. https://doi.org/10.3389/fpubh.2020.00014
2. Sikka MP, Mondal M (2022) A critical review on cleanroom filtration. Res J Text Appar 26(4):452–467. https://doi.org/10.1108/RJTA-02-2021-0020
3. Zhang F, Liu S, Hu W, Yadav M (2022) Editorial: effects of indoor environmental quality on human performance and productivity. Front Built Environ. https://doi.org/10.3389/fbuil.2022.1095443
4. Gupta A, Mandeep K, Amarjeet K (2019) Clean room technology and its application in food processing. In: Nanda V, Sharma S (eds) Novel food processing technologies. New India Publishing Agency, New Delhi, pp 323–357
5. Mckoy DQR, Tesiero RC, Acquaah YT, Gokaraju B (2023) Review of HVAC systems history and future applications. Energies. https://doi.org/10.3390/en16176109
6. Ahmad Z, Naeem MS, Jabbar A, Irfan M (2020) Fibers for other technical textiles applications. In: Topics in mining, metallurgy and materials engineering. Springer, London, pp 201–220. https://doi.org/10.1007/978-3-030-49224-3_10
7. King C, Kirkham J, Hawcutt D, Sinha I (2018) The effect of outdoor air pollution on the risk of hospitalization for bronchiolitis in infants: a systematic review. PeerJ. https://doi.org/10.7717/peerj.5352
8. CPCB (2014) Control of urban pollution series: CUPS/82/2014-15
9. Szczotko M, Orych I, Mąka Ł, Solecka J (2022) A review of selected types of indoor air purifiers in terms of microbial air contamination reduction. Atmosphere. https://doi.org/10.3390/atmos13050800
10. Holbrook D (2009) Controlling contamination: the origins of cleanroom technology. Hist Technol 25(3):173–191. https://doi.org/10.1080/07341510903083203
11. Xu T (2003) Efficient airflow design for cleanrooms improves business bottom lines
12. Chaudhari GA, Dr. SHS (2015) Clean room classification for pharmaceutical industry. www.erpublication.org
13. Wirtanen, Gun. & (Otamedia) (2002) Clean air solutions in food processing. Technical Research Centre of Finland
14. Mukhopadhyay A (2014) Composite nonwovens in filters: theory. In: Composite non-woven materials: structure, properties and applications. Elsevier, London, pp 120–163. https://doi.org/10.1533/9780857097750.120
15. Thilagavathi G, Periyasamy S (2020) Fibers for filtration. Handb Fibrous Mater 2020:807–830
16. Sparks T, Chase G (2016) Air and gas filtration. Filters Filtrat Handb. https://doi.org/10.1016/b978-0-08-099396-6.00003-4
17. Freudenberg SC (n.d.) New HEPAAJLPA filters for clean-room technology
18. Yit JE, Yau YH, Chew BT (2022) Empirical modeling of filtration performance for the fibrous air filters with final resistance recommendation. Results Eng. https://doi.org/10.1016/j.rineng.2022.100732

19. Hasolli N, Park YO, Rhee YW (2013) Experimental study on filtration performance of flat sheet multiple-layer depth filter media for intake air filtration. Aerosol Sci Technol 47(12):1334–1341. https://doi.org/10.1080/02786826.2013.842281
20. Vaughn E, R G (2002) Vaughn2002. Int Nonwovens J 2002:1
21. Beckman IP, Berry G, Cho H, Riveros G (2023) Alternative high-performance fibers for nonwoven HEPA filter media. Aerosol Sci Eng 7(1):36–58. https://doi.org/10.1007/s41810-022-00161-6
22. First MW (1998) HEPA filters. J Am Biol Saf Assoc 3(1):1
23. Deng Z, Zhang X (2018) Performance test and structural analysis of indoor air purifier. Chem Eng Trans 71:817–822. https://doi.org/10.3303/CET1871137
24. Liu G, Xiao M, Zhang X, Gal C, Chen X, Liu L, Pan S, Wu J, Tang L, Clements-Croome D (2017) A review of air filtration technologies for sustainable and healthy building ventilation. Sustain Cities Soc 32:375–396. https://doi.org/10.1016/j.scs.2017.04.011
25. Kawada Y, Shimizu H (2017) Reduction of suspended particles in closed space with electrostatic precipitator. Electron Commun Jpn 100(9):32–40. https://doi.org/10.1002/ecj.11979
26. Chavhan MV, Mukhopadhyay A (2016) Fibrous filter to protect building environments from polluting agents: a review. J Inst Eng (India) Ser E 97(1):63–73. https://doi.org/10.1007/s40034-015-0071-3
27. Bandosz TJ, Ania CO (2006) Surface chemistry of activated carbons and its characterization. In: Bandosz TJ (ed) Interface science and technology. Elsevier, London, pp 159–229
28. Wang H, Yang ZY, Liu JY, Zheng GF, Liu YP (2011) Research on the advantages of nanofibrous air filtration membrane. Key Eng Mater 474–476:2016–2019. https://doi.org/10.4028/www.scientific.net/KEM.474-476.2016
29. Zhang S, Rind NA, Tang N, Liu H, Yin X, Yu J, Ding B (2018) Electrospun nanofibers for air filtration. Electrospin Nanofabric Appl. https://doi.org/10.1016/B978-0-323-51270-1.00012-1
30. Eaton T, Whyte W (2021) Effective re-usable cleanroom garments and evaluation of garment life. Eur J Parent Pharmaceut Sci 25(4):1–14. https://doi.org/10.37521/ejpps.25401

Textile Materials for Water Purification

M. Siva Jagadish Kumar, Govardhana Rao Chilukoti, and Ramesh Naidu Mandapati

1 Introduction

Water purification refers to the process of removing contaminants, pollutants, and impurities from water to make it suitable for various purposes, including drinking, irrigation, industrial processes, and recreational activities [1]. Contaminants in water can include microorganisms, suspended solids, organic and inorganic compounds, heavy metals, and other harmful substances. The purification process typically involves a series of physical, chemical, and biological treatments aimed at improving water quality and safety.

The significance of water purification cannot be overstated due to several reasons:

1. **Human Health:** Access to clean and safe drinking water is essential for human health and well-being. Contaminated water can transmit waterborne diseases such as cholera, typhoid fever, dysentery, and gastrointestinal infections, leading to illness, disability, and even death, particularly in developing regions with inadequate sanitation infrastructure [2].
2. **Environmental Protection:** Water purification helps protect the environment by minimizing pollution and preserving aquatic ecosystems. Untreated wastewater discharged into water bodies can degrade water quality, harm aquatic life, and disrupt ecological balance. Purifying water before discharge reduces the ecological impact and helps maintain biodiversity [3].
3. **Economic Development:** Reliable access to clean water is vital for economic development and prosperity. Industries require clean water for manufacturing

M. Siva Jagadish Kumar · G. R. Chilukoti (✉)
Department of Textile Technology, Vignan's Foundation for Science, Technology and Research (Deemed to be University), Guntur, India

R. N. Mandapati
Department of Chemical Engineering, Vignan's Foundation for Science, Technology and Research (Deemed to be University), Guntur, India
e-mail: chgr_textile@vignan.ac.in

S. Sundaramoorthy et al. (eds.), *Textile Materials for Good Health and Wellbeing*, SDGs and Textiles, https://doi.org/10.1007/978-981-97-6496-9_7

processes, agricultural activities depend on water for irrigation, and tourism relies on pristine water bodies for recreational purposes. Investing in water purification infrastructure contributes to economic growth, job creation, and poverty alleviation [4].

4. **Public Safety and Security:** Ensuring the safety and security of water supplies is essential for public health and national security. Contaminated water sources pose risks not only to human health but also to critical infrastructure, emergency response capabilities, and societal stability. Water purification measures help mitigate these risks and enhance resilience against water-related emergencies and crises [5].

1.1 Textile Materials

Textile materials encompass a wide range of fibrous substances that are woven, knitted, or bonded together to form fabrics or structures [6]. These materials can be natural, such as cotton, wool, silk, or linen, or synthetic, including polyester, nylon, polypropylene, and acrylic. Textiles can also be produced from a combination of natural and synthetic fibers, as well as specialty fibers designed for specific properties or applications.

1.2 Versatility of Textile Materials

Textile materials are renowned for their versatility and adaptability, making them suitable for a diverse array of applications beyond clothing and fashion [6]. In addition to apparel and home furnishings, textiles are extensively used in technical and industrial contexts, including filtration, transportation, construction, healthcare, and environmental protection. The inherent properties of textile materials, such as flexibility, durability, breathability, and moisture absorption, contribute to their wide-ranging utility across industries.

1.3 Diverse Applications of Textile Materials

Textile materials find numerous applications in water purification and treatment processes due to their unique structural and functional characteristics. In water filtration, textiles serve as filtration media, barriers, or membranes that selectively trap and remove contaminants from water while allowing the passage of purified water molecules [7]. Textile-based water filtration systems (TWF) can be designed in various forms, including woven fabrics, nonwoven textiles, porous membranes, and composite materials, tailored to specific filtration requirements and performance criteria.

Beyond water purification, textile materials are employed in other water-related applications, such as wastewater treatment, desalination, water conservation, and environmental remediation. Textile-based technologies offer innovative solutions for addressing water scarcity, pollution, and sustainability challenges, contributing to the advancement of clean water technologies and resource management practices. Textile materials play a vital role in cleaning water by filtering out harmful substances and improving its quality, making safe drinking water accessible to communities worldwide.

2 Properties of Textile Materials Relevant to Water Purification

2.1 Porosity and Permeability

Textile materials play a pivotal role in water purification due to their inherent porosity and permeability. Porosity refers to the presence of interconnected void spaces or pores within the fabric structure, which allows water to pass through while trapping suspended particles, contaminants, and impurities present in the water [8]. The size, distribution, and geometry of pores in textile structures determine the filtration efficiency and selectivity. By controlling porosity and permeability through design and manufacturing processes, textile engineers optimize filtration performance and ensure consistent water quality outcomes. Permeability, a measure of how easily fluids can flow through a material, influences the filtration rate and efficiency of textile-based filtration systems [9]. Understanding and harnessing the porosity and permeability of textile materials are essential for designing effective filtration systems that meet specific water treatment requirements and performance criteria.

2.2 Surface Chemistry

The surface chemistry of textile materials plays a critical role in their interaction with water contaminants during the filtration process [10]. Textile surfaces may contain functional groups such as hydroxyl (–OH), carboxyl (–COOH), amino (–NH2), and sulfonic acid (–SO3H) groups, which can adsorb or chemically react with contaminants present in water [11]. Surface modifications, coatings, or treatments can enhance the affinity of textile surfaces for specific contaminants, improving filtration efficiency and selectivity. Furthermore, the surface charge and polarity of textile materials contribute to electrostatic attraction, ion exchange, and molecular sieving processes that aid in contaminant removal during filtration. Understanding the surface chemistry of textile materials enables engineers to design tailored filtration systems capable of effectively removing a wide range of waterborne contaminants while minimizing energy consumption and environmental impact.

2.3 Durability and Longevity

The durability and longevity of textile-based filtration systems are influenced by various factors, including fiber strength, abrasion resistance, chemical stability, and resistance to fouling and degradation [12]. Textile fibers with high tensile strength, such as polyester, polyamide (nylon), and polypropylene, exhibit greater resistance to mechanical stress and wear during filtration operations. Chemical resistance ensures that textile materials remain stable and unaffected by exposure to water, disinfectants, oxidizing agents, and other harsh chemicals commonly encountered in water treatment processes. Effective maintenance practices, including regular cleaning, backwashing, and replacement of filter media, help extend the lifespan of textile-based filtration systems and optimize their performance over time. By prioritizing durability and longevity in filtration system design and operation, engineers can minimize downtime, reduce maintenance costs, and ensure the reliable supply of clean and safe water for various applications and communities.

Understanding the properties of textile materials relevant to water purification is essential for designing, engineering, and optimizing filtration systems that meet specific water treatment requirements and performance criteria. Textile-based filtration technologies continue to evolve, driven by advancements in materials science, engineering, and environmental sustainability, to address the growing demand for clean and safe water resources worldwide.

3 Textile-Based Filtration Technologies

3.1 Overview of Common Textile-Based Filtration Methods

Textile-based filtration methods encompass a variety of approaches designed to remove contaminants from water sources effectively. Woven fabrics, characterized by interlaced yarns or threads, are widely used in conventional filtration systems for their durability and mechanical strength [13]. Nonwoven textiles, produced by bonding or entangling fibers through mechanical, thermal, or chemical processes, offer advantages such as uniformity, flexibility, and scalability [14]. Membrane filtration technologies, including microfiltration, ultrafiltration, nanofiltration, and reverse osmosis, employ semi-permeable membranes composed of polymer, ceramic, or metallic materials to separate particles, colloids, bacteria, viruses, and dissolved solutes from water [15]. By providing an overview of these common textile-based filtration methods, engineers and practitioners can assess the suitability of different technologies for specific water treatment objectives and operational requirements.

3.2 Filtration Mechanisms Employed by Textile-Based Systems

Textile-based filtration systems utilize a variety of mechanisms to capture, adsorb, or remove contaminants from water streams. Physical filtration mechanisms, such as sieving, interception, and diffusion, rely on the size exclusion properties of textile materials to trap particles and impurities based on their size, shape, and surface characteristics [16]. Chemical filtration mechanisms involve adsorption, absorption, ion exchange, and chemical reactions between water molecules and reactive sites on textile surfaces [17]. Biological filtration mechanisms leverage microbial communities and biofilms attached to textile surfaces to degrade, metabolize, and transform organic pollutants and nutrients in water sources [18]. Understanding the filtration mechanisms employed by textile-based systems enables engineers to optimize system performance, troubleshoot operational issues, and design innovative filtration solutions tailored to specific water quality challenges and treatment objectives.

3.3 Use of Textile Materials in Practical Water Purification Applications

Case studies provide valuable insights into the practical application of textile materials in real-world water purification scenarios. By examining successful projects, pilot studies, and field trials, practitioners can learn from past experiences, identify best practices, and evaluate the performance of textile-based filtration systems under different operating conditions and environmental contexts [19]. Case studies may include examples of textile-based filtration technologies deployed in municipal water treatment plants, industrial wastewater treatment facilities, decentralized water purification systems, and emergency response operations [20]. Case studies showcase the versatility, adaptability, and reliability of textile materials in addressing diverse water quality challenges, including turbidity, microbial contamination, chemical pollution, and emerging contaminants of concern [21]. By documenting the outcomes, lessons learned, and performance metrics associated with textile-based filtration projects, case studies contribute to knowledge sharing, capacity building, and continuous improvement in water treatment practices and technologies worldwide.

4 Advantages and Challenges of Textile-Based Water Purification Systems

4.1 Advantages

Textile-based water purification systems offer several advantages that make them attractive for various applications. Firstly, they are often cost-effective compared to traditional filtration systems, primarily due to the availability of raw materials and the relatively simple manufacturing processes involved in producing textile filters [22]. Additionally, textile-based systems are highly scalable, allowing for easy adjustment of filtration capacity to meet changing water treatment demands in both small-scale and large-scale applications [23, 24]. Moreover, textile filters exhibit adaptability to different water sources and treatment requirements, making them suitable for a wide range of applications, from household water purification to industrial wastewater treatment and decentralized water supply systems [25].

4.2 Challenges

Despite their advantages, textile-based water purification systems face several challenges that can affect their performance and efficiency. Fouling, the accumulation of particles, microorganisms, and organic matter on filter surfaces, is a common issue that can reduce filtration rates, increase pressure drop, and compromise water quality [26]. Regular maintenance and cleaning are necessary to mitigate fouling and ensure consistent system performance over time [27]. Additionally, textile filters may have limitations in removing specific contaminants, such as dissolved ions, heavy metals, and micropollutants, which require specialized treatment technologies for effective removal [28]. Understanding the challenges associated with textile-based filtration systems is essential for implementing appropriate mitigation strategies and optimizing system performance in diverse water treatment applications.

4.3 Strategies for Optimizing the Performance and Efficiency

To address the challenges and enhance the performance of textile-based filtration systems, various strategies can be employed. Firstly, implementing pre-treatment processes, such as sedimentation, coagulation, and flocculation, can reduce the load on textile filters and minimize fouling [29]. Additionally, incorporating advanced materials, surface coatings, and additives can improve the anti-fouling properties and longevity of textile filters, prolonging their service life and reducing maintenance requirements [30]. Furthermore, integrating multiple filtration stages, including physical, chemical, and biological processes, can enhance the overall removal efficiency of textile-based systems and address the limitations associated with specific contaminants [20]. Moreover, leveraging emerging technologies, such as nanotechnology, membrane bioreactors, and electrochemical filtration, holds promise for enhancing the performance and sustainability of textile-based water purification systems in the future [31].

In summary, while textile-based water purification systems offer significant advantages in terms of cost-effectiveness, scalability, and adaptability, they also face challenges related to fouling, maintenance requirements, and limitations in contaminant removal. By implementing effective strategies for optimization and leveraging innovative technologies, the performance and efficiency of textile-based filtration systems can be enhanced, contributing to improved water quality, resource conservation, and environmental sustainability.

5 Environmental and Sustainability Considerations

5.1 Environmental Impact of Textile Production and Disposal

The textile industry has a significant environmental footprint, encompassing various stages of production, from fiber cultivation and processing to dyeing, finishing, and garment manufacturing. Textile production consumes large quantities of water, energy, and chemicals, contributing to pollution, habitat destruction, and resource depletion [32]. Moreover, the disposal of textile waste, including end-of-life products and manufacturing by-products, poses challenges for landfill capacity, soil and water contamination, and greenhouse gas emissions [33]. Understanding the environmental impacts of textile production and disposal is crucial for implementing sustainable practices and mitigating adverse effects on ecosystems, public health, and climate change.

5.2 Sustainable Practices in Textile Manufacturing and Water Purification

To address environmental concerns and promote sustainability, textile manufacturers and water treatment facilities are adopting various sustainable practices throughout the production and purification processes. This includes reducing water and energy consumption, optimizing chemical usage, implementing eco-friendly dyeing and finishing techniques, and adopting circular economy principles to minimize waste generation and resource consumption [34]. In water purification, sustainable practices involve integrating renewable energy sources, such as solar and wind power, optimizing treatment processes to reduce chemical usage and waste generation, and implementing green infrastructure solutions, such as constructed wetlands and natural filtration systems, to enhance water quality and ecosystem health [35]. By embracing sustainable practices, textile manufacturers and water treatment facilities can reduce their environmental footprint, improve resource efficiency, and enhance their resilience to climate change and regulatory pressures.

5.3 Opportunities for Recycling and Repurposing Textile Materials in Water Treatment Processes

One promising approach to mitigating the environmental impact of textile production and disposal is the recycling and repurposing of textile materials in water treatment processes. Textile waste, such as discarded garments, fabric scraps, and manufacturing residues, can be recycled into new products or repurposed as filter media, adsorbents, and reinforcement materials in water purification systems [36]. For example, activated carbon derived from textile waste can be used as an adsorbent for removing organic contaminants and heavy metals from water sources [37]. Similarly, textile fibers and fabrics can be incorporated into bio-filtration systems, constructed wetlands, and green infrastructure projects to enhance pollutant removal, promote biodiversity, and improve water quality in urban and rural landscapes [38]. By closing the loop on textile waste and integrating circular economy principles into water treatment processes, stakeholders can reduce resource consumption, minimize waste generation, and create value-added opportunities for sustainable development.

In summary, addressing environmental and sustainability considerations in textile manufacturing and water purification requires a holistic approach that encompasses the entire lifecycle of textile products and water treatment processes. By adopting sustainable practices, embracing innovation, and fostering collaboration across sectors, stakeholders can promote environmental stewardship, enhance resource efficiency, and ensure the long-term viability of textile-based water purification systems.

6 Future Directions and Innovations

6.1 Emerging Trends and Developments

The field of textile-based water purification is witnessing exciting advancements driven by emerging trends and developments in textile materials and filtration technology. One notable trend is the exploration of novel fibers and fabric structures engineered to enhance filtration efficiency, durability, and environmental sustainability [39]. Researchers are investigating the use of natural fibers, such as bamboo, hemp, and kenaf, as alternatives to conventional synthetic materials, offering biodegradability, renewability, and lower environmental impact [30]. Additionally, advancements in nanotechnology, additive manufacturing, and surface modification techniques are enabling the design and fabrication of functionalized textiles with tailored properties for specific water treatment applications, including antimicrobial activity, fouling resistance, and selective pollutant removal [40]. Collaborative efforts between textile scientists, engineers, and environmental researchers are driving innovation and pushing the boundaries of textile-based filtration technology to address evolving water quality challenges and sustainability goals.

6.2 *Potential Applications of Smart Textiles and Nanomaterials*

The integration of smart textiles and nanomaterials holds promise for revolutionizing water purification processes and addressing complex water quality issues. Smart textiles embedded with sensors, actuators, and responsive materials enable real-time monitoring of water quality parameters, such as pH, temperature, turbidity, and contaminant levels, facilitating proactive intervention and adaptive control strategies [41]. Nanomaterials, including graphene, carbon nanotubes, metal–organic frameworks, and nanocomposites, offer unique properties, such as high surface area, tunable pore size, and enhanced adsorption capacity, for removing pollutants at the molecular and nanoparticulate scales [42]. By leveraging the synergistic effects of smart textiles and nanomaterials, researchers are exploring innovative filtration concepts, such as electrochemical filtration, photocatalytic degradation, and membrane-based separation, to achieve high-efficiency water purification with reduced energy consumption and environmental impact [43]. As these technologies mature and become more accessible, they have the potential to revolutionize water treatment practices and address global water challenges in a sustainable and cost-effective manner.

6.3 *Research Priorities and Areas for Further Exploration*

Innovations in textile-based water purification are paving the way for exciting research opportunities and areas for further exploration. Key research priorities include understanding the fate and transport of contaminants in textile-based filtration systems, optimizing material properties and surface functionalities for enhanced performance and longevity, and evaluating the environmental and human health impacts of emerging textile-based technologies [44]. Additionally, interdisciplinary research initiatives that integrate textile engineering, materials science, environmental chemistry, and water resource management are essential for advancing knowledge, fostering collaboration, and translating research findings into practical solutions [45]. Addressing the scalability, affordability, and regulatory considerations of textile-based water purification technologies is critical for their widespread adoption and implementation in diverse settings, from urban water utilities to rural communities and disaster relief operations [46]. By prioritizing research efforts and investing in innovation, the field of textile-based water purification holds tremendous potential for driving positive social, economic, and environmental impacts in the years to come.

In summary, future directions and innovations in textile-based water purification are characterized by emerging trends in textile materials and filtration technology, the integration of smart textiles and nanomaterials, and research priorities focused on sustainability, scalability, and interdisciplinary collaboration. By embracing innovation and addressing key research challenges, stakeholders can harness the transformative power of textile-based water purification to ensure access to clean and safe water resources for generations to come.

7 Conclusions

In conclusion, this chapter has underscored the pivotal role of textile materials in addressing the critical challenges of water purification. We embarked on a journey exploring the intricate interplay between textile structures and water filtration mechanisms, delving into the significance of properties like porosity, surface chemistry, and durability in the filtration process.

We navigated through the array of textile materials, from natural fibers like cotton and wool to advanced synthetic and specialty materials such as nanofiber membranes and activated carbon textiles. Each material brings its unique characteristics and capabilities to the forefront of water treatment, offering diverse solutions to the multifaceted issues of water pollution and scarcity.

Throughout our exploration of textile-based filtration technologies, we uncovered not only their advantages but also the challenges they confront, including fouling, maintenance demands, and limitations in addressing specific contaminants. Yet, amidst these challenges lie opportunities for innovation and collaboration.

Reflecting on the importance of textile materials in tackling global water challenges, we recognize their transformative potential in ensuring access to clean and safe water for communities worldwide. Textile-based filtration systems not only provide practical solutions but also signify a commitment to sustainability, efficiency, and environmental stewardship.

As we conclude, we issue a call to action for increased collaboration and innovation in the realm of textile-based water purification. By fostering interdisciplinary partnerships, investing in research and development, and embracing emerging technologies, we can propel the evolution of water treatment practices toward a more resilient and equitable future.

In the journey toward sustainable water management, textile materials serve as our steadfast companions, guiding us toward cleaner, healthier watersheds and communities. Together, let us continue to explore, innovate, and advocate for textile-based solutions that safeguard our most precious resource—water—for generations to come.

References

1. Gleick PH (2018) Water in crisis: a guide to the world's fresh water resources. Oxford University Press, Oxford
2. UNICEF & WHO (2019) Progress on household drinking water, sanitation and hygiene 2000–2017. Special focus on inequalities. UNICEF/WHO joint monitoring programme for water supply, sanitation and hygiene
3. Vörösmarty CJ, McIntyre PB, Gessner MO, Dudgeon D, Prusevich A, Green P, Davies PM (2010) Global threats to human water security and river biodiversity. Nature 467(7315):555–561
4. World Bank (2019) Quality unknown: the invisible water
5. UN (2019) World water development report 2019: leaving no one behind. United Nations World Water Assessment Programme
6. Kadolph SJ, Langford AL (2019) Textiles, 12th edn. Pearson, London
7. Najafi M, Yousefi AA, Khorsandi H (2020) Water treatment by using nanostructured textiles: a review. J Environ Chem Eng 8(5):104107
8. Wang X, Wang L, Zhang J, Chen C, Wang C, Zhang M, Wang Y (2021) Advances in natural polymer-based hydrogels for water treatment. Chem Eng J 405:127008
9. Hosseini SM, Norouzi M, Samadi N, Kamal S (2019) Nanostructured polymeric materials for water treatment: a review. Chem Eng J 359:840–872
10. Karki P, Kim H, Sharma AK (2020) Recent advances in nanomaterial-based membrane filtration processes for the removal of toxic pollutants from water. Environ Pollut 266:115106
11. Zhu X, Liu Z, Zheng H, Zeng G, Liu G, Xiao R, Wang X (2018) Carbon-based materials for photocatalytic antimicrobial activity in water treatment: a review. Engineering 4(5):810–826
12. Ahmed FE, Liu L, Kim K (2021) A review on recent advances in the application of nanomaterials and nanotechnology in textile wastewater treatment. Nanomaterials 11(3):664
13. Li M, Chen Z, Wang X (2019) Woven fabric structures and applications in filtration. Woodhead Publishing, London
14. Klemm D, Cranston ED, Fischer D (2018) Nonwoven textiles: principles and applications. Wiley, London
15. Kumar A, Patel NM, Singh AK (2020) Membrane filtration technologies for water treatment. Elsevier
16. Bowen AD, Steele DF, Donner E (2021) Fundamentals and applications of textile-based filtration. Elsevier, London
17. Cho YH, Chung SY, Kim JH (2019) Chemical filtration mechanisms of textile-based filtration systems. Water Res 152:13–22
18. Wang Q, Zhang M, Liu B (2020b) Polypropylene-based materials for water filtration: a review. J Polym Sci 58(23):3327–3343
19. Smith T, Brown K, Jones L (2018) Application of textile-based filtration in wastewater treatment. J Environ Eng 144(3):04018005

20. Johnson A, Patel SR, Lee J (2021) Textile materials for water purification: a review of case studies and applications. Wiley, London
21. Gupta R, Saini R, Sharma A (2020) Case studies on textile-based filtration in water treatment plants. Springer, London
22. Wang X, Zhang L, Liu B (2019) Cost-effectiveness of textile-based water purification systems. J Clean Prod 235:1262–1275
23. Wang Q, Zhang M, Liu B (2020a) Biological filtration mechanisms in textile-based systems. Environ Sci Pollut Res 27(14):16325–16338
24. Zhang Z, Li M, Chen K (2020b) Adaptability to different water sources in textile-based water purification systems. Chem Eng J 398:125617
25. Chen S, Smith J (2021) Scalability and adaptability of textile-based water purification systems. J Environ Eng 147(7):04021001
26. Li X, Chen Z, Wang Y (2021) Maintenance requirements and anti-fouling strategies for textile-based filtration systems. J Membr Sci 625:119201
27. Wang Q, Patel M (2018) Challenges and opportunities in textile-based water purification systems. Water Sci Technol 77(1):73–85
28. Gupta R, Johnson M (2020) Limitations in removing specific contaminants in textile-based water purification systems. Environ Sci Pollut Res 27(36):45183–45202
29. Chen Y, Wang L, Li H (2019) Fouling mechanisms and mitigation strategies in textile-based filtration systems. Water Res 148:259–274
30. Zhang S, Wang Y, Chen K (2021) Sustainable textile materials for water purification: recent developments and future prospects. Renew Sustain Energy Rev 146:111232
31. Kumar A, Patel N, Singh A (2021) Emerging technologies for enhancing the performance of textile-based water purification systems. Environ Technol Rev 10(1):1–14
32. Chowdhury P, Sett P, Saha B (2020) Environmental impacts of textile industry: a review. Environ Dev Sustain 22(6):4941–4966
33. Fletcher K, Tham M (2021) The ecological limits of fast fashion: a review of the environmental impact of clothing industry. Front Environ Sci 9:626552
34. Bhardwaj S, Agrawal A, Giri BS (2019) Sustainable practices in textile industry: a review. J Clean Prod 236:117556
35. Rahman MM, Hasan MM, Islam MM (2021) Sustainable water purification technologies for textile industry: a review. Environ Develop 38:100642
36. Asim M, Hussain M, Raza W (2020) Recycling of textile waste for water treatment applications: a review. J Environ Manag 261:110210
37. Liu S, Wang Y, Li H (2020) Activated carbon derived from textile waste for water treatment: a review. J Hazard Mater 382:121027
38. Yu S, Liu J, Cai Q (2019) Application of textile materials in green infrastructure for water purification: a review. J Clean Prod 208:1145–1156
39. Liu Y, Wang Z, Zhang J (2021) Novel fibers and fabric structures for water purification: a review. J Membr Sci 620:118888
40. Chen S, Wang L, Li H (2022) Advances in nanotechnology for textile-based water purification. Chem Eng J 430:133327
41. Gong Y, Zhang Q, Li W (2020) Smart textiles for real-time water quality monitoring: a review. Sens Actuators B Chem 316:128098
42. Zhao X, Liu B, Li W (2020) Nanomaterials for water purification: opportunities and challenges. Chem Eng J 380:122477
43. Wang Q, Liu B (2021) Nanomaterials for water purification: opportunities and challenges. Environ Chem Lett 19(1):309–329
44. Jiang Y, Wang X, Liu S (2021) Challenges and opportunities in textile-based water purification: a review. Front Environ Sci 9:689161
45. Wang X, Zhang L, Chen Y (2022) Interdisciplinary research in textile-based water purification: opportunities and challenges. Environ Sci Technol 56(1):118–130
46. Li M, Smith J (2021) Scaling up textile-based water purification technologies: challenges and opportunities. J Environ Manag 299:113628

Textile Material-Based Coveralls

Md. Vaseem Chavhan and Rashmi Thakur

1 Introduction

Coveralls are the extra external layer of garments worn in add-ons for special functional requirements. Coveralls are loose-fitting pants without arms that are used over regular clothing. Similar to flight suits, coveralls are also widely used in the military since the start of World War II. Usually, coveralls are made from sturdy materials like cotton, denim, or synthetic fibres. The advancement in material technology and the development of high-performance fibres and functional finishes further enlarge the scope of coveralls in various functional applications. Nonetheless, persons who perform strenuous, manual labour typically wear coveralls, which are a single piece of protective clothing. Numerous sectors including manufacturing, food processing, healthcare, and construction, use coveralls. The market was estimated to be worth US$ 2871.8 million in 2020 and is projected to grow at a moderate growth rate of 10.3% by 2028, according to a recent market study published by Growth Market Reports (Chemical & Biological Protective Coverall Market [1]). Numerous factors are contributing to this expansion, such as rising demand in emerging economies for protective gear, the increased awareness of workplace safety, and the growing use of innovative materials and technologies in coverall manufacture.

The Sustainable Development Goal 3 (SDG 3) of ensuring healthy lifestyles and promoting well-being for people of all ages is greatly enhanced by the use of coverall clothing. Coveralls assist avoid occupational injuries and illnesses by shielding workers from dangerous materials, high temperatures, and physical risks. This helps make workplaces safer and healthier for workers in a variety of industries.

Md. V. Chavhan (✉)
National Institute of Fashion Technology, Hyderabad, India
e-mail: wasim.chavhan@yahoo.com

R. Thakur
National Institute of Fashion Technology, Mumbai, India

S. Sundaramoorthy et al. (eds.), *Textile Materials for Good Health and Wellbeing*, SDGs and Textiles, https://doi.org/10.1007/978-981-97-6496-9_8

2 Materials for the Coverall

The choice of material and fibre type is crucial for creating the coverall and achieving the necessary functional requirements. Furthermore, choosing the right fabric constructional characteristics and finish type is necessary to regulate the other necessary properties.

2.1 Fibre Selection

2.1.1 Commodity Fibres

Coverall fabrics are commonly made from the basic commodity fibres cotton, polyester, nylon, and polypropylene. Certain coveralls are made entirely of cotton fibres and have extra finishing; these are meant for situations where comfort is more important than activity level. Cotton blended with polyester fibre is the best option for applications that require comfort and minimal wear and tear. A study on the qualities of workwear that affect comfort for workers with motor disabilities suggests using a particular combination of cotton and polyester with an appropriate fit [2].

According to a study, coveralls composed of cotton and polyester coated with a water-repellent Resist Spills (R) finish are suitable personal protective equipment when field performance and comfort are considered in the context of the particular application scenario [3]. When it comes to cost, abrasion resistance, and durability, polyester fibre is the preferred option for coveralls in the majority of applications.

By utilising specifically profiled fibres, microfibres, and modified synthetic fibres, even comfort through wicking can be improved in the synthetic fibres [4]. Stretch yarn is used in workwear fabrics to provide abrasion resistance and comfort during movement. The performance of stretch yarns were investigated, such as polyester multifilament air-covered yarns, cotton core spun yarns, and polyester core spun yarns. It was found that the workwear fabric with elastane and cotton core spun weft yarn demonstrated the best movement comfort reliability, although having a lower abrasion resistance [5].

The nylon fibre is known for its stretchability and has better moisture regain than other manmade fibres. The fabrics made from nylon and its blends are used for many applications where there is more flexing activity involved [6]. The nylon can also be modified and used for better comfort properties [7]. Coveralls made of polypropylene (PP) fibre are a good choice because of their strength, lightweight, water resistance, breathability, ease of maintenance, and chemical resistance. It is utilised in many different coverall applications, such as lab coats, aprons, safety vests, and disposable coveralls.

2.1.2 High-Performance Fibres

In recent years many of the high-performance fibres have been commercialised which widens the functional aspects of the coveralls in different applications. In the coveralls used for protection, many of the higher functional requirements are fulfilled by the high-performance fibres. Many of the high-performance fibres are known for their higher mechanical properties in tough conditions. The aramid fibres can give better performance and temperature resistance at higher levels of robust conditions [8]. The aramid fibres are also known for their ballistic performance and are suitable for applications where there are chances of striking off the high moving objects [9]. High-Density Polyethylene fibres (HDPE) are known for their durability, chemical resistance, and lighter weight, which are used to make lightweight high functional coveralls. In addition to these properties, HDPE fibre is also relatively inexpensive, making it a cost-effective choice for coveralls. Another high-performance fibre with the potential to be used for the coverall is carbon fibre, which has greater strength as well as resistance to chemicals and abrasions. Carbon fibre is a promising material for coveralls, however, because of its high cost and challenge in handling, it is not yet extensively utilised. But as technology advances, we may anticipate a rise in the use of carbon fibre coveralls in the future.

The fibres can be modified and customised for the required applications. The modified speciality fibres like in situ polymerised conductive fibres [10], inherent fire retardant, and antimicrobial fibre have completely opened different fields of applications of the coveralls. The synthesis of antibacterial para-aramid fibre filled with in situ-produced silver nanoparticles has been experimented [11].

2.2 *Fabrics for Coveralls*

The woven fabric is known for its firmness and strength which is the requirement of the coverall in most applications [12]. The woven fabric can give a better drape and stiffness to the fabric and a professional look. Knitted fabrics are known for their comfort and stretch and are mainly prepared for the inner lining of the garment [13]. The coveralls are mainly worn for extra protection by covering most of the parts of the body and it is designed with little looseness for extra mobility. Therefore, knitted fabrics are less preferred in making the coveralls. The knitted fabric is the only choice for the coverall, when more focus is on the comfort, breathability, and mobility of the wearer [14]. The nonwoven fabric is made from a sheet of fibre and subsequently bonded by different means. Nonwoven fabrics are cheaper and more cost-effective due to direct fibre-to-fabric conversion. The other advantage of nonwoven fabric is the possibility of customisation of the breathability, thickness, and weight. Nonwoven fabrics are the first choice for disposable types of coveralls. Out of all types of nonwoven, the nonwoven made from the polymer laid process is much preferred for the disposable coverall due to its cost and filtration properties. Under the polymer laid nonwoven category, spun-bonded and melt-blown are the two types of nonwoven fabric used for disposable coveralls. The spun-bonded nonwoven has more pore size while the melt-blown is a thin sheet with less pore size. The spun-bonded (S) and melt-blown (M) fabrics are used in different combinations like SMS, SSM, and SSMS based on the requirements [15]. For the other application, the coated nonwoven is the choice to make the economically viable coveralls. The needle-punched nonwoven made from the multiple punching sequences with higher aerial density and subsequently coated with the required polymer can be used for the cost-effective coverall of medium durability fabric.

2.3 Functional Finish for Coverall

The fabric of the coverall provides the mechanical strength and the breathability to the coverall. In order to have extra protection from external agents light, bacteria, flash, oil, chemicals, heat, liquid, splash, etc., it is required to use the coat or treat the outer layer of the coverall with a suitable finish [16]. There are various finishes like antimicrobial, antiviral, UV resistance, waterproof, oil resistance, flame retardant (FR), and many other finishes are used [17].

One of the most important finishes required for the industrial coverall is the flame retardant (FR) finish. The workers are in direct or indirect contact with heat and radiation in some workplaces and the Standards and Regulations mandate the use of FR coverall. The other important finish for the general coverall is the stain or soil release finish. The coveralls come across oil, dust, and dirt during usage and they should not catch and retain the dust and stain. Even sometimes, the finish is used to reduce the friction between the fibres to improve the ballistic property of the fabric [18]. Most of the coveralls cover the full body and there is no or less ventilation for body heat and moisture to transfer. The heat build-up may lead to discomfort and hyperthermia. The thermoregulation finish is used in the interlining to absorb the heat generated on the inner side and to regulate the heat at microclimate [19].

3 Properties and Features of the Coverall

The coverall is used for different functions with the following properties

Protection: The main property expected from the coverall is protection from the external agent and protection from the external abrasive force, cut-marks, stretch, and chemical and biological agents [20].

Mechanical Strength: The strength is important for the coverall which is used for industrial applications. The tensile strength, multidirectional strength, tear resistance, etc., are to be maintained at the fabric level so as to get the same in the coverall. The joining point's strength is of the same importance as the fabric strength. The seam strength should be strong and optimum; the seam efficiency is expected to be around 90%.

Comfort: It is expected that the coverall should be comfortable so that the worker using the coverall can work efficiently and productively. The comfort is optimised based on the level of protection. The coverall used for the higher level of protection is fully coated and there is no ventilation, in such cases getting comfort from the coverall is the challenge. For the coverall having low level of protection, there is the possibility of higher breathability and ventilation and more comfort may be expected from such coverall [21].

Durability: It depends on the type and applications of the coverall. The coveralls used in the industry for routine work are expected to be durable. The coveralls that come under the contact of hazardous particles are of the disposable type.

Easy to Wash: The coveralls used for industrial applications and other routine works come under lot of dirt and dusts. It is expected to have easier washability of such coveralls.

Visibility: Most of the coveralls are used for a particular application. The person involved in such task or work needs to be identified easily. Sometimes even for the safety purposes, visibility is very important from the coverall. For visibility and for identity, the colour selection is important and for visibility under low light, the fluorescent material is used in the coveralls.

4 Coverall Types

The developments in material technology with the innovation in the special fibre, and fabric manufacturing technology as well as the innovation of the functional finish open up new dimensions of the coveralls. Also, industrial engineering with the care of work safety awareness and focusing on the efficiency of workers, the coverall is recommended at work site. Together this leads to the different varieties of the coverall for different applications. The coverall can be categorised based on the function and properties of the coverall and based on the applications of the coverall.

4.1 Broad Classification of Coverall

The coverall is broadly categorised into two types as reusable coveralls and disposable coveralls.

4.1.1 Reusable Coveralls

Most of the coveralls are used for industrial application and the daily routine work is of reusable coveralls. These coveralls are made up of woven fabrics and are easy to clean. Most of the coveralls designed for physical protection and mechanical stress fall under this category and being robust in construction can be easily washed and reused. The concern is to make sure the capability of the finishes applied on the fabrics to withstand the number of washes. In a few cases, the cleaning is done by simply putting it under the radiation. Using life cycle assessment technologies, the environmental sustainability of cleanroom coveralls is investigated for three different types of coveralls: a disposable flash spun-bonded high-density polyethylene (HDPE) coverall, a reusable woven polyester coverall, and a disposable SMS (spun bonded-melt blown-spun bonded) polypropylene (PP) coverall. In every category of environmental impact, it was found that the reusable cleanroom coverall system outperforms the two disposable cleanroom coverall systems [22].

4.1.2 Disposable Coveralls

The coveralls coming in exposer with the hazardous particles, chemical and biological agents and the radioactive isotopes must be disposed of after the use. Even for the low level of protection against these particles, it is better to dispose of the coverall. The nonwoven coveralls can be made at a very low cost and can be used as disposable coveralls. The largest volume of disposable coveralls is used in the pharmaceutical industry and in the medical field. Tyvek® is a popular brand of disposable coveralls that offers excellent protection against dangerous particles. It is constructed of synthetic flash-spun high-density polyethylene fibres. Tyvek is well-known for having characteristics including being waterproof against liquids yet enabling water vapour to pass through, as well as being hard to tear but easily cut. The study examined the physiological and subjective impacts of using various respirators while donning lightweight, disposable coveralls (Tyvek® 1412 polyolefin) that are frequently employed in the asbestos removal sector. These findings imply that, in certain low-workload asbestos abatement sector work conditions, heavier, more protective respirators may be linked to just minor increased physiological and subjective stress [23].

Decontamination is a crucial component of disposable coveralls, particularly for those come into touch with potentially dangerous particles. Many organisations offer guidelines for decontaminating such coveralls and disposing of them afterwards. Studies are also conducted regarding the effectiveness of the various decontamination techniques [24].

4.2 Types Based on the Function

Coveralls categorised based on the properties and functions are as follows.

4.2.1 Flame Resistance Coveralls

Flame resistance (FR) coveralls are the basic types of coveralls used in various industries in which there is a chance that workers come in contact with flame. With safety being a top priority, the coverall should be self-extinguishing and not easily combustible. Wool and glass fibres are the traditional fibres used to provide flame retardancy. For the coveralls where commodity fibres are used especially cotton that can catch the flame and propagate, it is mandatory to use the FR finishes over it. The FR finish can give flame resistance to these fibres. One such finish that is well-known for providing cotton with flame retardancy is PROBAN®. Cotton and materials containing cotton can be rendered flame retardancy with the PROBAN® process. In addition to providing protection against thermal risks, PROBAN® flame resistant (FR) apparel is designed for use as a protective workplace coverall. It was found in one investigation that the Proban® FR treatment was unaffected by washing; can be used in two-ply flame retardant (FR) "action coverall" (AC) and "anti-flash" (AF) hood and gloves [25]. The other option is to use the FR fibres for the coverall for the flame resistance property. The flame retardant high-performance fibre aramid is used in the coveralls as the fibre is inherently flame-resistant. Furthermore, some speciality fibres like poly(p-phenylene-2,6-benzobisoxazole) (PBO) fibre exhibit flame retardant qualities that surpass those of aramid fibres. Some of the commodity fibres coming with the FR property are FR viscose, FR cotton, FR acrylic, and FR polyester. By applying the right flame retardant finishes, nylon's flammability can be reduced. Multifunctional bio-based coatings have even been developed for this purpose [26].

FR coveralls are used for extremely high temperature with heat insulation property for the workers working at very high temperatures like mines. In such cases, the coverall should provide flame resistance as well as heat insulation. Fibres like carbon, glass, and asbestos can provide flame resistance at higher temperatures. For heat insulation purposes, the FR fabric with heat radiation retardation and reflective property is used.

4.2.2 Chemical Resistance Coveralls

The chemical resistance coveralls are very specific to the type and form of chemical. The chemicals are present in the form of gas, liquid splash, and liquid. In some of the chemical industries and agricultural farms, various hazardous chemical fumes may come across. In such cases, the full body coverall including the head cover has to be used, also for a higher level of protection, the coated type of coverall is recommended. The coverall may be incorporated with an adsorbent to adsorb the gases onto the surface like carbon powder [27]. In this type of coverall, there is a possibility of providing comfort by using non-coated fabric and providing a vent. For the chemical industries dealing with the liquid forms of chemicals, the coverall with coated fabric is used. The coverall is designed to be fully sealed in the area where the chance of liquid contact and penetration is there.

4.2.3 High-Visibility Coveralls

Sometimes the coveralls are just worn over the garment for visibility. The persons working under the site where the movement of the heavy load vehicles, cranes, and materials are frequent, it is required to highlight the worker for safety purposes. For the daylight and open area, wearing the coveralls of the bright colour may solve the purpose but for the dark site or low visibility area, the conditions are different. In such situations, it is required to wear the coveralls with reflective linings or a special coverall design. The present reflective interlinings are based on fluorescent fabrics and retro-reflective materials. For visibility, even certain cutting-edge technologies have been developed. In one study, tilt sensors with distress signalling capabilities were designed to promote safety during night time roadside work, rescue operations, and sporting activities. Belts with LED lights are also featured for improved visibility [28].

4.2.4 Special Coveralls

Some of the coveralls are designed for the very specific function and for specific conditions. The coverall for arch flash resistance, electromagnetic radiation protection, insect repellent, mosquito protection, extremely cold weather protection, and anti-dog bite jumpsuit are some examples of specially designed coveralls.

4.2.5 Arc Flash Coverall

Arc flash coveralls are specifically designed to offer maximum protection to workers from the intense heat and radiation that is produced by an electrical arc flash. Arc flashes are usually caused by a short circuit between energised conductors, and they have the potential to produce high temperatures and pressures that can be hazardous to workers. These coveralls are usually made of a combination of high temperature resistant materials, such as aramid fibres as discussed in the earlier section, as well as reflective materials, such as aluminized Mylar, that provide additional protection to workers who are exposed to the risk of arc flashes, such as electrical workers, linemen, and maintenance workers. However, it is also recommended that workers in other industries, wherever the risk of electrical arc flashes exists [29].

4.2.6 Electromagnetic Radiation (EMR) Protective Coverall

Electromagnetic Radiation coveralls are specially designed to protect workers from harmful EMR exposure such as X-rays, gamma rays, and microwaves. Prolonged exposure to EMR can lead to various health problems, including cancer, burns, and eye damage. These coveralls are typically made of materials that can absorb or reflect EMR, such as lead, tungsten, and silver. A coverall constructed of cotton woven fabric with a hybrid yarn composed of stainless steel wrap yarn can provide good electromagnetic radiation shielding effectiveness [30]. Workers exposed to high amounts of EMR, such as X-ray technicians, nuclear power plant workers, and microwave oven repair technicians, must wear EMR coveralls. Employees in other fields where there is a chance of exposure to EMR, like the medical, dental, and aviation industries are also advised to use them.

4.2.7 Insect Repellent Coveralls

The purpose of insect repellent coveralls is to shield workers from biting insects including fleas, ticks, and mosquitoes. Typically, the fabric used to make insect repellent coveralls has been treated with insect repellents, like permethrin or DEET. To help stop insects from biting through the coverall, they also have characteristics like tight cuffs and ankles, hoods, and long sleeves and pants. Workers who are at danger of exposure to biting insects, such as those in forestry, construction, and outdoor environments, must wear coveralls that repel insects. Workers in other fields where there is a chance of exposure to biting insects, like travel and tourism, agriculture, and public health, are also advised to use them.

4.2.8 Mosquito Protective Net Type of Coverall

Coveralls with mosquito protection nets are intended to shield workers from mosquito bites. Coveralls with mosquito protective nets are usually composed of a mesh material that is lightweight and treated with an insect repellent, such as permethrin. In addition, they have long sleeves, long legs, and a hood to assist keep mosquitoes from penetrating through the coverall. Workers who are at danger of mosquito exposure, such as those who treat malaria, those who work outdoors, and those who visit malaria-endemic regions, must wear coveralls with mosquito protective nets.

4.2.9 Anti-Dog Bite Coverall

The purpose of anti-dog bite jumpsuits is to shield employees from dog bites. Jumpsuits designed to ward off dog bites are usually composed of strong material that is impervious to tearing and punctures. Additionally, they have elements like sleeves and panels reinforced with Kevlar to assist keep dog bites from piercing the jumpsuit. Workers who are vulnerable to dog attacks, such as delivery drivers, postal workers, and animal control officials must wear anti-dog bite jumpsuits.

4.2.10 Extreme Cold Protective Coverall

Extreme cold suits are made to shield workers from bitterly cold conditions. Typically, extreme cold suits are constructed from a sturdy fabric and filled with a lightweight material, like synthetic fibres or down. In order to keep the wearer warm, the coveralls also have waist cinch, long sleeves, and pants. One example of these is the Iron-Tuff Insulated Coveralls by RefrigiWear Men, which offer protection down to – 50°F. With a ripstop outer shell, a warm and long-lasting inner lining, and 280 g of polyester insulation, the heavy-duty Freezer Edge® coveralls are comfortable down to – 30°F. These thermal coveralls' wind-tight, water-resistant outer shell helps keep out the bitter winter weather. Workers, who are exposed to extremely cold temperatures and those who operate in cold storage facilities, must wear extreme cold suits.

5 Application Domains of the Coveralls

There are many areas where the coveralls are used for different requirements and functions.

5.1 Civil Construction

During construction, the workers come in contact with lots of dirt and dust. Also, there is a higher probability of accidents at the construction site, especially if the construction sites are in public transport area. The coveralls are used at the construction site to protect the workers from dust, dirt, and as well as small debris. The coveralls designed for the construction workers are of general type and the visibility linings are used. The coveralls are made up of stronger fabric with soil release properties.

5.2 Manufacturing

Workers in the manufacturing industry are exposed to a variety of dangerous substances and compounds, such as acids, bases, and other chemicals. They are exposed to mechanical forces, abrasions, tears, and cuts. One crucial item of personal protective equipment (PPE) that can shield employees from these risks is a coverall. The coveralls required for the manufacturing industry are generally made with fabric with chemical and abrasion resistance. The coveralls designed for the manufacturing industry are generally used with other PPEs like gloves, goggles, etc. Each type of manufacturing industry has special requirements. The polypropylene-based coverall is a good option for manufacturing industry that involves general non-poisonous chemicals such as 3 M™ White Disposable Polypropylene Coveralls, 4500 Series. Hooded Coverall, DuPont™ Tychem™ 6000F orange colour provides outstanding chemical permeation protection against a wide variety of chemicals; it is constructed from strong, lightweight fabric that is laminated to heavy-duty Tyvek substrate with a proprietary barrier film. At food and beverage processing, wearing coveralls shields employees from exposure to substances that could contaminate food supplies. Here the focus of the coverall is to give good hygiene and to reduce the contaminant rather than the protection. The coveralls are designed along with the cap to cover all body parts including the hairs of the workers. At metal working industry, the workers are shielded from metal dust and other dangers by wearing coveralls. Other than dust cuts, protection is also important for the metal or glasswork industry workers. The coveralls with the hand gloves part with the cut-resistant fabric are recommended if work is related to the cutting operation.

5.3 Healthcare Coverall

Healthcare personnel can shield themselves from infections, bodily fluids, and other dangers by using healthcare coveralls, a type of personal protective equipment (PPE). Typically, disposable, chemical, and fluid-resistant nonwoven fabric is used to make healthcare coveralls. Cardinal Health™ coveralls are constructed from tri-layer SMS fabric. These coveralls are used in a range of environments, including clinics, nursing homes, hospitals, and labs. They are crucial for shielding medical personnel from a range of risks, including radiation, chemicals like disinfectants, infections, and human fluids. ChemoPlus™ a certified chemotherapy gown provides a barrier against 16 types of chemotherapy drugs.

5.4 Agriculture

To prevent exposure to pesticides, herbicides, and other chemicals, agricultural workers wear coveralls. A study emphasises the need to use personal protective equipment (PPE) to reduce skin contact with chemicals during farming [31]. The pesticides are toxic in nature and its exposure is harmful. It is recommended to use coverall with the hood to avoid exposure at the face area. They are also used to shield employees from additional dangers including dust and grime. The operator's exposure levels during fungicide applications in pepper produced in greenhouses were measured. It was found that the Hydrofoil® coverall significantly decreased the exposure when compared to the corresponding values for the cotton coveralls [32].

5.5 *Mining*

Coveralls are used for protecting miners from dust, filth, chemicals, sharp objects, high temperatures, falling objects, moving machinery, and electrical. The coveralls should have visibility for the miners working underground.

For the employees, in the military, waste management, auto repair, electrical work, food service, house renovation, janitorial work, landscaping, painting, plumbing, roofing, and welding, the coveralls are utilised.

6 Designing of Coverall

The coveralls are designed for the specific application with the required functions. The material selection such as fabric and finishes for the coverall is the first important step. The selection of the fibres, high-performance fibres, fabrics, and special finishes was discussed in the earlier sections. Apart from fabric selection, the garment design is important for the proper fit of the garment to wearers.

6.1 *Garment Fit*

A crucial factor is the coverall's size, which should match the average population size for coveralls to be sold in stores [33]. Most of the coveralls are designed with a loose fit for ease of movement and better cover and protection [34]. Also, from the comfort point of view, the loose fit for the coverall is preferred. Compared to Nomex® IIIA coveralls of the same style and size, the less drapable Kevlar®/PBI coveralls had bigger mean air gap widths.

6.2 *Vents*

From the comfort point of view, the breathability of the garment is important. Some of the coveralls with a lower level of protection come with breathable fabrics. For a higher level of protection, the coated fabric is used, where the issue of heat and moisture transfer arises and leads to discomfort. To provide relief from the heat and moisture built up at microclimate between the garment and the coverall, it is required to provide the vent. The coveralls are designed with open and closed vents based on the level of protection. The open vents are provided in the area where there is less exposure to the external agent. The close vents are closed by the zippers, so that whenever required zip can be opened for dissemination of the heat and moisture to the outside atmosphere. Opposite to the vent, the place at the coverall is required to be closed to avoid the direct flow of the outside particulate air. The opening at the cuff and the opening at the bottoms of the suits are to be closed at the time of the usage. The elastic bands are used for this purpose to provide a snug fit and to keep the coverall in place. The V01/V16 Series vented coverall from Red Wing is made for personnel in sectors where full-body coverage resistant to flames is necessary. To promote circulation, coveralls have mesh across the shoulders, arms, and across the back.

6.3 Donning and Doffing

While designing the coverall, the ease of wearing and removing should also be kept in mind. In situations, frequent donning and doffing of the PPE may be required, in such cases; the coverall is designed with a proper large opening with zippers and snaps.

6.4 Stitching and Seam Sealing

The coverall stitching is as equally important as the selection of the fabric. The stitching is done for the joining of different parts of the coverall together. For the woven fabric, the stitching can be done with the sewing machine and for the nonwoven fabric with seamless technology. After joining the panels, the stitch line strength and the permeability need to be checked. The seam line opening, seam strength, seam slippage, and seam efficiency have to be as per the standard. For the woven coverall, for the application where the mechanical forces and abrasion come into play, the seam should also bear the same without distortion. For the heat resistance coverall, the seam joints also should be stitched with the flame retardant type of the sewing thread. Wearing chemical protective gear (CPC), such as coveralls, is thought to be the final line of defence against airborne synthetic nanoparticles in the field of dermal protection (ENM). According to the study's findings, the penetration level increased noticeably in the presence of seams and, depending on how the seam was designed, might even reach 90% [35]. Coveralls are designed for a higher protection level with the requirements of the splash resistance, the seam sealing is important. Some coveralls also have features with sealed seams to help prevent molten metal from entering the coverall. The sealed seams and overlapping closures help to prevent EMR penetration.

Some of the portions of the coverall may have more exposure to external agents. Additional protection can be obtained by adding a protective pad at those regions. Finally, colour psychology and colour science also play an important role for a coverall; different colours are used for different workers in different professions.

7 Conclusions

In conclusion, this chapter explored the world of coverall garments, highlighting their essential role in protecting and aiding wearers across diverse work environments. By examining the potential of various textiles, high-performance fibres, and specialised finishes, coveralls can be tailored to meet specific functional requirements and enhanced performance. The chapter also delved into the ever-expanding market of coveralls, noting their varied applications and diverse range. Finally, the chapter explored key design considerations, including fit, vents, and seams, emphasising their crucial role in ensuring optimal comfort and protection for the wearer. This chapter underscores the significant contribution of coveralls to enhance workplace well-being and performance.

References

1. Chemical & Biological Protective Coverall Market | 2028. (n.d.). Retrieved December 7, 2023, from https://growthmarketreports.com/press-release/global-chemical-and-biological-protective-coveralls-market-expected-to-reach-us-62475-million-by-2028
2. Sledzińska K, Napieralska L, Jasińska I, Witczak E, Jarzyna V (2022) Comfort-related properties of workwear for employees with motor disabilities. Fibres Text Eastern Europe 30(2):47–54. https://doi.org/10.2478/FTEE-2022-0006
3. Machera K, Tsakirakis A, Charistou A, Anastasiadou P, Glass CR (2009) Dermal exposure of pesticide applicators as a measure of coverall performance under field conditions. Ann Occup Hyg 53(6):573–584. https://doi.org/10.1093/annhyg/mep032
4. Hooper DR, Cook BM, Comstock BA, Szivak TK, Flanagan SD, Looney DP, Dupont WH, Kraemer WJ (2015) Synthetic garments enhance comfort, thermoregulatory response, and athletic performance compared with traditional cotton garments. J Strength Cond Res 29(3):700–707. https://doi.org/10.1519/JSC.0000000000000783
5. Mukhopadhyay A, Shaw VP (2023) Reliability analysis of stretchable workwear fabric under abrasive damage: influence of stretch yarn composition. J Nat Fibers. https://doi.org/10.1080/15440478.2022.2134262
6. Kim HA (2016) Thermal comfort and tactile wearing performance of wool/nylon fabrics for tra-biz garment. Fash Text Res J 18(6):878–888. https://doi.org/10.5805/sfti.2016.18.6.878
7. Jin Z, Shen Q, Fan M, Tao J (2011) Research on thermal and moisture comfort of seamless knitted fabric made of triangle-hollow nylon. Appl Mech Mater 66–68:1311–1314. https://doi.org/10.4028/www.scientific.net/AMM.66-68.1311
8. Jassal M, Ghosh S (2002) Aramid fibres—an overview. Indian J Fibre Text Res 27(3):290–306
9. Chu Y, Liu Y, Fahaduzzaman M, Rahman MdR, Huang W, Li S, Chen C, Yang D, Chen X (2023) Ballistic performance of multi-layer fabric panels considering fabric structure and inter-yarn friction. J Ind Text. https://doi.org/10.1177/15280837231213411
10. Cruz-Estrada RH, Folkes MJ (2000) In-situ production of electrically conductive fibres in polyaniline-SBS blends. J Mater Sci 35(20):5065–5069. https://doi.org/10.1023/A:1004879614917
11. Kong X, Geng X, Geng S, Qu R, Zhang Y, Sun C, Wang J, Wang Y, Ji C (2022) Antibacterial para-aramid fiber loaded with in situ generated silver nanoparticles. Surf Interfaces. https://doi.org/10.1016/j.surfin.2022.101922
12. Kim HA, Kim SJ (2018) Flame-retardant and wear comfort properties of modacrylic/FR-rayon/anti-static PET blend yarns and their woven fabrics for clothing. Fibers Polym 19(9):1869–1879. https://doi.org/10.1007/s12221-018-1087-3

13. Vidya T, Prakash C, Babu VR, Rajwin AJ, Shah BA, Roy R (2023) Thermal properties of different kinds of polyester knitted fabrics. Indian J Fibre Text Res 48(3):295–301. https://doi.org/10.1016/j.ijheh.2010.08.007
14. Garzón-Villalba XP, Wu Y, Ashley CD, Bernard TE (2018) Heat stress risk profiles for three non-woven coveralls. J Occup Environ Hyg 15(1):80–85. https://doi.org/10.1080/15459624.2017.1388514
15. El-Ghoubary WA, El-Sayad GM, El Gamal FA, Abouzaid HAK (2023) Development of breathable and liquid/wet bacterial penetration barrier composite laminated fabrics for surgical gown applications. J Ind Text 53:15280837231170 2. https://doi.org/10.1177/15280837231170294
16. Zwane PEC, Masarirambi S (2011) Exploration of nano-finished non-wovens for potential use in protective clothing for agricultural workers in South Africa. Res J Environ Earth Sci 3(3):297–306
17. Aslam S, Hussain T, Ashraf M, Tabassum M, Rehman A, Iqbal K, Javid A (2019) Multifunctional finishing of cotton fabric. Autex Res J 19(2):191–200. https://doi.org/10.1515/aut-2018-0048
18. Gowthaman S, Srinu B, Balaganesan G (2023) Effects of ZnO nanowire and silica coatings on the inter-yarn friction and ballistic properties of Kevlar fabrics. Int J Impact Eng. https://doi.org/10.1016/j.ijimpeng.2023.104637
19. Arjun D, Hayavadana J, Prakash C (2023) Thermo-regulative finish on polyester blend knitted fabric. Indian J Fibre Text Res 48(3):247–252. https://doi.org/10.56042/ijftr.v48i3.6059
20. Nayak R, Kanesalingam S, Wang L, Padhye R (2019) Stab resistance and thermophysiological comfort properties of boron carbide coated aramid and ballistic nylon fabrics. J Text Inst 110(8):1159–1168. https://doi.org/10.1080/00405000.2018.1548800
21. Watson C, Troynikov O, Lingard H (2019) Design considerations for low-level risk personal protective clothing: a review. Ind Health 57(3):306–325. https://doi.org/10.2486/indhealth.2018-0040
22. Vozzola E, Overcash M, Griffing E (2018) Life cycle assessment of reusable and disposable cleanroom coveralls. PDA J Pharm Sci Technol 72(3):236–248. https://doi.org/10.5731/pdajpst.2017.007864
23. White MK, Hudnall JB, Hodous TK (1989) Physiological and subjective responses to working in disposable protective coveralls and respirators commonly used by the asbestos abatement industry. Am Ind Hyg Assoc J 50(6):313–319. https://doi.org/10.1080/15298668991374723
24. Slagley JM, Paschold H, Engler JM (2017) Evaluation of coverall field dry aerosol decontamination methods using a manikin. J Occup Environ Hyg 14(7):502–509. https://doi.org/10.1080/15459624.2017.1296235
25. House JR, Squire JD (2004) Effectiveness of Proban® flame retardant in used clothing. Int J Cloth Sci Technol 16(3–4):361–367. https://doi.org/10.1108/09556220410699552
26. Kulkarni S, Xia Z, Yu S, Kiratitanavit W, Morgan AB, Kumar J, Mosurkal R, Nagarajan R (2021) Bio-based flame-retardant coatings based on the synergistic combination of tannic acid and phytic acid for nylon-cotton blends. ACS Appl Mater Interfaces 13(51):61620–61628. https://doi.org/10.1021/acsami.1c16474
27. Sharma PK, Singh VV, Tripathi NK, Sathe M, Verma V, Sharma SP, Tomar LNS, Chaturvedi A, Yadav SS, Thakare VB, Acharya J, Gupta AK, Ganesan K (2019) Chemical protection studies of activated carbon spheres based permeable protective clothing against sulfur mustard, a chemical warfare agent. Defence Sci J 69(6):577–584. https://doi.org/10.14429/dsj.69.13958
28. Park S-J, Kim S-W (2020) Designing and fabricating of the high-visibility smart safety clothing. Korean Soc Emot Sensib 23(4):105–116. https://doi.org/10.14695/kjsos.2020.23.4.105
29. Doan DR, Floyd HL, Neal TE (2004) Comparison of methods for selecting personal protective equipment for arc flash hazards. IEEE Trans Ind Appl 40(4):963–969. https://doi.org/10.1109/TIA.2004.830796
30. Marciniak K, Grabowska KE, Stempień Z, Ciesielska-Wróbel IL (2019) Shielding of electromagnetic radiation by multilayer textile sets. Text Res J 89(6):948–958. https://doi.org/10.1177/0040517518760749

31. Protano C, Guidotti M, Vitali M (2009) Performance of different work clothing types for reducing skin exposure to pesticides during open field treatment. Bull Environ Contam Toxicol 83(1):115–119. https://doi.org/10.1007/s00128-009-9753-1
32. Tsakirakis A, Kasiotis KM, Arapaki N, Charistou A, Tsatsakis A, Glass CR, Machera K (2011) Determination of operator exposure levels to insecticide during bait applications in olive trees: study of coverall performance and duration of application. Int J Hyg Environ Health 214(1):71–78. https://doi.org/10.1016/j.ijheh.2010.08.007
33. Jeon E, Moon J (2013) Size analysis on disposable protective coveralls. J Korean Soc Cloth Text 37(5):655–666. https://doi.org/10.5850/jksct.2013.37.5.655
34. Adams PS, Keyserling WM (1995) The effect of size and fabric weight of protective coveralls on range of gross body motions. Am Ind Hyg Assoc J 56(4):333–340. https://doi.org/10.1080/15428119591016944
35. Vinches L, Hallé S (2017) Resistance of Type 5 chemical protective clothing against nanometric airborne particles: behavior of seams and zipper. J Occup Environ Hyg 14(12):939–946. https://doi.org/10.1080/15459624.2017.1368527

Personal Protective Gloves (Gloves for Medical Professionals and Others)

Vinay Kumar Chauhan, Ragini Dubey, Shiwanand Pandey, and P. Thennarasu

1 Introduction

Because of the COVID-19 pandemic, the concern for spreading of the infection through contact has been drastically altered. As a result, the use of personal protective devices (PPDs), which were previously reserved for health care workers (HCWs), is now in reach of general public, often without any discernible pattern. Providing care for COVID-19-infected patients has brought to light the need for improved accuracy and respect for novel guidelines, despite the fact that the use of medical gloves continues to be an essential component of infection control. It is possible that wearing gloves for extended periods of time when one is not actually providing medical care to a patient is unhelpful. There is a rising number of people are being diagnosed with dermatitis of the hands. Henceforth, the United States Food and Drug Administration has not given its clearance, approval, or permission for the use of any medical gloves for protection against COVID-19 or that prevent COVID-19 infection. The WHO has issued a warning regarding the gloves' inadequate protective capabilities. Incorrect usage poses a significant threat of infection and may actually

V. K. Chauhan (✉)
Department of Handloom and Textile Technology, Indian Institute of Handloom Technology, Chowkaghat, Varanasi 221002, India
e-mail: chauhanvkc12345@yahoo.com

R. Dubey (✉)
KVK PG College, Ghazipur 233001, India
e-mail: raginidubey91979@gmail.com

S. Pandey
Department of Fruit Science, College of Horticulture, Chandra Shekhar Azad University of Agriculture and Technology, Kanpur 208002, India

P. Thennarasu
Department of Handloom and Textile Technology, Indian Institute of Handloom Technology, Salem 636001, India

S. Sundaramoorthy et al. (eds.), *Textile Materials for Good Health and Wellbeing*, SDGs and Textiles, https://doi.org/10.1007/978-981-97-6496-9_9

promote the spread of severe acute respiratory syndrome coronavirus 2, (SARS-CoV-2). Wearing gloves on a regular basis might give you a false feeling of security and put you at risk for self-contamination. This might lead to involuntary facial touching or the transfer of fomites to inanimate surfaces like tables, phones, and keyboards. Viruses may live on gloves for up to four hours, according to research [1].

Medical workers and patients alike benefit greatly from the use of protective gloves against germs spreading. The Centres for Disease Control and Prevention in Atlanta state that medical practitioners should wear gloves to prevent the spread of disease [2]:

- Reduce the possibility of medical staff catching an infection from patients,
- Prevent the spread of germs from healthcare providers to patients.
- Reduce the risk of transmitting microorganisms from one patient to another through healthcare workers' hands.

According to CDC recommendations,

- Wearing protective gloves is necessary if there is blood or other potentially infectious material on mucous membranes or injured skin.
- After using gloves on a patient, they must be thrown away. It is not accepted practice to reuse gloves between patients.
- When moving from a filthy to a clean operation on the same patient, gloves must be changed.
- When handling more than one patient at a time, gloves should not be reused after being cleaned. The following are some more things that are included in WHO's suggestions:
- Just because you're wearing gloves doesn't mean you can skip washing your hands or using disinfectant.
- Gloves should be changed or discarded between handling the infected area and the rest of the patient's surroundings [2].

2 Gloves Purpose

Gloves are required if there is a risk of chemical, infectious agent, heat, cold, abrasive, or cutting object contact with the skin (www.research.uiowa.edu). The hands are especially susceptible to dangers such as chemicals, sharp objects, and harsh forces. Machine guarding on power tools, as well as self-sheathing needles and sharps-disposal boxes, are examples of protective engineering controls for the hands. Extremely few additional instances of hand-protecting engineering controls may be found. The use of proper gloves should support technical measures for hand protection because it is frequently not feasible to eliminate all hand exposure to dangers (www.lehman.edu).

3 Types of Protective Gloves (Sharman and Mahon [3], [19])

Gloves should be chosen depending on the kind of risk and the task at hand. Choosing the proper gloves for the job might be difficult due to the wide range of hand injuries that can occur on the job. There are typically four broad types of gloves (Sharman and Mahon [3], [16]).

1. *Leather, Canvas, or Metal Mesh Gloves*

   Protect your hands against abrasions and burns with a sturdy pair of gloves made of metal mesh, leather, or canvas. Sustained heat may also be avoided by using leather or canvas gloves.

   - Leather gloves can withstand mild heat, light impact, small chips, and abrasive materials.
   - A synthetic insert is needed for further warmth and cold protection in aluminized gloves, which reflect and insulate heat.
   - Aramid fibre gloves provide excellent insulation from extreme temperatures, resistance to cuts and abrasions, and long durability.
   - Gloves made from synthetic materials are resistant to heat and cold, can endure cuts and abrasions, and may even be resistant to certain diluted acids. These components are vulnerable to acids, bases, solvents, and alkalis.

2. *Fabric and Coated Fabric Gloves*

   - Cotton and other fabrics may be used to make fabric gloves, and gloves that are coated with fabric provide an additional layer of protection.
   - Dirt, slivers, chafing, and abrasions may all be avoided while using fabric gloves. When working with rough, sharp, or heavy materials, they do not provide enough safety. Some fabric gloves may be made more durable by adding a plastic coating.
   - Cotton flannel gloves often have a nap on one side and are composed of coated fabric. Fabric gloves may be made into all-purpose slip-resistant hand protection by covering the unnapped side with plastic. Chemical laboratory containers are handled using these gloves.

3. *Insulating Rubber Gloves*

   These gloves are used for protection against electrical hazards.

4. *Chemical and Liquid Resistant Gloves*

   Butyl rubber, neoprene rubber, natural rubber, nitrile rubber, and fluorocarbon (viton) rubber are only a few of the rubber and plastic materials used to create chemical-resistant gloves.

   Blending or laminating these materials may improve their functionality. Chemical resistance tends to increase with glove thickness, but gloves that are too thick might reduce grip and dexterity, which can be dangerous.

There are four primary varieties of disposable gloves used in healthcare, although they are all made and tested in much the same manner.

Types of Disposable Rubber Gloves (Sharman and Mahon [3], [21])

- Natural rubber is processed into latex gloves. Ammonia is used to conserve the sap from the rubber tree, and the sap is then filtered and concentrated to remove the water and contaminants. Latex may be used to make gloves after being processed with extra chemicals. These gloves are the strongest and most durable of their kind, as well as being biodegradable. They are also the best option for preventing viral infections. However, they may be allergenic for certain individuals, and prolonged contact with latex can cause sensitization.
- Comfortable to wear for long periods of time, natural (latex) rubber gloves are widely used for a variety of tasks. They have remarkable elasticity, tensile strength, and thermal stability. These gloves are designed to protect workers' hands against corrosive substances such as acids, alkalis, salts, and ketones that are often found in aqueous solutions. Not all workers should wear latex gloves because of the risk of allergic responses. Workers with latex allergies may be able to use alternatives such as hypoallergenic gloves, glove liners, or powder-free gloves.
- Vinyl gloves are not very effective in preventing the spread of germs, thus they should only be used in low-risk settings. These gloves are the most reasonably priced option, but they don't provide the same level of comfort as others. These gloves' rubber is made from plasticized polyvinyl chloride (PVC) that has been polymerized.
- Gloves made of the copolymer nitrile butadiene rubber (NBR) are known as nitrile gloves. This substance is the result of a copolymerization between acrylonitrile and butadiene. Nitrile gloves outperform latex gloves in durability and are more comfortable and flexible than vinyl gloves. They are anti-microbial and anti-chemical as well. Nitrile gloves, a kind of copolymer, can withstand exposure to high concentrations of chlorinated solvents like trichloroethylene and perchloroethylene without deteriorating. Nitrile gloves are designed for applications that need dexterity and sensitivity, yet they can withstand heavy usage and continuous contact with toxins that break down conventional gloves [15].
- Chlorine, carbon, and hydrogen are polymerized into neoprene gloves, which are then sulfur-crosslinked. These gloves are less popular yet have advantages over both latex and nitrile. They fit well, are long lasting, allergy free, and protect against chemicals and body fluids. They often outperform natural rubber alternatives in chemical and wear resistance.
- Butyl gloves are resistant to abrasion, oxidation, and degeneration brought on by ozone, and they are flexible even when they are cold. Butyl gloves are flexible even when cold and resistant to abrasion, oxidation, and ozone deterioration. When exposed to aliphatic hydrocarbons, aromatic hydrocarbons, or halogenated solvents, butyl rubber deteriorates rapidly (Sharman and Mahon [3], https://www.osha.gov).

4 Construction of Gloves [21]

Rubber gloves are constructed in much the same manner regardless of the material used. Ceramic or aluminium moulds made by hand are placed on a conveyor and dipped into wash tanks containing hot water as well as chlorine/bleach before being dried. This dipping process ensures that all traces of the previous batch of gloves have been removed. The moulds need to be coated once they have been cleaned so that the rubber will adhere to them. Since the materials need distinct treatments, here is where the procedures diverge. The length of time latex gloves spend in the tank determines their final thickness.

After the moulds are released, they are cleaned in hot water with chlorine to remove any remaining rubber and then spun to remove any remaining debris. Excess latex and chemicals are washed away in this procedure to mitigate allergic responses. After the moulds are filled, rubber molecules are cross-linked via a process called vulcanization to strengthen the gloves.

The gloves are next run through a succession of brushes that roll (also called beading) their ends into cuffs, after which they are given a final washing to eliminate any remaining extra latex. Then, air jets are used to blast them away. The gloves are inspected and packaged after they have been removed from the moulds [15].

Nitrile and vinyl gloves, in contrast to latex gloves, aren't crafted from natural rubber. Although these gloves are not made from natural latex, their manufacture is quite similar. The ceramic, hand-formed moulds are washed in bleach water to eliminate any residue before being used in the production process. Calcium carbonate and calcium nitrate are used to aid in the coagulation of synthetic materials around the formers after they have dried. Once again, the formers are dried out. The formers are dipped into the tanks containing NBR or PVC. The gloves are then formed by drying the materials at a high temperature. Nitrile gloves are either chlorinated or polymer coated to make donning them less of a hassle. Gloves may be chlorinated to make them more durable and slippery by exposing them to chlorine in the form of an acid combination or gas. The addition of a polymer layer to the glove surface makes it more slippery. Finally, during the stripping step, the gloves are blown off the formers [14].

5 Factors to Be Considered for Gloves Selection [19]

When choosing the right glove, there are number of things to take into account in addition to glove material.

- area that has to be protected (arm, forearm, hand only),
- grip requirements (oily, wet, dry) and dexterity,
- electrical protection,
- thermal protection,
- cut and abrasion resistance,

- size and comfort,
- nature of contact (vapour exposure, splash, total immersion),
- type(s) of chemicals handled,
- duration of contact,
- physical state of the chemical(s) (gas, liquid, solid).

Various glove materials offer protection against distinct chemical compound groups. In general, the better a glove is in protecting its wearer from chemicals, the thicker the material used to make it (https://www.lehman.edu).

- Fine dexterity may be achieved by wearing two pairs of gloves, one made of a less suitable material than the other; the outer glove should be removed and discarded at the first evidence of contamination. Wearing a tight-fitting glove over the loose glove may improve overall dexterity in particular situations.
- The mils or gauge thickness of the gloves is another factor to think about. The thickness of a 10-gauge glove is 10 mils or 0.01 inches. Thinner, lighter gloves may have faster breakthrough times but provide less touch sensitivity and flexibility. In most cases, the breakthrough time will increase by a factor of four if the glove's thickness is doubled.
- Determine the level of immersion for the arm or the likelihood of chemical splash when selecting a glove length. Gloves with a length of more than 14 inches provide greater protection against water contact.

It is possible that glove size matters, too. There is no one-size-fits-all solution. Work gloves should fit snugly without causing tiredness, but should not be so large that the wearer's fingers poke out. Glove sizes are often stated as a fraction of the hand's diameter in inches. For some chores, it's also important to choose up gloves with the right colour, cuff style, and lining [19].

6 Glove Properties [4]

6.1 Barrier Properties

Gloves' principal purpose is to prevent the spread of disease by acting as a reliable barrier for the wearer and the one being treated. They are necessary during invasive operations and each time skin, mucous membranes, or sterile areas are breached. As a result, criteria have been devised to ensure that all gloves work satisfactorily regardless of material [4, 5] and that leakage is kept to a minimum even when the gloves seem intact.

Multiple studies examining the longevity of barrier protection have shown that natural rubber latex (NRL) gloves provide fewer rates of perforation and lower virus leakage rates than vinyl gloves [6–8].

6.2 Fit and Comfort

As per the European Commission's Scientific Committee on Medicinal Products and Medical Devices, nitrile gloves have a little greater elastic modulus (or stiffness) but a slightly lower TS (Tensile Strength) than latex gloves. Nitrile, unlike latex, does not completely recover after being stretched because it has a greater permanent set. Since nitrile gloves are often made to fit more loosely than latex ones, the material and construction of these gloves may have an impact on the wearers' tactile sensitivity and fine motor control. Nitrile gloves that suited the fingers were too thin for the hands, and gloves that fit the hands were huge for the fingers, as participants reported in research [9].

6.3 Enhanced Barrier Performance by Means of Double Gloving

There is a danger of contamination as well as exposure to blood-borne pathogens during procedures since holes in gloves are sometimes not discovered [10, 11]. Two pairs of gloves should be used in high-risk surgical situations to prevent the spread of illnesses like HIV. Most studies on double gloving have focused on developing methods of puncture indication (i.e., utilizing various coloured gloves to signal the outer glove layer has ruptured) for use in surgical procedures [8]. Therefore, some surgeons have been using double gloving consistently ever since the early 90 s. Nevertheless, reports indicate that double gloving is less prevalent in the United Kingdom, the United States, and Europe than in other nations, with the exception of orthopaedic and maxillofacial surgery [12].

NRL gloves are often used for double gloving, although synthetic gloves may be used on the inside to lessen the likelihood of allergic responses. Single gloving provides enough protection for low-risk surgery, while double gloving greatly increases protection against perforations generally attained in medium-risk surgery, according to a recent evaluation of gloves utilized in surgical procedures. Glove liners, triple gloving, and knitted gloves offered the maximum degree of protection necessary for high-risk surgery, including orthopaedic, craniofacial, and cardiac procedures [13]. Further, the author suggests using high-quality low-allergen NRL gloves at all these levels to have the highest barrier performance possible when neither the patient nor the operating staffs has an allergy to NRL.

7 Glove Testing

Gloves are required to pass two distinct tests. To begin, medical-grade gloves must be verified as such by regulatory bodies before they may enter the market. After these conditions are satisfied, the quality of each batch of gloves is checked using a series of progressively less rigorous tests.

Gloves are initially tested with air and then subjected to the pinhole leak test in accordance with ASTM guidelines while being tested in batches. All gloves contain tiny holes, but this test makes sure the holes aren't large enough to let any liquid through. The gloves are tested for leaks by filling them with a litre of water and keeping an eye on them for two minutes. There is a 2.5% quality tolerance for batches of industrial gloves, which implies that for every 100 gloves tested, there can be no more than 2.5 that don't pass. If there are less than 1.5 defective gloves per hundred, then the gloves are of a high enough quality to be utilized in medical settings.

In addition to the testing that must be done on each batch of gloves, manufacturers must also pass regulatory tests to demonstrate that their gloves meet the standards necessary to get a medical rating. Each pair of gloves must measure up to strict specifications for length (from cuff to middle fingertip), breadth (palm), and thickness (thickness of material). Glove size and type affect these dimensions. In addition to tensile and elongation performance criteria, gloves must also pass sterility tests in accordance with the rules in the United States Pharmacopeia and ageing tests in which they are heated to 70 to 100° C and kept at 100° C for a certain amount of time. Powder per glove must be less than 2 mg. In addition, the protein content and antigenicity of latex gloves must pass further testing. When using disposable gloves, they must be discarded at the first indication of contamination. If it is planned to use reusable gloves for a lengthy amount of time, it should be washed often [14–21].

8 General Manufacturing Processes and Machine Types

There are two main types of machinery used in the production of gloves made from natural rubber or synthetic materials. Both batch-style and continuous chain/drag line machines fall within this category. Surgical gloves are often produced using batch machines (Fig. 1a), which may be seen as a collection of indexing formers moving sequentially through a set of dipping tanks. Each dipping function is constrained simply by the pace of the slowest dip that came before it (Fig. 1b) [8].

There is a lot more riding on the design of a continuous chain machine than there is with a batch machine (Fig. 1). There is a lot of interdependence between the chain's movement speed, the length of the dip tanks, and the distances between each dipping point. If, for any reason, there is a need to change the soak duration in a leach tank, it has to be changed everywhere. Physically modifying the route lengths or lengthening

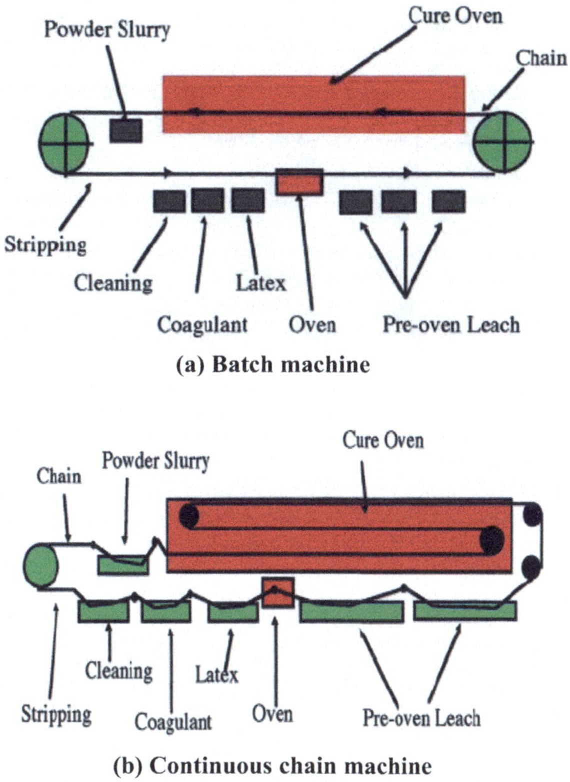

Fig. 1 Schematic representation of gloves manufacturing machine [8]

the leach tank is the only method to do this. Surgical gloves are often made on batch machines, whereas examination gloves are made on chain machines [8].

9 Conclusions

There has been significant development in the field of medical gloves. Care for patients with COVID-19 has highlighted the need for more precise and observance of innovative recommendations, but it remains a vital aspect of the infection control approach. It is critical that gloves be worn correctly and removed hygienically; moreover, protective gloves are considered disposable and should not be reused.

There are several varieties of protective gloves made from many different materials. The production process may be batch or continuous, depending on the requirements. Varying testing bodies had different criteria, therefore gloves produced had to be put through a variety of tests. Biohazard danger necessitates regular glove change; however, gloves may vary in protection, tactility, allergy risk, and expense.

References

1. Anedda J, Ferreli C, Rongioletti F, Atzori L (2020) Changing gears: medical gloves in the era of coronavirus disease 2019 pandemic. Clin Dermatol 38(6):734–736. https://doi.org/10.1016/j.clindermatol.2020.08.003
2. Garus-Pakowska A, Sobala W, Szatko F (2013) The use of protective gloves by medical personnel. Int J Occup Med Environ Health 26(3):423–429. https://doi.org/10.2478/s13382-013-0095-1
3. Sharman M, Mahon HSM (2023) Personal protective equipment. Infect Control Small Anim Clin Pract. https://doi.org/10.1079/9781789244977.0006
4. Palosuo T, Antoniadou I, Gottrup F, Phillips P (2011) Latex medical gloves: time for a reappraisal. Int Arch Allergy Immunol 156(3):234–246. https://doi.org/10.1159/000323892
5. Pratt RJ, Pellowe CM, Wilson JA, Loveday HP, Harper PJ, Jones SR, McDougall C, Wilcox MH (2007) Epic2: national evidence-based guidelines for preventing healthcare-associated infections in NHS hospitals in England. J Hosp Infect 65(Suppl 1):S1–S64
6. Klein RC, Party E, Gershey EL (1990) Virus penetration of examination gloves. Biotechniques 9:196–199
7. Rego A, Roley L (1999) In-use barrier integrity of gloves: latex and nitrile superior to vinyl. Am J Infect Control 27(5):405–410. https://doi.org/10.1016/S0196-6553(99)70006-4
8. Yip E, Cacioli P (2002) The manufacture of gloves from natural rubber latex. J Allergy Clin Immunol 110(2):S3–S14. https://doi.org/10.1067/mai.2002.124499
9. Sawyer J, Bennett A (2006) Comparing the level of dexterity offered by latex and nitrile SafeSkin gloves. Ann Occup Hyg 50(3):289–296. https://doi.org/10.1093/annhyg/mei066
10. Chan KY, Singh VA, Oun BH, To BH (2006) The rate of glove perforations in orthopaedic procedures: single versus double gloving. A prospective study. Med J Malays 61:3–7
11. Naver LPS, Gottrup F (2000) Incidence of glove perforations in gastrointestinal surgery and the protective effect of double gloves: a prospective, randomised controlled study. Eur J Surg 166(4):293–295. https://doi.org/10.1080/110241500750009113
12. St. Germaine RL, Hanson J, De Gara CJ (2003) Double gloving and practice attitudes among surgeons. Am J Surg 185(2):141–145. https://doi.org/10.1016/S0002-9610(02)01217-5
13. Tanner J (2008) Choosing the right surgical glove: an overview and update. Br J Nurs (Mark Allen Publ) 17(12):740–744. https://doi.org/10.12968/bjon.2008.17.12.30292
14. https://blog.ammex.com/how-nitrile-and-vinyl-gloves-are-made/ (Accessed 25 July 2023).
15. https://blog.ammex.com/how-latex-gloves-are-made/#.Xp4Kx8hKg2x. Accessed 22 July 2023
16. https://ehs.berkeley.edu/glove-selection-guide. Accessed 1 Aug 2023
17. https://www.lehman.edu/administration/environmental-health-safety/hazard-control-PPE-gloves.php. Accessed 3 Aug 2023
18. https://www.osha.gov/sites/default/files/publications/osha3151.pdf. Accessed 1 Aug 2023
19. https://ehs.princeton.edu/workplace-construction/workplace-safety/physical-safety/personal-protective-equipment-ppe/hand-and-arm-protection. Accessed 25 July 2023
20. https://ehs.research.uiowa.edu/ppe-glove-information. Accessed 1 Aug 2023
21. https://www.thomasnet.com/articles/other/how-to-make-medical-gloves/. Accessed 20 July 2023

Bandages (Compression, Retention, and Support)

Monica Sikka, Samridhi Garg, and Neha Kumari

Abstract In the context of medical treatment, bandages play crucial functions in compression, retention, and support. This chapter focuses on the diverse world of bandages for treating Chronic Venous Insufficiency (CVI). Throughout history, bandages, the hidden heroes of healthcare, have undergone substantial evolution, providing cutting-edge remedies for a variety of ailments. For academics, students, and healthcare professionals, this book chapter serves as a thorough reference that provides insightful information on compression bandages and stockings. It highlights the vital role that these seemingly straightforward medical instruments play in enhancing patient outcomes and quality of life. The role of bandages in medical treatment is still crucial, whether it be for compression, retention, or support.

1 Introduction

Chronic Venous Insufficiency (CVI) is a widespread illness that predominantly affects the lower extremities and represents a significant health concern for adults across the globe. CVI is estimated to affect 2–6 million people in the US alone, with nearly five lacs of them also suffering from painful venous ulcers. This condition's impact extends far beyond the U.S., with approximately 100 million individuals in China (accounting for 8.89% of the population) grappling with lower extremity venous disorders, resulting in a yearly morbidity rate ranging from 0.5 to 3.0%, with venous ulcers contributing to 1.5% of these cases. Furthermore, in Africa, varicose veins afflict around 3–8% of the female population, indicating a pervasive issue. The prevalence of CVI is also notably high in various European countries, including France, Germany, Poland, and Finland. This condition tends to affect individuals over the age of 50, with a higher incidence among women than men. It impairs the normal functioning of the venous system and results in various symptoms and complications.

M. Sikka (✉) · S. Garg · N. Kumari
Department of Textile Technology, Dr B R Ambedkar National Institute of Technology, Jalandhar, Punjab 144 008, India
e-mail: sikkam@nitj.ac.in

S. Sundaramoorthy et al. (eds.), *Textile Materials for Good Health and Wellbeing*, SDGs and Textiles, https://doi.org/10.1007/978-981-97-6496-9_10

The disease mainly affects the lower extremities and is caused by damaged or weakened vein valves that lead to improper blood flow. This, in turn, causes blood to pool in the lower limbs, leading to symptoms such as leg swelling, pain, skin changes, and the development of venous leg ulcers. To manage CVI, bandages, particularly compression bandages, play a vital role. Its goal is to minimise edoema and improve the venous and lymphatic return from the lower extremities. Recent advances in technological materials and product design have broadened the options for upgrading and optimizing compression therapy. Among these innovations, compression textiles have garnered increasing attention due to their distinctive structures and manufacturing methods, which allow for controlled compression tailored to various body shapes at low cost. Compression bandages (CBs) and medical elastic compression stockings (MECS) are two common forms of compression textiles used in the prophylaxis and management of CVI. These bandages help improve blood flow, reduce oedema, and alleviate discomfort by exerting controlled pressure on the affected limbs. Additionally, they promote tissue healing and the restoration of skin integrity, which is crucial in preventing the progression of CVI-related complications. Table 1 provides a succinct overview of Chronic Venous Insufficiency, covering its definition, prevalence, causes, symptoms, diagnosis, management, and the importance of early treatment.

This chapter explores compression therapy in detail, examining the different types of bandages, materials, and application techniques used in CVI treatment. Its ultimate goal is to better understand the processes of compression treatment and the usage of compression bandages or stockings in treating CVI.

2 Compression Therapy

Compression therapy is a prevalent treatment for various venous conditions, such as deep vein thrombosis, varicose veins, ulcers, and oedema. It effectively reduces swelling and enhances blood flow, but some patients find it difficult to tolerate. In such cases, intermittent compression therapy (ICT) emerges as an alternative. This method utilizes an air compressor for inflating and deflating an airtight limb bag, reinforcing venous function and preventing fluid leakage. Research suggests that ICT not only fosters healing and fibrinolysis but also has the potential to replace traditional techniques. Remarkably, rapid ICT application yields better results than slow inflation. However, it's crucial to note that ICT is not suitable for individuals with peripheral vascular disease or deep vein thrombosis. Furthermore, individuals with congestive heart failure require strict monitoring while introducing ICT into their therapy regimen. Overall, this therapy serves as a critical tool for addressing chronic venous and lymphatic issues in the lower extremities by mitigating the effects of gravity and enhancing blood flow while reducing the risk of blood clot formation and promoting fluid reabsorption. The required pressure varies depending on the patient's position, with lower pressure in a lying position and higher pressure in a standing position.

Table 1 Chronic venous insufficiency (CVI)

Aspect	Chronic venous insufficiency (CVI)
Definition	A chronic medical situation characterized by inefficient venous blood return from the lower limbs, leading to blood pooling in the legs
Prevalence	Affects millions of individuals worldwide
Etiology	Damage or malfunctioning of venous valves
	Weakened vein walls
	Inherited predisposition
	Prolonged sitting or standing, obesity, and other factors
Symptoms	Leg swelling (oedema)
	Leg pain or aching, exacerbated by prolonged standing or sitting
	Skin changes, such as hyperpigmentation
	Development of venous ulcers in advanced cases
	Varicose veins
Diagnosis	Clinical assessment of symptoms and medical history
	Doppler ultrasound to assess blood flow in the veins
	Venous duplex ultrasound to identify valve dysfunction
	Venography (less commonly used)
Management	Compression therapy (use of compression stockings)
	Lifestyle modifications (leg elevation, regular exercise, avoiding prolonged inactivity)
	Medications to reduce inflammation and improve vein function
	Minimally invasive procedures (endovenous ablation)
	Surgical intervention (in severe cases)
Importance of early treatment	Early detection and management are crucial to prevent complications and enhance the patient's quality of life

2.1 *Classification of Compression Therapy*

The compression therapy is a popular method of treating cardiovascular disease. Bandages, stockings, and intermittent pneumatic compression are the three main modalities used for this therapy [1]. Each procedure has pros and cons, and selecting among them depends upon the patient's condition, duration of therapy, and patient compatibility. The most popular and effective solutions for long-term therapy of lower limbs in the early stages of CVI are compression stockings or tights, whereas sleeves are chosen for the upper limbs. Compression bandages are more appropriate for more severe stages of CVI, particularly if there is a venous leg ulcer with a dressing present or a need for a substantial decrease in venous or lymphatic oedema.

Table 2 Classification of compression bandages

Class	Compression level	Pressure range at ankle (mmHg)	Bandage examples
3a	Light	14–17 mmHg	Elset® (Mölnlycke)
3b	Moderate	18–24 mmHg	Velkomp (DattMediproductsPvt. Limited)
3c	High	25–35 mmHg	Setopress (Mölnlycke), Elodur® forte (BSN medical)
3d	Extra high	Up to 60 mmHg	Various extra-high-compression bandages

2.1.1 Compression Bandages

These are classified as retention, support, or compression based on their efficacy in standard testing. Furthermore, compression bandages are classified based on the force needed to stretch them and, thus, the maximum amount of compression they may exert on a limb. In addition, it is essential to consider that the efficacy of a bandage in clinical settings may differ from its laboratory performance. This variability is contingent upon factors such as training of the operator and application method, including the number of layers applied, the extent of extension utilized, and whether the bandage is presented in a spiral or figure eight configuration. The following are common compression systems utilized to treat venous leg ulcers [2] (Table 2).

Bandages have a variety of uses, such as compression, support, and retention.

2.1.2 Bandages for Retention

These are specifically designed to secure and hold medical dressings, wound coverings, or other medical devices in place. They are essential for maintaining the position and integrity of these materials, ensuring they stay in the right location and perform their intended functions effectively. Retention bandages are often made from materials like gauze, elastic, or cohesive materials and are commonly used in wound care, post-surgical dressing changes, and various medical procedures. They provide stability and prevent shifting or displacement of dressings or devices, promoting proper healing and patient comfort.

2.1.3 Support Bandages

These bandages offer retention and stop a mass of tissue from developing a deformity or changing shape due to sagging or swelling. Support bandages come in various forms, including elastic bandages, wraps, braces, and sleeves. The type of bandage to be used depends on the particular requirements of the person, the nature of the injury or ailment, and the amount of support needed. It is crucial to follow the instructions of a healthcare professional while using these bandages to ensure their efficiency and safety.

Leg ulcers and varicose veins represent two prevalent medical conditions that are effectively managed with the use of *compression bandages*. Leg ulcers, often found on the lower legs, are open sores that can result from a variety of underlying causes, with impaired circulation being a common factor. Compression bandages play a central role in the treatment of leg ulcers. By applying consistent pressure to the affected area, these bandages help enhance blood flow and reduce swelling, which are essential factors in the healing process. This compression therapy not only promotes wound healing but also alleviates discomfort and minimizes the risk of infection. Varicose veins, on the other hand, are swollen and twisted veins that frequently appear on the legs, often due to weakened vein valves. The pooling of blood in these veins can lead to discomfort, pain, and aesthetic concerns. Compression bandages or stockings are employed to address the symptoms associated with varicose veins. They work by supporting the venous system, improving blood circulation, and reducing the appearance of varicose veins. This compression therapy not only offers relief from discomfort but also aids in preventing the worsening of varicose veins. In both cases, the application of compression bandages is a cornerstone of conservative treatment prescribed by healthcare professionals. Properly fitted and maintained, these bandages effectively contribute to the management and alleviation of symptoms associated with leg ulcers and varicose veins, ultimately enhancing the quality of life for individuals affected by these conditions.

In the late nineteenth century, rubber-based elastomeric compression bandages were initially employed. But lately, bandages composed of Lycra or other elastane fibres have taken their place—they're stronger, lighter, more comfortable, and machine-washable. These days, bandages are made of knit or woven materials and are intended to give certain compression levels based on performance criteria.

2.1.4 Short-Stretch Bandages or Low-Elasticity Bandages

Short-stretch bandages are a valuable resource in treating venous leg ulcers, effectively reducing edema and improving venous return. They are characterized by their low elasticity and high working pressure, with a recommended resting pressure range of 40–60 mmHg [3]. However, this pressure diminishes relatively quickly, necessitating frequent checks and replacements, ideally daily. To prevent pressure injuries and ensure even pressure distribution, padding with cotton or foam is essential. These bandages play a pivotal role in the decongestion phase of treatment, alongside non-elastic and adaptive compression bandages, and multicomponent systems. They contribute to the reduction of swelling and the facilitation of venous leg ulcer healing [4].

In practical terms, short-stretch bandages display low resting pressure but exhibit significant working pressure, particularly during a patient's active moments. This dynamic contrast results in a pronounced massaging effect on the calf muscles, reinvigorating the muscle pump's efficiency by facilitating the movement of venous blood from the superficial network to the deep network. As a consequence, these bandages prove highly effective in the management of advanced stages of chronic venous disease (CVD), most notably in venous leg ulcers and severe edema. Their capacity to enhance venous return and lymphatic drainage is particularly advantageous for patients with adequate mobility. These bandages can be worn continuously, day and night, due to their low resting pressures. However, it's worth noting that, as indicated in the literature [5], the interface pressure decreases notably a few hours after application, potentially leading to the need for frequent adjustments to maintain the necessary sub-bandage pressure. This challenge is particularly evident at the outset of compression therapy when venous edema is in the absorption phase, and significant reductions in leg volume occur. The unique behaviour of short-stretch bandages, which adapt their compression irregularly based on a patient's movement and mobility, means that, at rest, they function as a passive system with minimal compression. When a patient is active, the calf muscles exert pressure on the bandage, creating robust resistance due to increased muscle volume, resulting in elevated working pressure [6]. As patients walk, this dynamic interplay of low and high pressures generates a pronounced massaging effect, significantly enhancing venous return and contributing to the reduction of edema.

2.1.5 Long-Stretch Bandages or Elastic Bandages

Long-stretch bandages are characterized by their high extensibility, primarily owing to the presence of elastic fibres. These bandages can be either dry or cohesive. Their elastic properties result in a consistent level of pressure, which remains nearly constant whether the wearer is at rest or engaged in physical activity. As a consequence, they exhibit a low static stiffness index (SSI) and do not provide a significant massaging effect. This type of bandage is particularly suitable for individuals who are immobile or have limited mobility, especially those who are unable to activate their calf muscle pump effectively [3]. Elastic bandages may need to be removed during nighttime as the pressure they exert can be uncomfortable for sleeping. The requirement for daily reapplication can sometimes pose a significant challenge in terms of patient compliance with the treatment. Furthermore, applying elastic bandages, especially when lacking a visual compression guide, can be somewhat challenging for individuals without prior training, as they need to be stretched to approximately 50% of their original length, which necessitates regular practice and knowledge of proper bandaging techniques. These elastic systems deliver substantial pressure while at rest, thanks to the recoil force generated by the elastic materials they incorporate. During physical activity, the elastic nature of the bandage reduces the resistance to muscle contraction. In this scenario, two forces come into play: the recoil force of the bandage and the radial force generated by muscle contraction. These opposing forces partially negate each other, resulting in minimal differences in the pressure applied to the leg between rest and activity, hence maintaining a low static stiffness index (SSI). As a result, there is little to no massaging effect while walking. This is why long-stretch (or elastic) bandages are primarily recommended for individuals who are immobile or have limited mobility.

2.1.6 Multicomponent System

Multicomponent multilayer compression systems, formed by amalgamating multiple bandages with distinct and complementary characteristics, harness the benefits of both long-stretch and short-stretch bandages. These systems deliver effective compression, featuring noticeable pressure peaks during ambulation, which results in a substantial massaging effect on the calf-muscle pump. When the wearer is at rest, these systems maintain consistent intermediate pressure, ensuring the uninterrupted flow of venous return. Remarkably, this combination of bandages is well-tolerated during nighttime use, allowing for continuous application for up to 7 days without any loss of pressure or slippage. Traditionally, multicomponent multilayer systems were comprised of four bandages [7, 8].

2.1.7 Adaptive Compression Bandages

Adaptive compression bandages (Fig. 1) feature Velcro strips for easy adjustability. They are employed in the treatment of venous leg ulcers, effectively reducing swelling and enhancing blood circulation. These bandages are characterized by their low elasticity and their ability to exert significant working pressure. Users can self-apply these bandages and fine-tune the pressure to their comfort. They are well-suited for both initial treatment and ongoing maintenance. Research studies have demonstrated their efficacy and user-friendly nature [9].

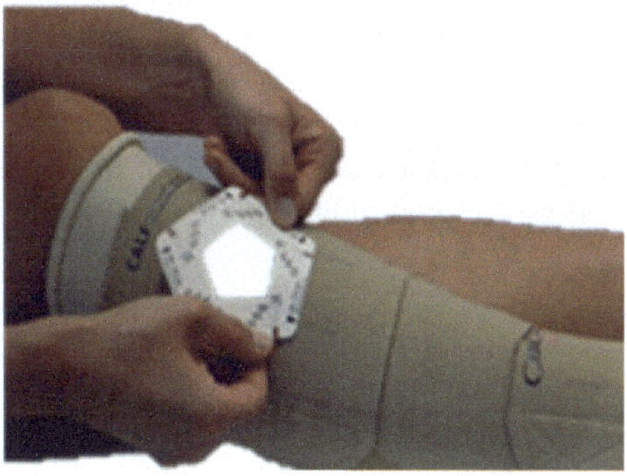

Fig. 1 Adaptive compression bandages

3 Compression Stockings

These can treat open ulcers and lower the chance of recurring after healing. Stockings are classified similarly to bandages based on the amount of compression given to the limb. Additionally, compared to bandages, stockings are less susceptible to operator variability (Table 3).

Compression stockings treat venous leg ulcers caused by poor blood circulation. After the ulcer heals, stockings prevent recurrence and oedema. They come in different materials and compression classes. To ensure effectiveness, measure and fit correctly. Proper washing and care maintain quality [10]. MCS for legs constructed as per standard, also known as ready-made, series, or standard stockings, are available in 4 models: calves, knee-length stockings, thighs, and compression support tights. Arm stockings can also be used to treat upper-limb problems, with a shoulder cap or without a shoulder cap and with an integrated glove or without an integrated or separate glove. There are various knits used to make medical compression stockings.

1. Machine-made, ***flat-knit garments*** with at least one knitted thread and one elastic thread inserted every other course. These stockings can be made with a high degree of compression and an extremely tight fit. Commercially available knit MCS are often stiffer, but because of their increased rigidity, it is more difficult to slide them across tissue folds. Circular-knit stockings lack these characteristics (inflexibility, increased stiffness, and difficulty slipping on). You can add or remove individual courses from flat-knit stockings. As a result, they can be modified to fit atypical leg circumferences, like those caused by severe lymphedema. The stocking has a seam at the rear of the leg and is knitted as a single piece that is sewn together.
2. Machine-made ***circular knits*** with seams on both single and double surfaces, at least one knitted thread, and one inserted elastic thread every other course. A knitting cylinder with a predetermined number of needles is used to create these MCS. It is not possible to add or remove more courses. Only by adjusting the thread tension or the course (loose or tight knitting) can the stocking be made to fit the contour of the leg. Therefore, the shaping of circular-knit MCS is limited. Circular-knit stockings cannot fit limbs with extremely small circumferences, drastic circumference variations, or deep tissue folds.

Table 3 Classification of compression stockings

Class	Support level	Pressure range (mmHg)	Intended use
1	Light support	14–17	Treatment of varicose veins
2	Medium support	18–24	More severe varicosities, prevention of venous leg ulcers
3	Strong support	25–35	Severe chronic hypertension, severe varicose veins, and prevention of venous leg ulcers

4 Ulcer Stocking System

Ulcer stocking systems (Fig. 2) are used to treat chronic venous leg ulcers caused by poor blood circulation. They are divided into two sections: the understocking and the compression stocking that is worn below the knee. The understocking protects the wound dressing while the compression stocking provides the necessary pressure for healing. They are easy to apply and remove, available in different sizes, and can improve patient comfort and quality of life. These systems prevent oedema and the recurrence of ulcers and are recommended during the maintenance phase of compression therapy [11] (Table 4).

Compression therapy products are predicted to reduce leg and foot volume, venous reflux, and hypertension, driving demand. The rise in orthopaedic treatments, the ageing population, and sports medicine are driving the industry (Compression therapy market size, share and trends, n.d.). Compression bandages and stockings are essential medical devices that apply controlled pressure to improve circulation, reduce swelling, and accelerate wound healing. With various options available, such as short-stretch and long-stretch bandages, non-elastic bandages, adaptive compression bandages, and medical compression stockings, these devices cater to diverse patient needs and clinical situations. The table provides a list of commercially available bandages and stockings (Table 5).

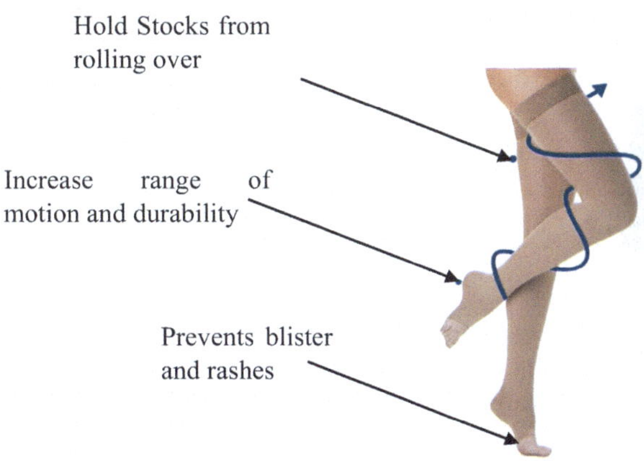

Fig. 2 Ulcer stocking system

Table 4 Class of ulcer stocking system

CCL	Intensity	Pressure range (mmHg)	Pressure range (kPA)
I	Slight	18–21	2.4–2.8
II	Medium	23–32	3.1–4.3
III	High	34–46	4.5–6.1
IV	Very high	49 and above	6.5 and above

Table 5 Commercially available compression bandages and stockings

Types	Commercial name
Short-stretch bandages	Comprilan®, Rosidal K®, BiaForm®, Durelast®, Pütter®,
Long-stretch bandages	Dauerbinde® K, Lastodur®, Perfekta®
Non-elastic bandages	Varicex®, Varolast®, ALEGRO®
Multicomponent system	UrgoKTwo®, Coban 2®, Ultra Four®, K-Four®, Compri2®, Profore®, UrgoK2® (lite)
Ulcer stocking system	JobstUlcerCare®, mediven ulcer kit®, Tubulcus®, ULCER X®, Venotrainulcertec®
Adaptive compression bandages	Circaidjuxtacures®, ReadyWrap®
Medical compression stockings	JOBST UltraSheer/Classic®, Mediven plus®, Venosan®, VenoTrain®

Intermittent Pneumatic Compression is an additional alternative for patients who cannot tolerate compression bandaging, suffer from decreased ankle mobility or calf muscle function, or have peripheral artery conditions where other forms of compression are unsuitable. This method involves sequentially inflating and deflating the limbs to mimic the movement of the venous foot and calf muscles as well as normal circulatory function. However, this technique can be expensive, heavy, noisy, and power-required [12].

Smart Compression's Memory Materials: Compression treatments using bandages or stockings have always faced challenges in attaining and sustaining the prescribed compression level and gradient. Additionally, these devices cannot provide dynamic, massaging compression, which is often favoured, particularly for elderly and less active patients, to enhance blood flow. Moreover, issues like reduced mechanical strength of fibres, stiffness, and complications such as active ulcers make it impossible to over-bandage. As a result, modern approaches are gaining prominence with the introduction of functional polymers for effective compression management, addressing the limitations of conventional methods. Memory polymers can be programmed with stress, stored at a temperature lower than their thermal transition temperature (Tg), and then retrieved reversibly, when necessary, via an external thermal stimulation above Tg. This allows for real-time control of material stress, essentially enabling in-situ material stress management. By integrating these memory polymer fibres into fabric structures, it becomes possible to adjust the internal stress of the material to compensate for pressure deviations from the initially designated level. This innovation holds the potential to maintain the required pressure on the leg and deliver dynamic compression advantages [13].

5 Shape Memory Alloy in Compression Management

Shape memory alloys (SMAs) are temperature-sensitive materials that return to their original shape when subjected to heat or an electric current stimulus. This memory effect is due to a transformation between a weak, asymmetrical parallelogram crystal structure in the martensitic phase and a highly elastic bcc (body-centred cubic) crystal lattice in the austenitic phase. SMAs are widely used in different applications because of their high corrosion resistance, excellent fatigue performance under cyclic loading, and high strength-to-weight ratio. However, their cost has limited their extensive use. Active compression devices seek to mimic nature's mechanisms for intelligent actuation. Several compression prototypes have been designed using smart and shape memory alloys' actuation capabilities, such as NiTi (nitinol) wire. This wire may exhibit significant recoverable strain or force in response to temperature or load. Coiled actuators offer substantial displacement but relatively low recoverable force. One of the significant advantages of these alloys is their ability to generate high force per unit area (200 MPa) through Joule heating and electric current. Pressure is controlled by adjusting the electric current applied to the SMA wires. Precisely applying stress in shape memory materials to achieve dynamic compression is crucial.

6 Shape Memory Polymer in Compression Management

One essential aspect of pressure bandages is maintaining the appropriate gradient pressure between the ankle and knee to treat venous ulcers. Advances in smart materials, especially shape memory polymers, have made them a popular choice for this purpose. These polymers can shift from a deformed shape to a permanent shape when stimulated. Their flexibility, high recovery rates, and ease of processing make them more desirable than memory alloys. Kumar et al. have developed a prototype bandage that includes a memory polymer fabric actuator and a heating layer for thermal therapy and memory polymer actuator control. The dynamic compression of the system is regulated by an external heating device programmed for this purpose [14].

Similarly, Small et al. used shape memory polymer to remove arterial blood clots that could lead to ischemic stroke. Their shape memory polymer micro-actuator device, powered by laser therapy, effectively dislodged vascular obstructions. This technology offers a promising method for restoring normal blood flow. Smart stockings are trending for venous ulcer compression management [15]. A smart stocking has been developed using a blend of nylon yarns and SMP, where temperature changes regulated external or control pressure levels, enabling over 50% additional pressure in the wrapped position due to the memory polymer [16]. Narayana et al. integrated shape memory filaments into knitted fabrics to achieve maximum memory stress and provide a massage effect for maintaining memory stress when the temperature changes [17]. Stockings made from memory material hold the potential for adjustable compression therapy and smart wound care.

7 Compression Textiles: Properties and Behaviours

Compression textiles, widely used in medicine, sports, and everyday attire, possess unique properties and behaviours that make them valuable tools for enhancing performance and promoting health. These garments are characterized by their elasticity like spandex, allowing them to comfortably conform to the body's contours while maintaining consistent pressure [18]. The level of compression varies across different garments, with measurements often expressed in millimetres of mercury (mmHg). Gradient compression, a common design feature, ensures that pressure is highest at the extremities and decreases gradually towards the core, facilitating better blood circulation and minimizing swelling risk. Moisture-wicking properties, frequently achieved through nylon and polyester, enhance comfort during extended wear by keeping the body dry [19]. Additionally, some compression textiles incorporate antimicrobial treatments and breathable materials to maintain hygiene and prevent overheating. Durability is a crucial consideration, and materials like nylon, spandex, and polyester are chosen for their ability to withstand frequent stretching and washing. Seamless construction minimizes friction and irritation, while certain textiles offer UV protection and thermal regulation. Moreover, emerging technologies are introducing smart features, including sensors and conductive threads, to monitor physiological parameters and enhance overall performance [20]. Understanding these properties and behaviours is essential for selecting the right compression textile for specific purposes, whether for therapeutic benefits or performance enhancement. Here's an overview of these aspects:

7.1 Material Selection and Fabrication Technology

Stretchability is important in the product; therefore, it is essential to understand the factors that cause stretch generation. The three critical factors are as follows:

(i) stretchability of fibres,
(ii) type of yarn (textured, core-spun "Lycra™", rubber thread)
(iii) fabric with the right constructional geometry

The materials used for compression bandages and stockings include nylon, polyester, cotton, viscose, spandex, rubber, and non-latex materials. These materials can be combined in different ways to produce fabrics with different properties and structures. Compression textiles can be made using knitting or weaving techniques (Fig. 3). Seamless knitting technology is used to make many compression garments,

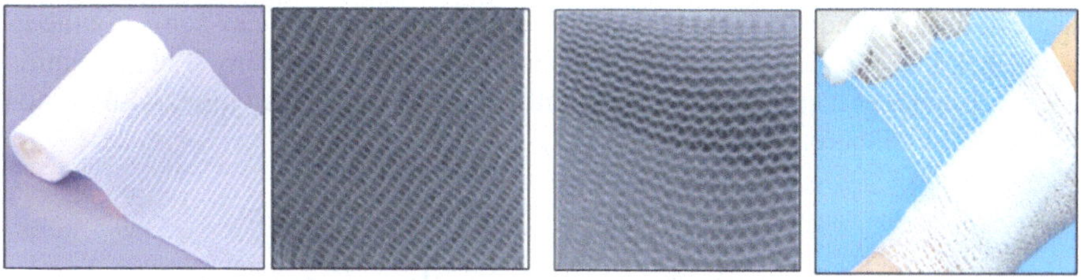

(a) (b)

Fig. 3 Structure of bandages **a** woven; **b** knitted

which reduces the risk of chafing and pressure points. Advanced fabrication techniques are used to create gradient compression garments, which promote efficient blood circulation and reduce swelling. 3D printing allows for precise customization based on an individual's anatomy. Smart textiles are also used to monitor physiological parameters or enhance performance. The materials are usually warp or weft-knitted. Weft-knitted constructions include rib, single jersey, and pique. Nylon and fine spandex yarn are used to produce loops, and spandex or rubber threads are inlaid to make the fabric stretchable [21, 22].

Compression bandages with high levels of pressure and minimal stretch are crucial for treating certain ailments. These bandages need to be constructed/chosen based on their capacity to maintain pressure over time. The constructional parameters, like fabric structure, the number, and the denier of elastic yarns, all affect the bandage's ability to provide the desired pressure levels. As the high-stretch inlay yarn in the fabric coarsens, the breaking load of the knitted and woven fabric samples increases. Furthermore, bandage samples with coarser yarns exhibit low fabric growth and excellent elastic recovery values. The sample with the most elastic yarns and a close structure generates significantly more tension force at low stretch values. Knitted bandage materials with smaller amounts of high-stretch threads tend to apply high pressure with minimum stretch, while woven specimens require a larger number of inlay elastic yarns. However, the recovery behaviours of the knitted and woven samples are similar [23].

7.2 Material Construction

Bandages are constructed based on their required stiffness, which is an important criterion in selecting them. Highly stiff bandages include inelastic short-stretch, multiple-component short-stretch, and multiple-component multilayer bandages. On the other hand, long-stretch ones have lower stiffness. One of the popular highly stiff bandages is the multiple-component four-layer bandage (4LB). It consists of a padding bandage, a crepe bandage, an elastic bandage, and an outer layer made of an elastic cohesive bandage. Commercial multicomponent bandage systems use the P-L-A-C-E ("Pressure level-layers-components of material-elasticity") classification scheme. Bandages can be of two types: knitted or woven construction, with knitted ones providing inherent stretchability [24].

8 Compression Textile Market and Future Scope

The compression textiles industry is expected to grow rapidly in the future years. According to Market & Market Research Pvt Ltd, the compression therapy market is projected to grow significantly, with an estimated revenue of $3.7 billion in 2022 and an anticipated increase to $4.9 billion by 2027, at a compound annual growth rate (CAGR) of 5.5%. The growth of this market is driven by the increasing prevalence of sports injuries and accidents, along with the largest target patient population (Compression therapy market, n.d.). From 2023 to 2030, the market for compression therapy is expected to develop at a CAGR of 6.2%, reaching a value of USD 4.14 billion. Bandages and stockings are increasingly being utilized to treat venous problems, propelling market expansion. The future of the compression bandage market is likely to be shaped by ongoing technological advancements, an ageing population, and a continued focus on patient comfort and personalized care. The "Medical-Grade Compression Garments Market" global report reveals a consistent and robust growth trajectory in recent years, anticipated to extend positively until 2030. A notable trend in this market is the growing demand for eco-friendly and environmentally sustainable products. Additionally, there is a noteworthy shift towards incorporating advanced technologies, including artificial intelligence, machine learning, and blockchain, to enhance the quality and efficiency of products, resulting in innovative solutions surpassing traditional options. The 2023 Medical-Grade Compression Garments Market Research Report accentuates trends, growth prospects, and potential scenarios up to 2030. As healthcare systems seek cost-effective and efficient ways to manage chronic wounds and venous disorders, compression bandages will remain a critical component of treatment plans. Additionally, as the market evolves, there may be further developments in smart textiles and wearable technology integrated into compression garments to monitor patient health and improve treatment outcomes. It's important to note that the market for compression bandages is subject to ongoing changes and developments. It is advisable to consult recent industry reports, market analyses, and news sources to obtain the most current information and insights [25, 26].

9 Conclusions

This chapter includes the dynamic and diverse world of bandages, with a particular emphasis on their role in treating Chronic Venous Insufficiency (CVI). Throughout the annals of medical history, these unassuming yet essential tools have undergone remarkable evolution, consistently delivering cutting-edge solutions for a wide spectrum of medical conditions. It will serve as a testament to the enduring relevance and significance of bandages in modern medicine, shining a light on their multifaceted contributions to patient well-being. It is an invaluable resource for those seeking a deeper understanding of the pivotal role that these seemingly unassuming tools play in the pursuit of improved medical outcomes.

References

1. Farah R, Davis M (2010) Venous leg ulcerations: a treatment update. Curr Treat Options Cardiovasc Med 12:101–116
2. Thomas S (1995) Compression bandages. In: Leg ulcers: nursing management. s.l.:s.n.
3. Hanna R, Bohbot S, Connolly N (2008) A comparison of interface pressures of three compression bandage systems. Br J Nurs 17:16–24
4. Göstl K, Obermayer A, Hirschl M (2009) Pathogenese der chronisch venösen Insuffizienz durch Adipositas—Aktuelle Datenlage und Hypothesen. Phlebologie 38:108–113
5. Thomas S (2003) Laboratory-based evaluation of a compression bandaging system. J Biomimet Biomater Tissue Eng 99:24–28
6. Mosti GV, M RP, M M (2009) Compression therapy in the treatment of leg ulcers. Acta Vulnol 7(3):1–43
7. Ashby RL, Gabe R, Ali S (2014) Clinical and cost-effectiveness of compression hosiery versus compression bandages in the treatment of venous leg ulcers (Venous leg Ulcer Study IV, VenUS IV): a randomised controlled trial. Lancet 2014:1
8. Moffatt C, Kommala D, Dourdin N, Choe Y (2009) Venous leg ulcers: patient concordance with compression therapy and its impact on healing and prevention of recurrence. Int Wound 6(5):386–393
9. Stücker M, Link K, Reich-Schupke S, Altmeyer P, Doerler M (2013) Compression and venous ulcers. Phlebology 28:68–72
10. Nelson HS, Agashe SN, Mahesh PA, Katial R (2014) Cochrane Datab Syst Rev 5:1
11. Mosti G, Cavezzi A, Partsch H, Urso S, Campana F (2015) Adjustable velcro® compression devices are more effective than inelastic bandages in reducing venous edema in the initial treatment phase: a randomized controlled trial. Vasc Endovasc Surg 50(3):368–374
12. Vicaretti M (2010) Compression therapy for venous disease. Aust Prescr 33(6):186–190
13. Kumar B et al (2022) Shape memory textiles for smart compression management for chronic venous disorders—a review. Indian J Fibre Text Res 47:131–145
14. Kumar B, Hu J, Pan N (2016) Smart medical stocking using memory polymer for chronic venous disorders. Biomaterials 75:174–181
15. Small W, Metzger MF, Wilson TS, Maitland DJ (2005) Laser-activated shape memory polymer microactuator for thrombus removal following ischemic stroke: preliminary in vitro analysis. IEEE J Sel Top Quant Electron 11(4):892–901
16. Kumar B, Hu J, Pan N, Narayana H (2016) A smart orthopaedic compression device based on a polymeric stress memory actuator. Mater Des 97:222–229

17. Narayana H, Hu J, Kumar B, Shang S, Ying M, Young RJ (2018) Designing of advanced smart medical stocking using stress-memory polymeric filaments for pressure control and massaging. Mater Sci Eng C 91:263–273
18. Xiong Y, Tao X (2018) Compression garments for medical therapy and sports. Polymers 10(6):663
19. Lim CS, Davies AH (2014) Graduated compression stockings. CMAJ Can Med Assoc J 186(10):E391
20. Gulati R, Sharma S, Sharma RK (2022) Antimicrobial textile: recent developments and functional perspective. Polym Bull 79(8):5747–5771
21. Ghosh S, Mukhopadhyay A, Sikka M, Nagla KS (2008) J Tissue Viab 17:82–94
22. Vickery A, Vickery BC (2004) Int J Cloth Sci Technol 16(2):173–183
23. Ghosh S, Mukhopadhyay A, Sikka MKN (2008) Pressure mapping and performance of the compression bandage/garment for venous leg ulcer treatment. J Tissue Viab 17(3):82–94
24. Kankariya N, Laing RM, Wilson CA (2021) Textile-based compression therapy in managing chronic edema: complex interactions. Phlebology 36(2):100–113
25. Compression therapy market size, share and trends, *Compression therapy market size, share and trends analysis report,* GVR report cover
26. Compression therapy market: markets and markets

Textile-Based Wound Dressings

S. Kubera Sampath Kumar, S. Subramanian, and K. Harish Kumar

1 Introduction

A wound is defined as disruption of epithelial tissue integrity, which may extend deeper and involve sub-epithelial tissues that are dermis, fascia, and muscle. Wounds can result from accidental incidents, intentional actions, or be part of a disease process. The historical journey of wound care is from prehistoric times to modern medicine, progressing from basic wound dressings such as vinegar-soaked dressings, through the use of topical antibiotics, to the application of growth factors. Wounds can be categorized based up on factors such as (i) chronic or acute nature and (ii) their cause, distinguishing between open wounds, where skin is torn, cut, or punctured due to physical trauma, and closed wounds, resulting from blunt force trauma causing contusion. The primary goal of wound healing is to achieve homeostasis and restore tissue integrity. This process is intricately orchestrated and complex, beginning with the injury of tissue and ending with regeneration. The term healing is classified into several categories based on the expected nature of the recovery process.

Although the fundamental process remains the same for all wounds, the rate of healing is affected by factors such as the immune system, the wound and age, and the specific healing conditions. Treatment is not only based on the medication, but it also depends on the use of appropriate dressing methods and appropriate dressing materials. An appropriate wound dressing will create an optimal environment for

S. K. S. Kumar (✉)
Department of Textile Technology, Vignan's Foundation for Science, Technology and Research (Deemed to be University), Vadlamudi, Guntur, Andhra Pradesh, India
e-mail: sampathtext@gmail.com

S. Subramanian
Department of Textile Technology, A. C. Tech., Anna University, Chennai, Tamil Nadu, India

K. H. Kumar
Department of Fashion Design, Footwear Design and Development Institute, Raidurgam, Hyderabad, India

S. Sundaramoorthy et al. (eds.), *Textile Materials for Good Health and Wellbeing*, SDGs and Textiles, https://doi.org/10.1007/978-981-97-6496-9_11

healing, with the goal of reducing hospitalization stays by requiring a couple of dressing replacements. The latest dressing theory advocates the ideal dressing that minimizes trauma during changes, promotes exudate absorption, insulates the wound from extreme temperatures, acts as a bacterial barrier, and remains non-adherent to the wound, ensures comfort on the patient, facilitates gas exchange, conforms to the contours of the body and prevents eschar formation [1, 2]. Various wound dressings are available on the market, such as gauze cloth with monolayer, thin films, hydrogels, polyurethane foams, alginates, hydrocolloids, the activated char cloth with silver, and chitosan, which offer some advantages but also limitations, failing to fully meet the criteria of an ideal wound dressing [2, 3]. Cotton gauze, being hygroscopic, adheres to the wound surface upon absorbing exudates, causing disturbance to newly formed tissues during redressing. Thin film dressings also face the issue of adhesion to the wound surface. Hydrogel dressings, while absorbing exudates, tend to swell and move away from the wound site, creating potential air spaces for bacterial proliferation. Hydrocolloid and hydrogel dressings, effective up to a certain exudate level, are not suitable for highly exuding wounds. Hydrocolloids may dissolve and contaminate the wound upon absorbing large quantities of exudates. The challenge lies in achieving all requirements of the wound dressing within a single-layer medical device. Current research globally is focused on the development of multilayer wound dressings, in which each layer is designed to meet the criteria of the ideal wound dressing.

2 Wound Healing Processes

In the wound healing process, inflammation is the initial phase, characterized by vasoconstriction followed by vasodilation, increased blood vessel permeability, and immune cell recruitment to the site of injury. The early inflammatory phase aims to eliminate the cause of injury, paving the way for tissue repair. Proliferation follows, involving the activation of fibroblasts and other cells that contribute to tissue repair. During this phase, collagen is deposited, supporting the restoration of normal tissue structure and function. Epithelialization is a critical component where new epithelial cells migrate and cover the particular wound, forming a protective barrier. As healing progresses, the remodeling phase begins, focusing on rebuilding and enhancing the recently formed tissue. The collagen fiber undergoes, realignment and maturation, enhancing the tensile strength of the healing tissue. This phase is essential for restoring tissue integrity and function. Finally, scar maturation occurs, refining the characteristics of the formed scar. The process involves continued collagen remodeling and the reduction of vascularity within the scar tissue. While the scar may never fully replicate the original tissue, scar maturation contributes to improved strength and flexibility. A successful wound-healing process involves the orchestrated progression through these phases, ensuring effective tissue repair and minimizing functional impairment [4–6] (Fig. 1).

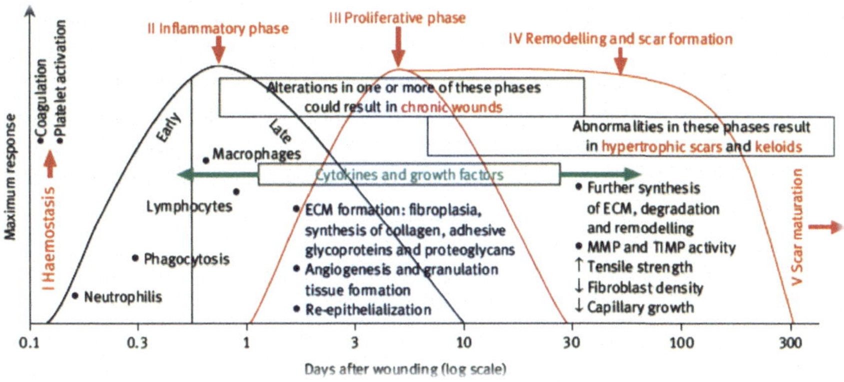

Fig. 1 Phases of wound healing [4]

2.1 *Factors Which Are Affecting the Wound Healing*

In many instances, the wound healing process is a natural and uncomplicated process that culminates in the restoration of tissue integrity. However, certain wounds present challenges and resist healing, evolving into complex medical issues that demand specialized care and treatment. The intricate process of wound healing hinges on the coordinated interaction among various cells, mediators, and growth factors. Creating an optimal microenvironment free from cytotoxic factors is imperative for successful wound healing. Both the local factors, such as infection, poor surgical technique, and tissue ischemia, and systemic factors, such as aging, vitamins, underlying diseases, and particular medication have been implicated in the delayed wound healing. The success of any wound repair requires timed and balanced activities between connective tissue, epithelial cells, vascular and inflammatory, and the respective mediators. Establishing a conducive environment for wound healing, addressing the root cause of the wound, and correcting associated abnormalities are crucial aspects of effective wound management [7]. Wound dressings fall into passive, interactive, or active categories. Passive dressings primarily provide coverage, whereas interactive, or active dressings are believed to modify the physiology of the wound environment. Examples of interactive dressing include foams, alginates, hydrocolloids and hydrogels. The ideal dressing must compulsorily maintain a moist environment at the wound interface and act as a barrier against microorganisms.

2.2 *Evolution of the Wound Dressings*

Since sixteenth century, various periods of the enlightenment have shaped the field of wound management. In the context of World War I, the responsibility of changing wound dressings rested with physicians and medical students, marking an interesting journey for wound care. Reflecting on the historical progress, the field has evolved significantly from the days when Ambrose Pare, a barber-surgeon with the French army, scalded wounded soldiers' injuries with boiling oil to counteract the perceived poisonous effects of gunpowder. Pare's improvisation during the Battle of Turin, using a mixture of turpentine, oil of roses, and egg yolks inadvertently became one of the first clinical trials in wound care, showcasing the soldiers' improved healing with less pain. This might be considered one of the earliest instances of an interactive moist wound dressing [8].

A defining development in moist wound healing was presented in initial work that highlighted the value of the moist wound environment. The concept of fluid wound healing only gained serious attention in the 1970s and 1980s, Scab formation, and the rate of epithelialization of the superficial wounds in the skin of young domestic pigs greatly advanced the understanding of wound healing functions [9]. The study revealed that wound healing in particularly moist conditions preceded 50% faster than in dry, outdoor conditions.

Before this shift in understanding, dressings were primarily perceived as absorbers of exudate, maintainers of dry wounds, and protectors against external pathogens and further injury, all while ensuring comfort. Subsequent advancements in comprehending the various stages of healing and the specific requirements for diverse wounds have led to the availability of a wide array of dressings in the market. The extensive range, while beneficial, also introduces the potential for confusion in selecting the right type. This was crucial to assess the wound and its healing stage to identify specific objectives; For instance, for a sloughy wound, the focus should be on de-sloughing through exudate absorption. In contrast, a clean and granulating wound aims to create a moist environment conducive to healing. In some cases where the particular wound has a large cavity, appropriate dressing choices should involve plugging considerations.

3 Material for Wound Dressings

Biopolymers such as alginates, collagen, chitosan, gelatin, and silk have found applications as wound dressing materials. These dressings are available in diverse forms, including thin films, foams, hydrogels, and hydrocolloids, as well as woven, nonwoven, and knitted fabrics. Silk, renowned for its compatibility with both biological and hemostatic processes, proves to be a valuable scaffold in promoting wound healing. This effectiveness is attributed to the presence of the amino acids such as aspartic acid, glycine and arginine, which facilitate the reconstruction of the epithelial tissues. The Eri silk fibroin (ESF) scaffold exhibits superior cell attachment, spreading of L6 fibroblast cells, and binding compared to mulberry silk fibroin. Furthermore, cell viability is enhanced on ESF scaffolds [10].

Poly L-lactic acid (PLA) is a biopolymer that demonstrates remarkable wicking ability, rapid water spreading, and swift drying, contributing to highly favorable moisture management characteristics within its fibers. In addition to these attributes,

PLA offers, water vapor transport properties, thermal insulation, breathability, and superior buffering capacity for the liquid sweat compared to polyester and cotton [11]. Recognized as a biodegradable material with low toxicity, in vivo bio-absorbability, and exceptional biocompatibility, PLA is widely utilized in biomedical applications, particularly in the case of sustained drug delivery systems and the implants for orthopedic devices due to their low hydrophilicity [12].

3.1 Essential Criteria for Wound Dressings

Certain prerequisites define the characteristics of an optimal wound dressing, including

- Absorption of secretions and harmful substances from the surface of the wound with high efficiency.
- Maintaining optimal moisture levels at the wound dressing interface.
- Provides thermal insulation and Supports gas exchange.
- Protection of wound from bacterial invasion.
- It presents non-toxic properties and allows easy removal without causing any trauma to the patient.

4 Types of Wound Dressings

Various wound dressings are currently available on the market, offering a variety of options, including polyurethane foams, single-layer gauze, thin films, silver-activated carbon cloth, hydrocolloids, hydrogels, chitosan, alginate, and alginate collagens.

A multilayer wound dressing system was introduced by John et al. [13], comprising a layer that is skin-facing, a backing layer, and an absorbent layer situated in between them, all forming a cohesive unit. The skin-facing layer incorporates an opening for direct wound contact, facilitating rapid absorption of exudate by the absorbent layer.

It is important that the backing layer is free of adhesives, ensuring vapor permeability while maintaining liquids impermeable. Kim et al. [14] have ingeniously devised a technique to create antimicrobial moist wound dressings with silver. This innovation includes a structure with an adhesive layer that is pressure sensitive, and intermediate layer containing a compound of silver-carboxymethylcellulose (silver-CMC), and an outer protective membrane layer—all are bonded and laminated together. In addition, Eckstein has developed an absorbent multilayer wound dressing system specifically tailored for the healing of moderate to severe exuding wounds.

This wound dressing incorporates three layers that have a carrier layer, an absorbent layer, and a wound contact layer which is hydrophilic and is bonded to an absorbent layer. Specifically, the wound contact layer incorporates hydrophilic polyurethane [15].

4.1 Advancements in Wound Dressing Technologies

Pepper et al. [16, 17] introduced a groundbreaking wound dressing assembly crafted with nanomaterials using needleless electrospinning. This assembly comprises single or multiple layers of nano-electrospun materials, each possessing unique characteristics. Patrick et al. [18] innovated polyethylene glycol (PEG)-based multiblock thermoplastic polyurethanes infused with polyhedral oligomeric silsesquioxane (POSS), showcasing enduring antimicrobial effects, reduced hospital expenses, enhanced fluid management, and simplified application and removal processes, thereby alleviating patient discomfort. Additionally, this device mitigates lateral wicking, thereby preventing skin peeling and moisture accumulation at the edges of highly exuding wounds, ultimately reducing swelling. It facilitates effortless adjustment of hydrogel properties and resultant antimicrobial activity. Lin and Yeh [19] developed a dual-layer wound dressing featuring a polymer layer containing anti-infective antibiotics or traditional Chinese herbs on the upper stratum and a porous carbon layer capable of housing skin-related epithelial cells on the lower stratum. Flach [20] devised an electrically conductive wound dressing employing a nonwoven layer utilizing a conductive plastic film.

Nadi and Peter [21] introduced a multi-polymer composite wound dressing system tailored for highly exuding wounds. This system includes a contact layer, an absorbent layer, and additional optional layers such as a backing sheet, adhesive, and a moisture vapor transmission rate control layer. Pepper et al. [16, 17] has unveiled a wound dressing assembly comprised of individually separated layers of nanomaterials, fabricated through a continuous needleless electrospinning process. These assemblies consist of single or multiple layers, each seamlessly joined together to form a cohesive dressing assembly. Reibel et al. [22] innovated a nonwoven fabric containing gelatine fibers imbued with antimicrobial properties. Bright, et al. [23] devised a nano-silver wound dressing incorporating multiple layers serving various functions, including a skin-contact layer, a disinfectant anti-toxic layer, a blood-absorbing and astringent layer, and an isolation layer. Buglino et al. [24] developed a non-adherent wound dressing featuring an absorbent layer infused with superabsorbent material. Arnold [25] introduced a multilayer molecular filtration membrane for wound dressing purposes. Robert et al. [26, 27] designed a flexible, antibacterial, and antifungal wound dressing comprising layers of woven fabric embedded with silver, absorbent material, and optionally breathable or waterproof material.

Liu and Valerio [28] devised a silicone-based antimicrobial wound dressing, presented as a transparent, self-adhesive gel sheet. Patrick et al. [29] introduced a durable polymer hydrogel containing thermoplastic polyurethane fibers, polyhedral oligo silsesquioxane, and polyethylene glycol. Helmut and Diefeng (2013) developed a wound care system utilizing polymeric cotton, silk, or porous media coated with a substance that hampers bacterial growth. Bright [23] formulated a multi-component dressing equipped with a wound-covering element and attachment points for material evacuation. Kubera Sampath Kumar et al. [30] developed a wound dressing coated with natural extracts of *Calotropis Gigantean*, *Eucalyptus Globulus* and Buds of *Syzygium Aromaticum* solution enhanced with rhEGF.

Bechert and Peter [31] innovated a dressing material featuring an absorbent matrix and an antimicrobial mineral substance. Lucie and Pang [32] created a multilayered polyurethane foam dressing infused with therapeutic agents. Alvarez [33] designed a multilayer wound dressing comprising a low oxygen-permeable outer layer, an oxygen-permeable inner layer, and an adhesive, applied to the wound to foster a hypoxic environment during the initial phase. Meyer et al. [34] crafted an occlusive dressing comprising a barrier layer in contact with the skin and an elastic support layer containing elastomeric and hydrocolloid phases for water absorption [13].

4.2 Biodegradable Polymers

Biodegradable polymer refers to solid polymeric devices and materials that undergo macromolecular degradation, leading to their dispersion in vivo. It is essential to note that this definition excludes degradation caused by environmental factors, fungi, or bacteria. Biodegradable polymeric systems or devices are susceptible to attack by biological elements.

As an outcome, the integrity of the system, as well as in some cases, the macromolecules themselves, can be affected, leading to the production of fragments or other degradation by-products. While all of these fragments have the potential to be removed from their original site of action, it is important to emphasize that they may not necessarily exit the body.

Bio-absorbable polymer materials dissolve in body fluids without experiencing polymer chain scission or molecular mass reduction. The gradual dissolution of water-soluble implants within the body fluids is an example. When dispersed macromolecules of bio-absorbable polymer are excreted, it is called a bio-resorbable polymer [35]. For various biomedical applications, biodegradable polymeric materials that include both synthetic and biologically derived (natural) polymers have been extensively studied. Biodegradation is a process causing cleavage of bonds within the polymer, leading to polymer erosion [36]. Polymeric biomaterials are categorized as hydrolytically degradable and enzymatically degradable polymers based on their mode of degradation. Depending on the implantation site and enzyme availability, the natural polymers exhibit varying degradation rates in vivo. Altering the chemical composition of these polymers can greatly impact their rate of degradation. While natural polymers offer advantages such as bioactivity, presence of receptor binding to cells, susceptibility to cell-mediated proteolytic degradation, and natural remodeling, they also have disadvantages such as strong immunogenic response, purification complexities, and potential risks of disease transmission [37].

Chitin, the most abundant natural amino-polysaccharide, along with its derivative chitosan, is a remarkable natural polymer. Chitin is predominantly found in the exoskeleton of insects and various invertebrates. They possess extreme hydrophobicity and insolubility in water and most organic solvents. Chitosan, derived from chitin through chemical processing, is highly recommended for wound dressing due to its excellent properties, such as biodegradability, biocompatibility, non-toxicity, and adsorption capabilities [38]. Despite their potential for wound healing, materials containing chitosan face limitations due to their low mechanical strength under physiological conditions [39]. Collagen is an important biomaterial in medical applications due to its unique characteristics such as biodegradability and weak antigenicity [40]. Collagen, which makes up a quarter of the body's total protein, has a fibrous structure that provides strength and support to various tissues such as tendons, ligaments, and skin. These proteins are composed of polypeptide chains (α-chains) that form a unique triple helical structure, measuring 300 nm in length and 1.5 nm in diameter. There are more than twenty different types of collagen in animal tissues, with types I, II, III, V, and XI known to form fibers. These types of collagen typically self-assemble into periodic, transversely striated fibers that can reach centimeters in length and tens of microns in diameters. Type I collagen, predominant in fiber formation, is found in bones, skin, teeth, and tendons, while Type II collagen, the second most abundant, is present in cartilage tissue, the developing cornea, and the vitreous humor [41]. Despite its beneficial properties, nanofibrous collagen matrices have a significant drawback of losing structural integrity in aqueous environments, such as the human body. However, combining collagen with biocompatible polymers that enhance cellular interactions and accelerate tissue regeneration can improve the dimensional stability of scaffolds [42]. One such synthetic semi-crystalline biodegradable polymer polycaprolactone (PCL) is synthesized through the ring-opening polymerization of caprolactone. PCL degrades through the action of various bacteria and moths. The polymerization of ε-caprolactone produces a semi-crystalline polymer with a glass transition temperature of $-$ 60 °C and melting point between 58 and 63 °C. The repeating molecular structure of PCL homopolymer includes a single relatively polar ester group and five non-polar methylene groups, imparting unique polyolefin-like properties due to its high olefinic content [43]. At the same time, the presence of the hydrolytically labile aliphatic ester bond makes the polymer biodegradable. Due to its higher crystallinity and hydrophobic nature, PCL biodegrades more slowly in vivo compared to PLA; it is used in both soft and hard tissue engineering, as well as in drug delivery systems. A major drawback of PCL is its low cell adhesion due to its hydrophobic nature, which can be enhanced by blending PCL with starch, cellulose, and PLA polymers [44].

4.3 *Textile Structure for Wound Dressings*

There are various types of wound dressings available, including thin films, woven fabrics, nonwoven fabrics, knitted fabrics, braided fabrics, and electrospun fibrous mats.

Woven fabrics, such as gauze, are produced using a loom where continuous sheet of yarns, known as the warp, is interwoven with cross-threads, called the weft, at right angles. These materials typically feature a plain weave structure, offering dimensional stability and varying degrees of porosity. However, a challenge with woven fabrics is the tendency for threads to unravel at the cut edge. To mitigate this issue, a technique called ‘Leno’ has been developed, where two warp threads twist around a weft, reducing-edge unraveling demonstrated by Rajendran et al. [7].

Knitted fabrics provide greater flexibility and conformability compared to woven fabrics. The threads in knitted fabrics are less likely to unravel when cut, making them suitable for specific applications. However, knitted fabrics may exhibit higher porosity due to their loop structure. Knitted nets and bandages are particularly effective in areas requiring three-dimensional adaptability. Compression stockings, commonly used for treating venous stasis, are a notable application of knitted materials in wound care.

Nonwoven materials are essential in wound dressings due to their unique characteristics and versatile applications. They offer advantages such as ease of manufacturing, customization, and enhanced absorbency, making them suitable for various wound types and sizes.

These materials are manufactured by bonding or felting fibers together, often without the need for weaving. In wound care, nonwoven materials offer several advantages it absorbs wound exudate, flexibility allows nonwoven dressings to conform easily to various wound shapes and contours, ensuring better contact with the affected area, and provide a breathable barrier that allows air circulation while preventing bacterial infiltration.

Electrospinning is a versatile and innovative technique used in the fabrication of fibrous mats, and its application in wound dressings has gained significant attention. Here are some key aspects of the application of electrospinning fibrous mats in wound dressings:

4.3.1 Fiber Structure and Morphology

Nano to micro-scale fibers; electrospinning allows the production of fibers ranging from nanometers to micrometers in diameter, mimicking the scale of natural extracellular matrix components. High surface area (nano structure), The fibrous structure provides a high surface area to volume ratio, promoting cellular interactions and facilitating enhanced absorption of wound exudate.

4.3.2 Controlled Release of Bioactive Agents

Incorporation of Drugs: Electrospinning allows the encapsulation of bioactive agents, growth factors, or antimicrobial substances within the fibers, enabling controlled release for therapeutic purposes.

Localized Treatment: This feature supports localized drug delivery directly to the wound site, reducing the need for frequent dressing changes.

4.3.3 Wound Healing Properties

Promotion of Cell Proliferation: Electrospun fibrous mats encourage cell adhesion, migration, and proliferation, supporting tissue regeneration.

Modulation of Inflammation: The structure of electrospun mats can influence the inflammatory response, contributing to a controlled and beneficial inflammatory phase during wound healing [45].

4.3.4 Moisture Management

Electrospun fibers can be engineered to be hydrophilic, aiding in the absorption of wound exudate and maintaining a moist environment, which is conducive to optimal wound healing.

4.3.5 Prevention of Infection

Antimicrobial Properties: Electrospun mats can incorporate antimicrobial agents or be designed with inherent antimicrobial properties, helping to prevent infection in the wound area.

4.3.6 Customization and Multilayered Systems

Multilayered Dressings: Electrospinning allows the creation of multilayered wound dressings with distinct functionalities, combining different materials or drug-loaded layers for a comprehensive therapeutic approach [45]. Subramanian and Kubera Sampath Kumar [46] developed a nano material-based multilayer wound dressing system consisting of three layers as shown in Fig. 2

Fig. 2 Schematic diagram of multilayer wound dressing system **a** primary layer with antibacterial agent **b** absorbing secondary layer, **c** protective top layer

Tailored Mechanical Properties: By adjusting parameters during electrospinning, the mechanical properties of the fibrous mats can be tailored to meet specific wound care requirements. The application of electrospinning in wound dressings demonstrates its potential to address various challenges associated with traditional wound care methods, providing advanced solutions for improved healing outcomes.

5 Conclusions

A wide range of wound dressings are available on the market, such as sterile cotton pads with gauze cloth, thin films, polyurethane foams, and hydrocolloids. Each type of dressing has its own set of advantages and limitations in meeting the criteria for an ideal wound dressing system. Ongoing global research is focused on developing multilayer wound dressings, where each layer is tailored to fulfill specific requirements.

The primary layer, which comes into direct contact with the skin, is designed to absorb exudate, promote gas exchange, maintain a sterile wound environment, and minimize trauma during dressing changes. The secondary layer is responsible for absorbing and retaining exudates from the primary layer, ensuring an optimal moisture balance at the wound site. Finally, the tertiary layer acts as a protective barrier, preventing environmental contamination while allowing water vapor transmission and securing the dressing in place.

Effective wound bed preparation involves managing exudates to promote healing and prevent complications. Different dressing materials aid in debridement, autolysis, and wound cleansing, and support granulation and epithelialization processes. Maintaining moisture balance is crucial for wound healing, necessitating wound dressing materials with good water vapor and liquid transmission properties.

Surface tension, wicking, and wetting are important phenomena associated with fibrous materials, influencing water and liquid transport through the dressing structures. In this study, a fibrous sheet was created from eri silk fibroin and incorporated with superabsorbent polymer for use in wound dressings. Nonwoven mats were produced using compression and spun-lacing methods, characterized for properties essential to the secondary layer of wound dressing systems.

Commercially available protective layers like gauze cloth, adhesive tape, and polyurethane film contribute to modern wound dressing theory. An ideal wound dressing should meet various requirements, including exudate absorption, gaseous exchange, bacterial barrier, non-adherence, patient comfort, body contour conformity, and prevention of eschar formation. However, achieving all these requirements in a single-layer medical device is challenging, highlighting the need for continued development of multilayer wound dressing systems capable of meeting diverse demands.

References

1. Czaja W et al (2006) Microbial cellulose—the natural power to heal wounds. Biomaterials 27:145
2. Reed AM (1991) Sustained growth factor delivery in tissue engineering applications. J Biomater Appl 6(3):3
3. Lin FH, Chen TM, Chen KS, Wu TH, Chen CC (2000) An animal study of a novel tri-layer wound dressing material—non-woven fabric grafted with N-isopropyl acrylamide and gelatin. Mater Chem Phys 64:189–195
4. Enoch S, Leaper DJ (2008) Basic science of wound healing. Surg Infect (Larchmt) 26:31
5. Goldsmith AL (ed) (1991) Physiology, biochemistry, and molecular biology of the skin, 2nd edn. Oxford University Press, New York
6. Leaper DJ, Harding KG (1998) Wounds: biology and management. Oxford University Press, Oxford
7. Rajendran S (ed) (2009) Advanced textiles for wound care. Woodhead Publishing Limited, Cambridge
8. Moore Z, Avsar P, Conaty L, Moore DH, Patton D, O'Connor T. The prevalence of pressure ulcers in Europe, what does the European data tell us: a systematic review. J Wound Care. 2019 Nov 2;28(11):710–719. https://doi.org/10.12968/jowc.2019.28.11.710. PMID: 31721669.
9. Winter GD (1962) Formation of the scab and the rate of epithelization of superficial wounds in the skin of the young domestic pig. Nature 193:293
10. Andiyappan, Muthumanickam, Sundaramoorthy, Subramanian, Vidyasekar, Prasanna, Srinivasan, Natarajan Tirupattur and Verma, Rama Shanker. "Characterization of electrospun fibrous scaffold produced from Indian eri silk fibroin" International Journal of Materials Research, vol. 104, no. 5, 2013, pp. 498–506. https://doi.org/10.3139/146.110888
11. Blackburn R (ed) (2005) Biodegradable and sustainable fibres, vol 47. Taylor & Francis, New York
12. Wu Y et al (2009) Polymeric micelle composed of PLA and chitosan as a drug carrier. J Polym Res 16:11
13. J.S. Nielsen., et al., Kolte, M.I. and Gundersen, B., Coloplast AS, 2009. Multi-layer wound dressing formed as a single unit. US Patent 7619130B2
14. Kim YS, Kim JY (2013) Methods for producing silver-bonded antimicrobial moist wound dressings and moist wound dressings produced by the methods. U.S. Patent 8,409,608
15. Axel E (2009) Multilayer absorbent wound dressing with a hydrophilic wound contact layer. US Patent 20090216168A1
16. Chen JP et al (2008) Electrospun collagen/chitosan nanofibrous membrane as wound dressing. Colloids Surf A 313:183
17. Pepper CB, McCarthy S, Real K, Kimball J, Ruzickova J, Svobodova J, Dvorak R (2011) Nanomaterial wound dressing assembly. US Patent 20110111012A1
18. Patrick M, Jian W, Dacheng R, Shuya H (2011) Study on performance of different wound dressings on surgical non infected wounds. J Nat Fibers 18:161–174
19. Lin YS, Yeh CL (2012) Wound dressing. US Patent 20120064145A1
20. Flach N (2012) Multilayer wound dressing with conductive regions. WO World Intellectual Property 2012008903A1
21. Nadi E, Peter R (2013) Laminate Polymer composite wound dressings, their manufacture and their use. US Patent 20130023844A1
22. Reibel D, Schlesselmann B, Kerhault JF, Groten R (2013) Bioresorbable wound dressings. ES European Patent 2414961T3
23. Bright R (2008) Multi-Component dressing for wound treatment on the human or animal body with application of reduced pressure. US Patent 20080004559A1
24. Buglino D, Constantine B, Hudak JC (1999) Improved wound dressing. EP European Patent 0999858B2
25. Arnold PS (1998) Multi-layer wound dressing. US Patent 5759570A

26. Robert OB, Flick AB, Becker AJ (2006) Multilayer Wound dressing. US Patent 7005556B1
27. Robert et al. (2006) designed a flexible, antibacterial, and antifungal wound dressing comprising layers of woven fabric embedded with silver, absorbent material, and optionally breathable or waterproof.
28. Liu Y, Valerio D (2011) Antimicrobial silicone-based wound dressings. US Patent 20110313048A1
29. Patrick M, Jian W, Dacheng R, Shuyu H (2013) Antimicrobial nanostructured hydrogel web containing silver. US Patent 8431151B2
30. Kubera Sampath Kumar S, Prakash C, Ramesh P, Sukumar N, Balaji J, Palaniswamy NK (2021) Study of wound dressing material coated with natural extracts of *Calotropis gigantean*, *Eucalyptus globulus* and Buds of *Syzygium aromaticum* solution enhanced with rhEGF. J Nat Fibers 18(2):2270–2283
31. Bechert T, Peter S (2009) Wound covering. US Patent 7605298B2
32. Lucie M, Pang NS (2003) Multi-Layer dressing as medical delivery system. US Patent 20030149406A1
33. Alvarez OM (1989) Multi-layer wound dressing having oxygen permeable and oxygen impermeable layers. US Patent 4875473A
34. Meyer RA, Habib WW, Stupar JA (1988) Occlusive wound care dressing. US Patent 4738257A
35. Vert M et al (1992) Bioresorbability and biocompatibility of aliphatic polyesters. J Mater Sci Mater Med 3:432
36. Katti DS et al (2002) Toxicity, biodegradation and elimination of polyanhydrides. Adv Drug Deliv Rev 54:933
37. Nair LS, Laurencin CT (2007) Biodegradable polymers as biomaterials. Prog Polym Sci 32:762
38. Dutta PK et al (2004) Chitin and chitosan: chemistry, properties and applications
39. Ignatova M et al (2009) Electrospun non-woven nanofibrous hybrid mats based on chitosan and PLA for wound-dressing applications. Macromol Biosci 9:102
40. Zhang Z et al (2006) Physicochemical properties of collagen, gelatin and collagen hydrolysate derived from bovine limed split wastes. J Soc Leather Technol Chem 90:23
41. Kadler KE et al (1996) Collagen fibril formation. Biochem J 316:1
42. Yeo IS et al (2008) Collagen-based biomimetic nanofibrous scaffolds: preparation and characterization of collagen/silk fibroin bicomponent nanofibrous structures. Biomacromol 9:1106
43. Vroman I, Tighzert L (2009) Biodegradable polymers. Materials 2:307
44. Yu L et al (2006) Polymer blends and composites from renewable resources. Prog Polym Sci 31:576
45. Shanmugam KSK, Sundaramoorthy S (2015) Development and characterization of an electrospun mat from Eri silk fibroin and PLA blends for wound dressing application. RSC Adv 5:31352–31364. https://doi.org/10.1039/C4RA15268A
46. Subramanian S, Kubera Sampath Kumar S (2024) Nano material based multilayer wound dressing system. Indian Patent, 409857

Surgical Sutures: An Overview

S. Viju, G. Thilagavathi, and R. Brindha

Abstract Sutures tie blood vessels together and bind tissues together in surgical procedures. A variety of materials is used in the making of sutures, none of which is ideal. The surgeon chooses the most appropriate suture material for a specific application based on the circumstances. This chapter discusses the characteristics of the most commonly used polymeric sutures, both absorbable and nonabsorbable. In addition, recent developments in surgical sutures are discussed.

1 Introduction

The process of wound repair depends on the utilization of suture materials in order to provide support to healing tissues. The use of suture material extends beyond the closure of skin wounds, encompassing a variety of purposes such as the closure of hemostasis, fascia, enterotomy, urogenital tract surgery, intestinal anastomosis orthopedic surgery, ophthalmic surgery, plastic surgery, vascular surgery, and neurosurgery. When selecting a suture material, it is crucial to take into consideration the physical and biological properties of the material, as well as the characteristics of the wound, the healing rate of the tissue, and the physical condition of the patient. Often, surgeons' selection is based on their personal preferences, training experiences, and economic considerations. Technological advances in suture materials have led to a greater variety of sutures available. Surgeons must understand the variety available today and keep up with new technology [1, 2].

Suture materials possess physical properties, handling characteristics, and biological properties. The perfect suture material would possess favorable handling characteristics, excellent knot security and tensile strength, sterility, non-electrolytic, and non-ferrous properties, non-capillary nature, nonallergenic, and noncarcinogenic properties, as well as the absence of any bacterial growth medium. Additionally, it should exhibit resistance to shrinkage, efficient absorption, easy sterilization without altering its properties, and cost-effectiveness [3].

S. Viju (✉) · G. Thilagavathi · R. Brindha
Department of Textile Technology, PSG College of Technology, Coimbatore, Tamil Nadu, India
e-mail: vijutext@gmail.com

S. Sundaramoorthy et al. (eds.), *Textile Materials for Good Health and Wellbeing*, SDGs and Textiles, https://doi.org/10.1007/978-981-97-6496-9_12

Sutures are made from a variety of materials, none of which is ideal. The choice of suture material in a specific application is determined by the surgeon, taking into account the circumstances. Factors such as the patient's age, weight, health status, and the location of the wound can influence the selection of suture material. Variations in tissue thickness, elasticity, healing speed, and scarring tendency throughout the body further contribute to the decision-making process. Additionally, certain conditions like diabetes, heart disease, dermatitis, and the use of steroids can impact the healing process. Infection and wound characteristics also play a role in determining the appropriate suture material. In an ideal scenario, surgeons should opt for a material that possesses a high strength-to-diameter ratio, consistent diameter, sterility, flexibility, and optimal tissue acceptance [1, 2].

2 Characteristics of Sutures

2.1 Suture Diameter

Sutures are quantified in millimeters and denoted by multiples of zeros. Sutures with a smaller diameter possess a greater number of zeros. For instance, the diameter of a 6.0 nylon suture is lesser than that of a 4.0 nylon suture. The diameter of sutures varies depending on their chemical composition; 5.0 surgical gut exhibits greater significance compared to 5.0 polypropylene. It is advised to employ the smallest suture size that imparts sufficient tensile strength for repair.

2.2 Tensile Strength

As per the United States Pharmacopeia (USP), the weight needed to fracture a suture is divided by its cross-sectional area to ascertain its tensile strength. Consequently, the relationship is not linear. Furthermore, the evaluation of sufficient tensile strength is performed to ascertain the tensile strength of looped and knotted sutures. The tensile strength of knotted sutures is approximately one-third of that observed in unknotted sutures. Generally, synthetic materials possess greater strength in comparison to natural ones [4].

2.3 Configuration

Sutures can be classified as either monofilament or multifilament in nature. Multifilamentous sutures are braided, a characteristic that lends them ease of manipulation but may also heighten the risk of infection and tissue reactivity. The phenomenon of capillary penetration, which allows bacteria and other foreign elements to permeate the tissue, increases the risk of infection. Bacteria can find refuge in the crevices of braided sutures, thereby evading phagocytosis. Consequently, monofilament sutures composed of materials such as nylon or polypropylene may be more suitable for the closure of contaminated wounds. However, it is worth noting that monofilament sutures are comparatively more challenging to manipulate [5].

2.4 *Coefficient of Friction*

The coefficient of friction determines the ease with which a suture traverses the skin. When performing subcuticular sutures, it is common practice to employ a suture material with a low friction coefficient, such as polypropylene. As the coefficient of friction diminishes, the suture material gains increased slipperiness, thereby rendering the knot more susceptible to unraveling. It is customary to incorporate multiple additional throws when employing polypropylene [6].

2.5 *Knot Security*

The strength of a knot is ascertained by evaluating the amount of force required to induce slippage. It exhibits a direct correlation with the coefficient of friction of the suture. Enhanced knot strength leads to a decrease in wound dehiscence. Although a suture with a high coefficient of friction provides excellent knot security, it presents difficulties in handling and tends to drag across the skin [7].

2.6 *Elasticity*

The elasticity of a suture material refers to its capacity to restore its initial shape and length subsequent to being elongated during the occurrence of wound edema. Although it is improbable for polybutester (Novafil TM) to incise the skin amid edema, it nevertheless aligns the wound edges subsequent to the reduction of swelling [8].

2.7 *Absorption*

Absorption is the process by which the body degrades a suture over time. Sutures can be categorized into two groups: absorbable and nonabsorbable. A nonabsorbable suture, unlike its counterpart, does not undergo absorption and retains its tensile strength. Conversely, an absorbable suture experiences a significant loss of its tensile strength within 60 days following its implantation. The suture material absorbs due to its reactivity with the tissue. The most common use of absorbable sutures is in burying sutures that approximate the edges of deep wounds. Superficially, nonabsorbable sutures are most frequently employed, and they are subsequently extracted. In exceptional instances, they may be utilized solely for deep structures necessitating extended reinforcement [8].

2.8 Tissue Reactivity

When a suture is inserted into a wound, the degree of inflammation elicited by the exogenous object determines its tissue reactivity. Natural substances (such as silk and surgical gut) elicit significantly greater inflammation compared to synthetic substances [8].

3 Suture Materials

3.1 Synthetic Absorbable Suture Materials

3.1.1 Polyglycolic Acid

The melting point of polyglycolic acid (PGA) is 220–225 °C, which makes it a biodegradable artificial polymer. The glass transition temperature ranges from 35 to 45 °C, and its density is 1.5–1.7 g/cm^3. The solubility of this acid is limited to highly fluorinated organic solvents, and its crystallinity lies between 45 and 55%. The properties of polyglycolic acid are influenced by both its average molecular weight and molecular weight distribution. The polyglycolic acid polymer can be synthesized through various routes, including enzyme-catalyzed polymerization, ring-opening polymerization, azeotropic polymerization, and solid-state condensation of haloacetates. Hydrolysis serves as the principal mechanism by which polyglycolic acid undergoes degradation. In this process, carboxyl groups experience an increase, pH values observe a decrease, and the ultimate degradation product, such as ethanoic acid, is eventually eliminated through the urine. In vivo, the degradation of polyglycolic acid usually takes 1.5–3 months. As a result of its more rapid biodegradability and biocompatibility, polyglycolic acid has been widely used for surgical suture applications. There is also evidence that the acid degradation product of polyglycolic acid can cause sterile inflammatory reactions in vivo [9].

3.1.2 Polylactic Acid

There is a glass transition temperature of 50–80 °C for polylactic acid (PLA), a melting temperature of 173–178 °C, and a density of 1.2–1.3 g per cubic meter for thermoplastic polylactic acid (PLA). The chiral molecule lactic acid can be transformed into three stereoisomers, namely poly(D-lactic acid), poly(L-lactide), and poly(D,L-lactic acid). In general, poly (L-lactide) possesses high modulus and mechanical properties, albeit with reduced service temperature, toughness, processability, and augmented crystallinity as the poly (L-lactide) content increases. Most commercially available poly (L-lactide) polymers consist of poly (L-lactide) and a small fraction of poly (D,L-lactic acid), which induces a melting point for processing but decreases crystallinity significantly. As a result, appropriate tradeoffs must be made based on the actual requirements. In addition to this, unaltered poly (L-lactide) exhibits a significant level of hydrophobicity and a diminished rate of degradation, both of which have an impact on its application in practice [10].

3.1.3 Poly-Lactic-Co-Glycolic Acid

A copolymer of glycolic acid and lactic acid cyclodimers in a precise ratio is produced by polymerizing poly(lactic-co-glycolic acid). As a result of its rheological and thermodynamic characteristics, it can be thermally stretched and exhibit good mechanical properties. Adjusting the glycolic acid and lactic acid ratio can control the degradation and mechanical properties. However, its low hydrophilicity may have an impact on therapeutic uses. The strength of Polylactic-co-glycolic acid sutures soaked in synthetic saliva dropped to 79.3% after 7 days. After 14 days, it dropped to 64.2%, slightly higher than Polyglycolic acid sutures [11].

3.1.4 Poly(*p*-Dioxanone)

The poly(*p*-dioxanone) (PPDO) molecules also have ester bonds, which are biocompatible molecules with an absorbable molecular structure. The synthesis of PPDO can be achieved through ring-opening polymerization from polydioxanone (PDO). Monofilament sutures known as PDS, which are manufactured from PPDO, possess greater elasticity but exhibit lower tensile strength compared to PGA and PGLA. Although PPDO sutures are less likely to break down than PGA and PGLA sutures, Compared to PGA and PGLA sutures, they tend to retain tension and exhibit little absorption until 90 days. In addition, PPDO exhibits low toxicity, improved hydrolysis properties, and high knot stability. In many cases, PPDO stitches are used to repair damage with a long healing time. The fiber's low melting point, plasticity, and thermal decomposition can still be improved during the spinning process [12].

3.2 *Natural Nonabsorbable Suture Materials*

3.2.1 Silk

Silk, a naturally occurring fiber, undergoes a degradation process that typically spans 2 years within the living organism. The tensile strength of silk may gradually decline over time when it is in vivo, even though silk is not completely non-degradable. There are many advantages to using silk sutures over other types of sutures, including their strength, low rate of bacterial adhesion, and ease of handling. The adoption of silk sutures in the closure of soft tissues, ocular ligation, neurological procedures, and cardiovascular surgeries has become widely prevalent. In order to achieve a sustained antimicrobial and anti-inflammatory effect, the application of berberine to coat silk sutures has also been employed [13].

3.2.2 Cellulose

Cellulose has sufficient mechanical strength and is low cost, widely available, and reproducible. Cellulose materials include cotton, viscose, ramie, cellulose nanocrystals, bacterial cellulose, and cellulose nanofibrils. Natural cellulose fibers such as cotton have excellent mechanical properties even when moist. Surgical sutures have been used for a long time. However, its low hydrophilicity may have an impact on therapeutic uses. The strength of Polylactic-co-glycolic acid sutures soaked in synthetic saliva dropped to 79.3% after 7 days.

Surgical sutures can also be made from other cellulose-derived materials besides cotton and ramie. It has been reported that sisal and coconut cellulose fibers have similar mechanical properties to silk fibers [14]. Although the tensile strength of coconut fibers was lower, their extraordinary ability to expedite deterioration and encourage bacterial growth was likely due in part to their porous surface. Sisal fibers, on the other hand, had consistent interior space and structure. Additionally, sisal cellulose fibers produced weaker biofilms than other fibers, although both were unable to fend off microbes.

3.3 *Synthetic Nonabsorbable Suture Materials*

3.3.1 Polyamide

Polyamide, commonly acknowledged as Nylon, is a synthetic material that is nonabsorbable and contains an amide group, thereby inhibiting the body's reaction to foreign substances. In addition to its long tensile strength and minimal interference with cells, it possesses excellent mechanical properties, elasticity, and resistance to abrasion. The glass transition temperature of typical polyamide is 50 °C, and its density is 1.14 g/cm^3. Although polyamide sutures demonstrate poor knot fastness and operability, the application of a chitosan coating can reduce surface friction and provide antibacterial properties. Despite lacking biological activity or the ability to facilitate signal exchange between cells, PA fibers can still allow cell adhesion due to their hydrophilicity. To manage the negative effects of the adhesion of cells, polyamide sutures are primarily used in the outermost layer of the skin, which limits direct contact between sutures and proteins/cells [15].

3.3.2 Polypropylene

Polypropylene (PP) has a low density of only 0.91 g per cubic centimeter and is nonabsorbable. In addition, it is easy to synthesize. Polyamide, commonly referred to as nylon, is a synthetic material that is not capable of absorbing liquids. Within the realm of stereospecific polymers, which include syndiotactic, isotactic, and atactic, the isotactic polymer stands out as the most appropriate choice for creating surgical sutures due to its distinctive molecular structure, superior mechanical properties, and improved ease of manipulation. PP sutures maintain their strength for extended periods after being implanted in the body because they are non-biodegradable and bio-inert. Sutures composed of polypropylene exhibit exceptional elasticity and possess a prolonged lifespan, rendering them highly suitable for skin suturing. However, despite these advantageous characteristics, PP sutures do possess certain inherent limitations, including suboptimal knot fastening capabilities and decreased flexibility, thereby increasing the likelihood of incisional hernias post-surgery. PP sutures are commonly used in general surgery, cardiology, and orthopedic surgery [16].

3.3.3 Polyethylene Terephthalate

The density of polyethylene terephthalate (PET) is 1.39 g per cubic centimeter, while its glass transition temperature is 75 Degree Celsius. PET sutures exhibit relatively low knot stability and elicit tissue reactions more frequently than nonabsorbable sutures. Consequently, it is advisable to avoid using PET sutures in proximity to infection-prone wounds. To enhance the antibacterial and biocompatible properties of PET, alkyl, and hydroxyl groups have been introduced into its molecular chain through antiseptic wet chemical surface modifications. In order to meet performance requirements, a study has investigated the blending of PET with other materials.

4 Recent Developments in Sutures

4.1 Coatings for Surfaces

Exploration has been conducted to develop suitable alterations and strategies for functionalization in order to enhance the clinical application of surgical sutures. The technique of surface coating has gained recognition as a widely employed method for modification. It is well established that surgical site infections (SSI) play a pivotal role in influencing the outcomes of postoperative recovery. Surgical sutures must prevent or mitigate wound infections. Various methods have been utilized in order to confer antibacterial, anti-inflammatory, and other healing-enhancing attributes to sutures, encompassing processes such as dip coating/soaking, layer-by-layer deposition, grafting, and impregnation [17].

Coating modification is usually carried out with antibacterial agents. The use of silver nanoparticles with antimicrobial agents has not resulted in drug resistance. A study used Streptomyces sp microbial mediation to synthesize silver nanoparticles (AgNPs) and incorporate them into commercial sutures. In another study, polypropylene (PP) sutures were coated with metallic particles via radiation grafting. Silver-coated sutures showed remarkable resistance to bacteria and cell toxicity [18]. Viju and Thilagavathi [1, 2] incorporated chitosan into silk sutures and studied antimicrobial activity. Based on the results reported by Viju and Thilagavathi [1, 2], the values of the Zone of Inhibition for silk braided sutures treated with chitosan were found to be 13 mm and 12 mm in opposition to *S. aureus* and *E. coli*, respectively. On the other hand, untreated silk braided sutures did not exhibit any zone of inhibition against *S. aureus* and *E. coli*. When compared to the untreated sutures, the treated sutures displayed improved antimicrobial activity against *E. coli* bacteria (as depicted in Fig. 1) and *S. aureus* bacteria (as shown in Fig. 2).

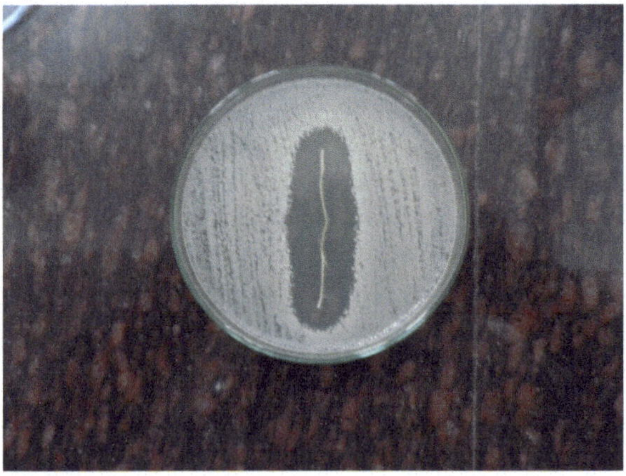

Fig. 1 Inhibition zone for chitosan-treated silk against *E. coli*

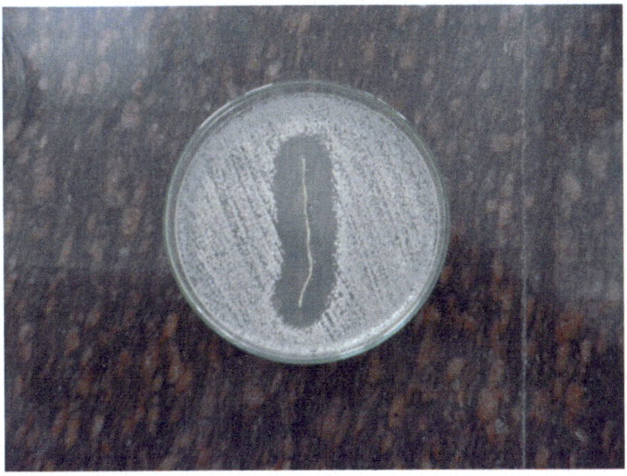

Fig. 2 Inhibition zone for chitosan-treated silk against *S. aureus*

4.2 Surface Grafting on Sutures

There needs to be more control over coating thickness with the coating techniques described earlier. The issues at hand can be effectively resolved by directly modifying the surface of the fiber. The technique of radiation grafting has gained widespread usage due to its ability to control the thickness, composition, and morphology of the grafted layer through the adjustment of reaction conditions such as monomers concentration, irradiation dose, proportion of monomers in copolymer grafting, temperature, and reaction time. First, the polymer suture is exposed to irradiation and subsequently grafted with a monomer. Consequently, these monomers transform into chemical groups that possess the capacity to interact reversibly with a specific drug. Various types of monomers have been successfully grafted onto different sutures. Silk and polymer-grafted polypropylene sutures can interact with antimicrobials or anti-inflammatory drugs either covalently or ionically. Additionally, the grafting of different monomers and the formation of a stimuli-responsive network can modify the drug release profile of the grafted coating. Grafting polypropylene sutures also has the potential to modify the crystallinity and, depending on the level of grafting, either enhance the mechanical properties by strengthening the sutures or diminish them by causing a separation between the polymer chains.

4.3 Drug Loading and Delivery

One promising strategy for drug loading differs from the surface coating technique by encapsulating functional additives directly into the fiber generation process. As the drugs are distributed both within and on the surface of the sutures, direct drug loading and delivery prevents drug burst release behavior and prolongs the duration of drug release [18]. The fabrication process of wet spinning is a well-known technique that is frequently used to produce filaments with functional additives. Through electrospinning, surgical sutures are developed with nanofibers that contain drugs. To achieve this, curcumin is dissolved in a polymeric solution composed of polylactic acid (PLLA), and the resulting solution is subjected to electrospinning, resulting in yarns that carry drug-loaded nanofibers. The ultimate stress of the synthesized yarn demonstrates a gradual decrease as the concentration of curcumin increases, and there is a slight increase in drug release. Although burst release persists for the initial 12-h period, the cumulative drug release can reach up to 78% within 72 h. It would prevent bacteria and inflammation and promote wound healing. It was demonstrated that electrospun nanofiber-constructed yarns loaded with levofloxacin decreased strength but extended the drug release period and controlled infections.

4.4 Sutures with Shape Memory and Self-Tightening Properties

It has been demonstrated that improving traditional sutures' knotting complexity can be achieved by developing intelligent sutures with shape memory and self-tightening mechanisms. Significantly, this can help fix the sutures more stably in a given position during minimally invasive surgical procedures. It is done while providing appropriately sized forces. In their study, the S2 suture was proposed by Phan et al. [19]. Self-tightening knots and barbed anchors could be employed on this S^2 suture to provide practical anchoring. Fluid pressure could be used to program this innovative suture to achieve a preset uniform tension. A pressure-locking mechanism maintained its length as well. Sutures with self-tightening capabilities return to their original form after release of pressure. S2 suture was useful in ex vivo experiments using porcine stomach and colon models.

Shape memory polymers are another emerging trend for generating shape memory sutures since they can return to their original shape after being temporarily deformed. Due to its low and easily controlled temperature response, shape memory polyurethane has attracted the most interest among shape memory polymers. As part of their research, A shape memory polymer was synthesized by Joo et al. [20] using polycaprolactone and polyurethane segments. A hot water bath at 40 °C led to a tight knot on the as-fabricated sutures within 3 s, which was a result of the polymer switching from its pre-stretched state back to its original state. Using hyper-branched polyurethane nanocomposites modified with starch-loaded reduced carbon dots, Rituparna et al. [21] demonstrated self-tightening sutures. Furthermore, the suture recovered its shape after 15 s of heating at 37 °C with 99.7% tightening. In addition, this intelligent suture showed excellent biodegradability, biocompatibility in vitro, and antithrombotic and antihemolytic properties.

4.5 *Sutures with Monitoring and Sensing Capabilities*

Monitoring the wound place after surgery is crucial, especially for severe wounds. Physiological parameters of the wound site are being monitored postoperatively with surgical sutures equipped with advanced sensing capabilities. Mostafalu et al. [22] report that conductive nanomaterial-infused threads can produce thread-based diagnostic devices that integrate physical and chemical sensors. Both in vitro and in vivo, this device is capable of measuring strain and pH values in the subcutaneous and gastric regions. It has been demonstrated by Owyeung et al. [23] that an all-thread device can be stitched transdermally and extracts biofluids online and analyzes them multiplexed, consisting of a sampling thread, a transistor, and an electrochemical sensor. Silk fibers were coated with multiple layers to accomplish multiple functions, including eluting and releasing drugs, monitoring tension, and detecting optical and electronic signals. The incorporation of thermal, strain, electrochemical, and optical sensors into surgical sutures has also been demonstrated in existing studies. The biocompatibility and mechanical properties of surgical sutures are still challenging when electronic devices are incorporated into them.

5 Conclusions

Using polymers in surgical procedures binds tissues together and ligates blood vessels. Polyglycolic acid, poly-lactic-co-glycolic acid, polylactic acid, and poly(*p*-dioxanone) are currently the most commonly used absorbable biomaterials out of a range of biomaterials. At the same time, nylon, polypropylene, and polyester are the most widely used non-absorbable biomaterials. Sutures have recently acquired additional antimicrobial, anti-inflammatory, anesthetic, and biocompatibility properties through technological advancements. The use of shape memory polymers, which can change their properties under small stimuli, was another improvement. The development of surgical sutures with advanced sensing capabilities has enabled the monitoring of physiological parameters following surgery. Despite the promising possibilities of bioactive, innovative, and electronic sutures, the clinical application still poses some challenges.

References

1. Viju S, Thilagavathi G (2013) Effect of chitosan coating on the characteristics of silk braided sutures. J Ind Text 42(3):256–268
2. Viju S, Thilagavathi G (2013b) Characterization of tetracycline hydrochloride drug incorporated silk sutures. J Text Inst 104(3):289–294
3. Viju S, Thilagavathi G (2012) Fabrication and characterization of silk braided sutures. Fibers Polym 13(6):782–789

4. Herrmann JB (1971) Tensile strength and knot security of surgical suture materials. Am J Surg 37:209–217
5. Gabrielli F, Potenza C, Puddu P, Sera F, Masini C, Abeni D (2001) Suture materials and other factors associated with tissue reactivity, infection, and wound dehiscence among plastic surgery outpatients. Plast Reconstruct Surg 107:38–45
6. Ammirati CT (2002) Advances in wound closure materials. Adv Dermatol Allergol 18:313–338
7. Bennett RG (1988) Selection of wound closure materials. J Am Acad Dermatol 18:619–637
8. Tajirian AL, Goldberg DJ (2010) A review of sutures and other skin closure materials. J Cosmet Laser Ther 12(6):296–302. https://doi.org/10.3109/14764172.2010.538413
9. Budak K, Sogut O, Sezer UA (2020) A review on synthesis and biomedical applications of polyglycolic acid. J Polym Res 27(8):208
10. Luo F, Li J, Ji F, Weng Y, Ren J (2021) Preparation of poly(lactic acid)-based shape memory polymers with low response temperature utilizing composite plasticizers. Polym Bull 79(7):4761–4781
11. Taysi AE, Ercal P, Sismanoglu S (2021) Comparison between tensile characteristics of various suture materials with two suture techniques: an in vitro study. Clin Oral Invest 25(11):6393–6401
12. Wu HW, Guo TT, Zhou F, Bu J, Yang S, Dai ZX, Teng C, Ouyang HW, Wei W (2022) Surface coating prolongs the degradation and maintains the mechanical strength of surgical suture in vivo. Colloid Surf B Biointerfaces 209:112–214
13. Qinting W, Chaoheng H, Xuchen W, Shujun Z, Li Z, Ruijuan X, Yi L, Xiaoqin W, Zhifen H, Zhaozhu Z, Gang L (2021) Sustainable antibacterial surgical suture using a facile scalable silk-fibroin-based berberine loading system. ACS Biomater Sci Eng 7(6):2845–2857
14. María Paula Romero G, Lilian S, Nelson Santiago V, Karla V, Alexis D, Daniel CW, Ralph SO, Frank A (2020) Natural cellulose fibers for surgical suture applications. Polymers 12(12):3042
15. Hufenus R, Yan Y, Dauner M, Kikutani T (2020) Melt-spun fibers for textile applications. Materials 13(19):4298
16. Aldhabaan SA, Hudise JY, Aldosari BF (2022) Comparison between using monocryl suture and polypropylene suture in closer of alar base excision in rhinoplasty. Plastic Surg Int 30(1):59–62
17. Feras A, Arul M, Omar P, Wenhui S (2016) Bioactive sutures: a review of advances in surgical suture functionalization. Curr Med Chem 24(2):215–223
18. Tuba B, Nurdan S, Aysel U, Inci Rana K (2018) Antimicrobial characteristics and biocompatibility of the surgical sutures coated with biosynthesized silver nanoparticles. Bioorg Chem 86:254–258
19. Phan PT, Hoang TT, Thai MT, Low H, Davies J, Lovell NH, Do TN (2021) Smart surgical sutures using soft artificial muscles. Sci Rep 11(1):22420
20. Joo YS, Cha JR, Gong MS (2018) Biodegradable shape-memory polymers using polycaprolactone and isosorbide based polyurethane blends. Mater Sci Eng C 91:426–435
21. Rituparna D, Yogendra PS, Prerak G, Biman BM, Niranjan K (2018) Smart self-tightening surgical suture from a tough bio-based hyperbranched polyurethane/reduced carbon dot nanocomposite. Biomed Mater 13(4):045004
22. Mostafalu P, Akbari M, Alberti KA, Xu Q, Khademhosseini A, Sonkusale SR (2016) A toolkit of thread-based microfluidics, sensors, and electronics for 3D tissue embedding for medical diagnostics. Microsyst Nanoeng 2:16039
23. Owyeung RE, Terse-Thakoor T, Nejad HR, Panzer MJ, Sonkusale SR (2019) Highly flexible transistor threads for all-thread based integrated circuits and multiplexed diagnostics. ACS Appl Mater Interfaces 11(34):31096–31104

Advancements in Implantable Medical Textile Materials

M. Indira, B. Sudarsini, and B. Sumalatha

Abstract The integration of textiles into medical devices has opened new avenues for enhancing patient care, surgical procedures, and therapeutic interventions. Implantable medical textiles exhibit a unique set of properties including biocompatibility, flexibility, and adjustable mechanical characteristics making them highly adaptable to the dynamic requirements of the human body. By examining the latest developments in materials science and engineering, this manuscript aims to provide insights into the multifaceted applications of these textiles, offering a glimpse into their potential transformative impact on the fields of vascular surgery, tissue regeneration, and various other medical domains. This manuscript seeks to unravel the multifaceted potential of these materials. The chapter emphasizes the significance of biomaterials used in the advancement of implants encompassing materials such as vascular grafts and tissue scaffolds. It also addresses the crucial aspects of biocompatibility assessment and explores the diverse biomedical applications across various fields. Additionally, the chapter offers insights into the prospects of these biomaterials providing a comprehensive overview of their evolving role in advancing medical science and technology.

Keywords Biocompatibility · Medical implants · Scaffolds · Tissue engineering · Vascular grafts

M. Indira (✉) · B. Sudarsini
Department of Biotechnology, Vignan's Foundation for Science Technology and Research, Vadlamudi, Andhra Pradesh, India
e-mail: indiramikkili@gmail.com

B. Sumalatha
Department of Chemical Engineering, Vignan's Foundation for Science Technology and Research, Vadlamudi, Andhra Pradesh, India

S. Sundaramoorthy et al. (eds.), *Textile Materials for Good Health and Wellbeing*, SDGs and Textiles, https://doi.org/10.1007/978-981-97-6496-9_13

1 Introduction

Medical textiles represent a burgeoning interdisciplinary field that bridges the gap between conventional textile technologies, such as weaving, embroidery, knitting, and other scientific fields such as medical sciences, chemistry, electrical engineering, and pharmaceuticals [1]. From a scientific perspective, medical textiles exist at the intersection of technical disciplines and life sciences [2].

On one front, the technical dimension encompasses textile engineering, material chemistry, testing, process control, and certification, all integral for the production of top-tier medical textile products [3]. Conversely, life sciences including medicine, microbiology, and related disciplines, play a pivotal role in shaping the functional characteristics of these products [3]. The intersection of these diverse scientific and technical domains fosters synergy with progress in one field frequently sparking innovations in others [4]. This field has expanded enormously leading to a wide array of innovative products, specialized high-tech products including blood filtration membranes, monitoring devices, protective textiles, modern baby diapers, adult incontinence products, surgical sutures, prostheses, implants, and other cutting-edge applications [5]. The synergy across disciplines and applications underscores the dynamic nature of medical textiles and their interconnectedness with broader scientific and technical realms [6].

The choice of materials for medical implants is contingent upon their specific application within the human body [7]. The acceptance or rejection of these materials is strongly influenced by their physicochemical properties. Four key physicochemical properties, surface functional group, surface topology, surface charge, and surface wettability play a pivotal role in governing the interaction between these materials and the human body [8]. The natural textile fibers can activate and adsorb plasma proteins, initiating interactions that impact biological responses including blood clotting and other tissue reactions [9]. On the other hand, synthetic textile fibers exhibit a high affinity to tightly bind with proteins resulting in foreign body reactions and adhesion to tissue materials [10].

The engineering of functional tissues relies significantly on the fabrication of constructs with precisely controlled mechanical properties, microstructure, and cellular distribution [11]. Scaffolds crafted from either synthetic or natural biomaterials are frequently employed to furnish mechanical support and establish three-dimensional (3D) environments conducive to cellular growth and function [12]. Conventional techniques for scaffold creation such as particle leaching, solvent casting, and freeze-drying produce porous constructs characterized by interconnected pores. These porous structures are well suited for facilitating the effective delivery of nutrients to the cells [13].

In the last few years, there has been a transformative shift at the intersection of material science and healthcare, leading to revolutionary breakthroughs in implantable medical textile materials [5]. These materials designed to seamlessly integrate with biological systems, have emerged as transformative agents in various medical applications [14]. This manuscript embarks on a comprehensive exploration of the latest advancements in implantable medical textiles with a keen focus on their applications in vascular grafts, tissue engineering scaffolds, and an array of other medical contexts.

2 Biomaterials

Biomaterials play a crucial role in advancing medical interventions, leveraging biological materials for human use, with significant potential for organ regeneration, wound healing, and tissue repair [15]. In earlier times, exogenous materials were predominantly used for wound healing and organ repair, but recent progress in biomaterial development has significantly enhanced these applications [16]. Biocompatibility, the interaction between biological systems and biomaterials is critical for successful integration and mitigating risks such as implant rejection and inflammation initiation [17]. Distinguished by their ability to elicit specific host responses, biomaterials contribute to biocompatibility during critical conditions [18]. Biomaterials have made a profound impact on tissue engineering and biomedical devices by facilitating the regeneration of human tissues or organs, thereby restoring normal bodily functions [19]. Their significance is further underscored by their sustainability and widespread application across various medical disciplines, compensating for traditional materials [20]. The selection of biomaterials considers factors such as the implantation site, the functions and structures of target tissues, durability, and the patient's medical history. This meticulous approach ensures optimal outcomes and underscores the critical role of biomaterials in advancing modern medical practices [21].

The most frequently utilized biomaterials for implant purposes encompass a diverse range of materials such as (1) Metals, (2) Ceramics, (3) Polymers, and (4) Biocomposites [18] as outlined in Fig. 1.

2.1 Metals

Individuals grappling with issues such as compromised joint function, weakened bones, bone trauma, and surgical procedures involving joints like knees, elbows, hips, and ears often necessitate implants. The utilization of biomaterials has seen a surge in demand within the medical field to address these challenges [22]. Among the metals employed as biomaterials for implants, notable examples include: stainless steel, titanium-based alloys, and cobalt and chromium-based alloys [23].

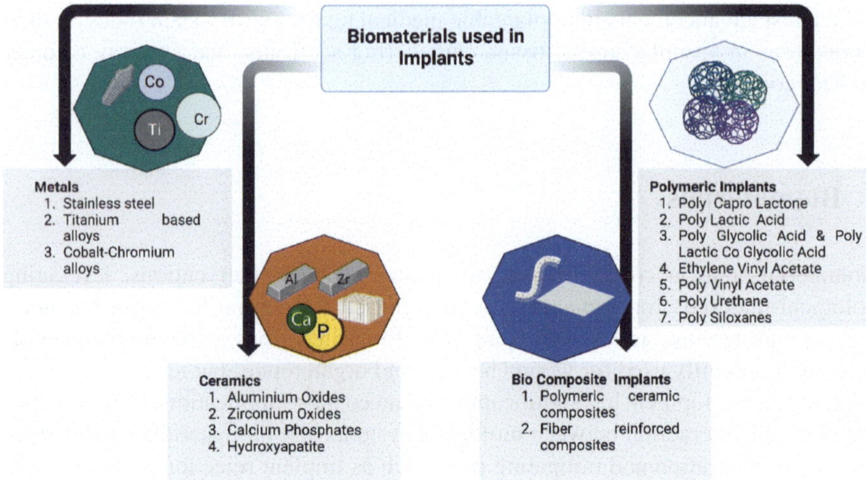

Fig. 1 Types of biomaterials used in implants

2.1.1 Stainless Steel

Substituting the prevalent use of vanadium steel in implants, stainless steel has emerged as a compelling alternative due to its robust nature and superior corrosion resistance compared to other materials with the SS302 variant being particularly prevalent [24]. This biomaterial has become ubiquitous globally, recognized for its exceptional strength and durability.

When alloyed with other materials, it gains enhanced malleability, a critical attribute for effective implant development [25]. Stainless steel biomaterials, especially when coated with chromium form an oxidized layer that renders them highly resistant to degradation and corrosion when exposed to bodily fluids and environmental factors [26].

Beyond its corrosion resistance, stainless steel exhibits biocompatibility and holds the distinction of being the first metal used in implants offering a cost-effective solution compared to other metals. For temporary usage, stainless steel finds application in the fabrication of implants such as intramedullary nails, bone screws, and bone plates, while for permanent applications, orthinox stainless steel is specifically employed in shoulder joint replacements and widely used in hip replacements. This versatility and adaptability make stainless steel a pivotal biomaterial in the realm of implantology [27].

The primary selection of stainless steel for implant devices is attributed to its remarkable durability and capacity to withstand high loads, making it an optimal choice for applications demanding stability and tolerance [28]. Despite its commendable advantages, it is noteworthy that stainless steel can release toxic elements such as nickel and chromium when exposed to a chlorine environment. Consequently, researchers have explored the substitution of nickel with nitrogen considering its potential harm to the human body [29].

To further enhance the performance of stainless steel implants, incorporating 3D-printing customizations has the potential to augment fixation in bone fractures [30]. However, it is essential to acknowledge that in certain conditions, these implants may be susceptible to corrosion by osteoclasts leading to inflammation and potential infections. Addressing this challenge is crucial for achieving enhanced biocompatibility in stainless steel implants for human usage [31].

2.1.2 Titanium-Based Alloys

Titanium, a transition metal existing in both liquid and solid states is widely utilized in implants with a preference for its α form, either in pure form or as an alloy owing to its exceptional resistance to corrosion [32]. An imperative criterion for effective implant functionality is the biomaterials' ability to mimic the natural structure of the affected bones within the human body. Titanium especially in its alloyed forms fulfills this requirement enabling the surrounding tissues to seamlessly integrate with the implant [33]. Renowned for their high biocompatibility, titanium, and its alloys find extensive use in various biomedical applications supported by advantageous characteristics such as low specific weight, Low modulus of elasticity, good ratio of strength and weight, extraordinary resistance to corrosion, and good tribological properties [34].

The exceptional attributes of titanium, characterized by its corrosion resistance, lightweight construction, and impressive toughness establish its alloys as a top choice for manufacturing both temporary and long-term external devices and fixations. This spectrum of applications includes the creation of artificial limbs and orthopedic calipers [35]. The stability of titanium in biological systems is a key attribute as it forms a passive oxide layer akin to the ceramic nature of bones, sharing an inorganic calcium phosphate composition. This characteristic supports stable osseointegration, a biological phenomenon that establishes a direct structural and functional connection between load-bearing implant surfaces and living bones [36].

Titanium's nontoxic and inert nature in the bodily environment further enhances its appeal for implant usage [37]. Its pivotal role extends to medical devices such as cardiovascular stents, where its crucial attributes of corrosion resistance and biocompatibility ensure prolonged effectiveness within the body's vascular system [35]. The indispensable role of titanium alloys in the biomedical field is underscored by their exceptional combination of optimal, mechanical, tribological, and biological properties, extensively utilized in myriad biomedical applications [38].

2.1.3 Cobalt and Chromium Alloys

Throughout the twentieth century, the use of metal alloys as biomaterials gained considerable attention with cobalt-chromium alloys emerging as prominent choices in the medical field due to their exceptional mechanical attributes such as stiffness, strength, and corrosion resistance, contributing to remarkable biocompatibility [39].

These cobalt-chromium alloys find widespread applications as components in modular prosthetic devices particularly in hip or knee joints, including ball and socket joints as well as in various other implants like tribal trays, acetabular cups, dental components, pacemaker lead casings, and cardiovascular stents [40]. Despite their widespread use and biocompatibility, a notable drawback of cobalt-chromium alloys is the potential for patient allergies or sensitivity to these implants [41]. Ongoing research and interest in biomaterial-based implants drive progress in manufacturing methods and surface alterations, continually enhancing the biocompatibility and functionality of cobalt-chromium implants. This ensures their continued efficiency and role in advancing implant technology [42].

2.2 Ceramics

The advent of ceramics in orthopedics dates back to around 1960, marked by the emergence of bioactive ceramics, often referred to as bio-glass or Calcium Phosphates (CaPs) in bone cement. These ceramics offer several advantages including injectability, minimal tissue necrosis, and excess solubility [43].

Initially, ceramics served as alternatives to metal hip prostheses and femoral heads, employing materials such as aluminum oxides, zirconia, and calcium phosphates. These ceramics formed through heating and pressing with variations in temperature and pressure exhibit biocompatibility, small size, high density, and favorable mechanical properties [44]. Hydroxyapatite, an inorganic form of ceramics is widely used in implants for various biomedical purposes owing to its outstanding properties like biocompatibility, bioactivity, osteoconductivity, and the ability to develop a direct bond with bones [45].

Porous aluminum oxide, due to its color similarity to dentin, Osseointegration, and allowance for bone penetration was widely used but faced challenges in mechanical strength leading to damage to the surrounding muscular and vascular cavities, necessitating amputation [46, 47]. Zirconium dioxide later replaced aluminum oxide implants offering refined properties especially stabilized zirconia-3 mol% yttrium oxide (yttria) becoming a primary choice for dental implants [48]. It maintained key characteristics like mechanical properties, biocompatibility, and exceptional fracture resistance [49].

To address the limitations of aluminum and zirconium oxides, a new material i.e., a combination of Al_2O_3-ZrO_2, emerged yielding Zirconia Toughened Alumina (ZTA) and (Alumina Toughened Zirconia (ATZ). These combinations offered higher strength, toughness, and hardness essential for maintaining resistivity to aging and ensuring optimal mechanical properties of the implant [50].

Calcium phosphate ceramics consisting of bioactive crystalline materials such as hydroxyapatite (HA) and β-Tricalcium phosphate (β-TCP) are used in implants.

Their degradation depends on factors like porosity, type, degree of crystallinity, and surface area. Examples include Noarin, Bone Source, and Mimix [51].

As an alternative approach, coating implants with Hydroxyapatite, a crystalline structure has been studied for its effectiveness in healing orthopedic and dental implants due to its osteoconductive and biocompatible behavior [52]. Commercially available coatings of implants once dominated by Calcium Phosphate bioceramics are now being replaced by hydroxyapatite drawing interest from researchers for its attributes in fighting infections, osteoinduction, addressing osteoporosis and exhibiting therapeutic effects [53].

2.3 Polymeric Implants

The contemporary landscape of biomedical implants has shifted toward polymeric biomaterials driven by their versatile nature, extensive applications in biomedical fields, and enduring biocompatibility [54]. Polymeric implants classified into biodegradable and non-biodegradable categories have garnered significant attention [55] which includes:

2.3.1 Polycaprolactones (PCL)

Polycaprolactones have recently gained prominence in biomedical implant development, offering an ideal material with a low melting temperature conducive to easy manufacturing. PCL is an emerging 3D printable biodegradable material, exhibiting considerable potential [56]. Implanted as coatings or 3D scaffolds, PCL degrades into glycolic acid, lactic acid, and hydroxyl hexanoic acid finding applications in biodegradable stents and bone fracture healing [57].

2.3.2 Polylactic Acid (PLA)

Polylactic acid derived from natural sources exhibits biodegradability and bioabsorbability, showcasing its versatile applications in various medical contexts [58]. Polylactides including PLA, are employed in tissue engineering scaffolds, covering membranes, and various medical implants and devices such as fixation rods, plates, pins, screws, and sutures [59].

2.3.3 Polyglycolic Acid (PGA) and Polylactic-Co-Glycolic Acid (PLGA)

Polyglycolic acid, a biodegradable polymer finds applications in implants owing to its biocompatibility and ability to degrade into nontoxic byproducts [60]. PLGA, a copolymer of PLA and PGA, is employed in various medical applications including sutures, stents, screws, plates, meshes, scaffolds, and synthetic skin generation [61].

2.3.4 Ethylene Vinyl Acetate (EVA)

EVA, a thermoplastic copolymer is a frequently employed non-degradable polymer in implantable applications holding FDA approval [62]. It is insoluble and nontoxic, finding applications in intravaginal rings, subcutaneous implants, and ocular implants. However, these non-biodegradable devices require retrieval post-use [63].

2.3.5 Polyvinyl Acetate (PVA)

FDA-approved PVA is used in implant coatings, valued for its minimal cytotoxicity favoring cell proliferation and facilitating tissue regeneration [64, 65].

2.3.6 Polyurethane (PU)

Polyurethanes are favored for their biocompatibility and ease of modifying chemical and mechanical properties. They are widely utilized in medical applications including tubings, dressings, catheters, surgical drains, cardiovascular devices, and coatings for devices such as breast implants [66]. Ongoing developments in PU scaffold development highlight their potential for both soft and hard tissue engineering in regenerative medicine, indicating a promising future for successful clinical trials [67].

2.3.7 Polysiloxane (PS)

Polysiloxanes or silicones are employed in implants due to their biocompatibility, flexibility, and biostability [68]. Polymerized polysiloxanes are used in the formation of nanofibers for stent implants and various other implants including breast implants, dental implants, and abutment screw holes [68–72].

2.4 Biocomposites

Biocomposites are increasingly employed in the medical field as sustainable, renewable, and ecofriendly alternatives, often incorporating reinforcing agents such as ceramics or polymers [73]. These materials characterized by biocompatibility, biodegradability, and natural polymeric components find applications in various medical implants [74].

2.4.1 Polymer Ceramic Composites

Polymer ceramic composites stand out as preferable choices for biodegradability, ease of processing, and biocompatibility. They are commonly utilized for applications such as the fixation of plates, treatment of jaw fractures, sutures, interference screws, and meniscus repair [75].

2.4.2 Fiber-Reinforced Composites (FRC)

Fiber-reinforced composites are utilized in patient-specific implants of standard shapes [76]. Biocomposites made from fibers exhibit characteristics such as lightweight, environment friendliness and renewability [77]. These materials find applications in diverse implants including dental implants, craniofacial implants, and other medical devices [76, 78–81].

3 Implantable Materials

Implantable materials play a crucial role in the repair and restoration of affected body parts. Medical textiles encompass both implantable and non-implantable materials utilized for tissue or organ regeneration, treatment, and growth [82]. Examples of implantable materials include vascular grafts, sutures for wound closures, artificial joints, and ligaments. Non-implantable materials, on the other hand, comprise bandages, plasters, wound dressings, orthopedic belts, and pressure garments [83]. Implantable materials find applications in replacing heart valves, joints, arteries, bones, and in bone fixation procedures. They are instrumental in the development of vascular grafts and tissue engineering scaffolds, contributing significantly to medical advancements [12].

3.1 Vascular Grafts

Vascular grafts also known as vascular prostheses find applications in cardiovascular reconstruction, hemodialysis blood access, trauma repair, and aneurysm repair [84]. Coronary artery bypass grafting (CABG) is a commonly employed procedure for bypassing blocked coronary arteries, reinstating proper blood flow to the heart muscle [85]. While healthy autologous vessels are the preferred choice, their availability is often limited. This scarcity has led to a substantial demand for tissue-engineered vascular grafts as a viable alternative [86]. However, the challenge lies in producing clinical-grade implantable tissue-engineered vascular grafts that can effectively integrate into the host, ensuring long-term viability. Overcoming this hurdle remains a significant challenge in the field [87].

Peripheral vascular bypass (PVB) grafts are used to bypass blocked arteries in the legs and arms [88]. Hemodialysis grafts (HG) are used to create access to dialysis in patients with kidney failure [89]. Aneurysm repair grafts (ARG) are used to reinforce or replace weakened blood vessel walls [90].

Key prerequisites for success in these applications include ensuring that the graft meets the handling requirements of surgeons, does not induce hyperplasia, avoids thrombosis, and maintains mechanical integrity [84]. Vascular grafts provide a conduit for blood to flow through, bypassing a blocked or damaged blood vessel [91]. The materials used for preparing vascular grafts are:

3.1.1 Types of Vascular Grafts

(a) **Synthetic Grafts**

These grafts are crafted from materials like expanded polytetrafluoroethylene (ePTFE) or polyethylene terephthalate (Dacron). The materials used in synthetic grafts are durable and resistant to infection [87].

(b) **Biologic Grafts**

These grafts are derived from human or animal tissues such as veins or arteries. They may undergo treatment to remove cells and prevent rejection [92].

3.1.2 Advances in Vascular Graft Design

Advancements in fabrication technology, coupled with progress in vascular tissue engineering and regenerative medicine, enable biological tissue-engineered vascular grafts to transform into living grafts [87]. These grafts have the ability to integrate, remodel, and repair host vessels responding to both mechanical and biochemical stimuli in their surroundings. Consequently, they hold the potential to address the shortage of available vascular grafts [93].

Over the past decade, significant strides have been made in the design of vascular grafts, aiming to enhance their performance and longevity. Innovations in materials science have led to the development of grafts with improved biocompatibility, flexibility, and mechanical strength [91].

The utilization of advanced biomaterials like ePTFE and Dacron has been instrumental in driving these advancements [94]. These modern materials not only offer durability but also mimic the natural properties of blood vessels, promoting better integration with the host tissue [95]. Additionally, novel polymers have been developed, enhancing the versatility and capabilities of vascular tissue engineering approaches [94]. In the realm of vascular tissue engineering, recent innovations have introduced advanced methodologies such as electrospinning, 3D printing, decellularization, and lyophilization. Moreover, there has been a shift toward more sophisticated graft architectures [96].

The incorporation of bioactive coatings and surface modifications has been explored to enhance cell adhesion, reduce thrombogenicity, and promote endothelialization, nanostructured surfaces, and smart polymers are being integrated to impart stimuli-responsive properties, allowing grafts to adapt to changing physiological conditions [97]. These design innovations hold promise for improving the overall performance and clinical outcomes of vascular grafts, addressing challenges such as restenosis and thrombosis associated with traditional graft designs [98].

3.2 *Tissue Engineering Scaffolds*

In the domain of tissue engineering, scaffolds serve as essential elements in establishing a favorable environment for cell attachment, differentiation, proliferation, and migration, both in in-vitro culture and in-vivo implantation [99]. Crafted from polymeric biomaterials, these scaffolds provide structural support for cell attachment and tissue development [100].

3.2.1 Scaffolds Interaction with Cellular Components

In engineered tissues, a triad of cells, scaffolds, and growth-stimulating signals collaborates to replace faulty tissues, restoring, or substituting damaged regions [101]. Functioning as temporary structures, scaffolds guide and facilitate the regeneration of damaged or lost tissues, playing a crucial role in regenerative medicine [100]. They create a supportive framework for the growth and development of new tissues and organs, playing a significant role in regenerative processes [102]. Therefore, the current paradigm in tissue engineering scaffolding aims to replicate at least to some extent, the functions of the native extracellular matrix (ECM) [103].

Scaffolds offer various advantages in tissue engineering includes:

It must offer void space to support vascularization, promote the formation of new tissue, and facilitate remodeling, ensuring effective integration with host tissue upon implantation [13]. The biomaterials need to undergo processing to create a sufficiently porous structure, allowing efficient transport of nutrients and metabolites without compromising the scaffold's mechanical stability significantly [104]. The biomaterials demonstrate a degradation rate upon implantation that corresponds to the formation of a new matrix by the developing tissue [23]. Scaffolds need to offer support for cells, whether they are added externally or come from within, allowing them to attach grow, and differentiate [100]. This applies to both laboratory cell cultures when the scaffolds are implanted inside the body. Scaffolds actively interact with the cellular components of engineered tissues, guiding and regulating their activities [101]. Additionally, the scaffold can function as a carrier or storage for external growth signals such as growth factors to expedite the regeneration process. Scaffolds offer mechanical stability and shape to the tissue defect

[105]. The inherent mechanical properties of the biomaterials used for scaffolding or their post-processing properties should align with those of the host tissue [13].

3.2.2 Role of Implantable Textiles in Tissue Engineering

Implantable textiles have become essential elements in the realm of tissue engineering, serving a pivotal role in coordinating the regeneration and repair of damaged or lost tissues [106]. The textiles act as three-dimensional frameworks, guiding the formation of new tissue and influencing its architecture. Textiles have various applications including wound closure, body wall repair, cardiovascular implants, and tissue engineering and regeneration [1, 107]. The choice of materials for implantable textiles is pivotal in determining their effectiveness in tissue engineering applications. The use of advanced polymers such as polydioxane (PDO) and poly (L-lactic acid) has gained prominence due to their favorable biocompatibility and mechanical properties [108]. In addition to their role as structural supports, implantable textiles often serve as carriers for bioactive molecules and growth factors [109]. Functionalized textiles can release these bioactive agents in a controlled manner influencing cell proliferation, differentiation, and tissue maturation [110]. As a result, implantable textiles contribute not only to the physical reconstruction of tissues but also actively participate in the biochemical signaling crucial for successful tissue engineering outcomes [19]. The integration of these textiles in tissue engineering strategies holds great promise for advancing regenerative medicine and addressing complex clinical challenges [1].

4 Biocompatibility Assessment

Ensuring optimal biocompatibility is paramount for the clinical implementation of biomaterials in implants [33]. Biocompatibility denotes the ability of biomaterials to fulfill their intended function without causing detrimental effects on biological systems while inducing an appropriate host response [111]. The ideal biomaterial should possess both high biocompatibility and exceptional functional properties. Various factors encompassing the chemical, mechanical, and structural characteristics of biomaterials, along with their interactions with the biological environment, and evaluation methodologies can profoundly impact biocompatibility [112]. The comprehensive biological evaluation of biomaterials encompasses a diverse array of in-vitro and in-vivo tests addressing cytotoxicity, hemocompatibility, genotoxicity, carcinogenicity, pyrogenicity, developmental toxicity, degradation assessment, and in-vivo animal models [113]. These assessments adhere to various international standards reflecting the multifaceted nature of ensuring the safety and efficacy of biomaterials intended for implant applications [111].

4.1 *In-Vitro Studies*

For the assessment of biomaterial biocompatibility, in-vitro studies play a crucial role as they provide insights into the interactions occurring between biomaterials and the immune system ensuring their safety and efficiency [114]. Serving as the initial step in any research, these in-vitro studies help assess the biocompatibility and cytotoxicity of various cells particularly their impact on stimuli generated through blood vessels [115]. These studies are instrumental in comprehending the relative compatibility of biomaterials checking for positive responses with host cells [116]. To assess the biocompatibility of the implants in-vitro, various tests can be conducted including cytotoxicity, hemocompatibility, genotoxicity, carcinogenicity, pyrogenicity, developmental toxicity, and degradation assessment [117]. These tests collectively contribute to a comprehensive understanding of how the implants interact with biological systems and ensure their suitability for clinical applications.

4.1.1 Cytotoxicity Testing

The safety and quality of biomaterials can be analyzed in the early stages of their development, even before conducting animal studies through cytotoxicity tests, which help assess the reactivity of biomaterials [118]. Cytotoxicity testing, a mandatory assessment for checking the biocompatibility of implants serves as a screening test for implants that come into contact with the patients [119]. The approval of each cytotoxicity test ensures consistency among different implant materials, allowing for the assessment of harmful or toxic substances before progressing to in-vitro or human evaluations [120].

4.1.2 Hemocompatibility Testing

Biomaterials containing blood are designed to maintain a non-inflammatory and non-thrombogenic surface achieved through exposure to various tissue types and shear forces within the body, ensuring compatibility with blood; this property is termed hemocompatibility [121]. Biodegradable implants, upon being implanted in the body, gradually degrade over time, facilitating the growth of new tissues around their placement [122]. Given the adverse effects of biomaterials interacting with blood components, it is crucial to focus on their impact. Biomaterials have the potential to disrupt blood components and trigger platelet activation upon exposure leading to undesirable thrombosis issues. Therefore, the analysis of platelet activation becomes a vital aspect of hemocompatibility testing [123]. Many artificial implants have the potential to induce thrombosis and inflammation upon contact with blood, underscoring the essential nature of hemocompatibility testing. This testing helps evaluate immune responses, particularly platelet activation, which is pivotal in understanding and preventing thrombogenesis [124].

4.1.3 Genotoxicity Testing

There is a focused effort on assessing the genotoxic effects of implants driven by concerns related to their biomedical applications which could potentially impact the regulatory components of DNA or even the entire DNA structure [125]. This scrutiny becomes particularly crucial when dealing with implants coated with nanoparticles as a significant number of them have been identified as genotoxic causing harm to the genome and potentially leading to mutations and in some cases the development of cancers [126]. Therefore, it is imperative to thoroughly evaluate the genotoxicity of implants as their adverse effects may include gene mutations, chromosomal damage, and DNA injury, thereby contributing to the induction of carcinogenesis [111].

4.1.4 Carcinogenicity Testing

Carcinogenicity testing is typically conducted when a specific implant material exhibits genotoxic effects aiming to identify the presence of any carcinogenic substances within the implant materials [127]. In-vitro carcinogenic tests are crucial for assessing epidemiological and clinical data related to carcinogenicity and toxicology, providing essential insights into the safety of implant materials [128]. These studies often involve animal models as they have been observed to exhibit similar responses to humans facilitating the evaluation of the implant material's carcinogenic potential [129].

4.1.5 Pyrogenicity Testing

Implants made from pyrogenic substrates of microbial origin have been observed to induce fever and may result in undesirable consequences [130]. These pyrogens can stimulate fever and immune responses, and trigger inflammatory reactions that may lead to complications. This underscores the critical necessity for comprehensive safety testing in blood containing medical devices to monitor long-term effects in patients [131].

4.1.6 Developmental Toxicity

Teratogenicity testing evaluates the materials that may pose toxicity risks during an organism's development with indicators such as embryonic or fetal death, retardation, functional impairment, and malformation serving as markers of developmental toxicity [132]. Long-lasting implants should therefore undergo testing for reproductive and developmental toxicity to study their toxicological effects [133]. To assess developmental toxicity, trials are conducted on pregnant animal models by subjecting them to implants or their components [134].

4.1.7 Degradation Assessment

Regardless of their composition, biodegradable materials used in implants are expected to gradually break down over time, serving as scaffolds or aiding in the healing process [135]. Factors such as the composition of the implant materials and environmental conditions can influence the degradation rate of the implants [136, 137]. Ongoing research in biodegradable materials is focused on methodologies for controlling impurity levels, enhancing coatings and alloying components, functionalizing implant materials, refining the rate of biodegradation at the interface between the implant and tissue, and exploring novel degradable materials [23].

4.2 *In-Vivo Studies*

The toxicity of biomaterials used is initially assessed through in-vivo experiments before their clinical applications, aiming to examine host responses and ensure efficiency [138]. The interactions of these biomaterials with human response are studied using animal models [139]. When selecting an animal model for clinical trials that resemble humans, factors such as the length and diameter of conduits, time of implantation, blood flow rate, and accessibility should be carefully considered [140].

Recent studies involving preclinical and clinical trials on in-vivo animal models have been increasing particularly in assessing the alternative biomaterial implants [141]. Before commencing scientific enquiries, researchers are obligated to adhere to animal welfare standards to ensure the meaningful use of animals in biomedical studies [142]. Over the past few decades, a diverse range of animals from small creatures like mice, rats, and rabbits to larger ones including dogs, sheep, cows, and pigs have been employed for preclinical assessments of burdened bone defects. Considerations for their selection encompass the cost of purchase, maintenance of animals, the animal's tolerance to captivity, ease of handling, and ethical or societal acceptance for their experimental use [143]. Additionally, it is crucial to maintain sterile conditions while handling animal models, exercising utmost care during surgeries [144].

It is imperative to regularly monitor the animals selected for experiments to assess their exposure to pain, distress, or suffering, particularly instances of moderate or severe pain and procedures that could potentially impair their conditions [145]. This monitoring is essential as it allows for the identification of concerns and complications [146]. Molecular analysis tools such as MS-based proteomics, hold high potential for analyzing proteins to understand the interactions between biological systems and biomaterials [147]. Non-clinical implantation studies, crucial for assessing medical device safety, involve evaluations of implanted medical devices that require thorough microscopic and macroscopic pathology assessments due to the device's physical presence and the unique tissue response to its materials [148]. Thus, in-vivo clinical trials are essentially required to establish fundamental documentation that lays the groundwork for human clinical studies on biomaterial implants [19].

5 Biomedical Applications

The biomedical applications of wearable and medical sensors and devices ranging from disease prevention to monitoring have been highlighted as transformative in the medical era. Notable advantages include a reduction in healthcare costs and an enhancement of patient well-being [149] (Fig. 2). These biomaterials are well matched with the human body due to excellent properties such as nontoxicity, biocompatibility, and osteoconductivity leading to reliable health improvements for patients and contributing to the growing demand in the biomedical industries [150] (Table 1). Some of them include:

5.1 *Blood Vessel Prostheses*

Blood vessel prostheses, also known as vascular grafts are composed of synthetic biodegradable polymers and find significant utility as biological implants, particularly in replacing damaged or diseased blood vessels [151]. These prostheses play a crucial role in vascular surgeries addressing conditions such as atherosclerosis, particularly in arterial reconstruction procedures [152]. They come in various forms, including synthetic grafts made from commercially available materials like ePTFE

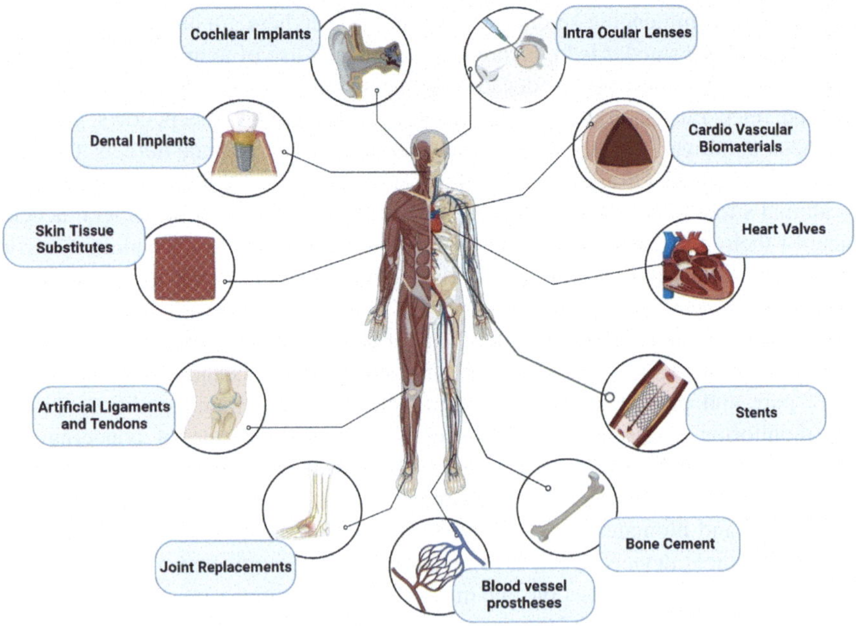

Fig. 2 Biomedical applications of implants

Table 1 Implant materials advantages and disadvantages

Name of the implant	Advantages	Disadvantages	References
Blood vessel prostheses	They resemble the natural blood vessels both in their function & structure highlighting the crucial role of stem cells in vascular tissue engineering	The development of these vascular grafts poses significant challenges underscoring the importance of carefully choosing a suitable animal model to ensure success	[194, 195]
Cardiovascular biomaterials	They are patient-friendly implants crafted from biomaterials demonstrating a notable enhancement in engraftment rates and overall efficiency. They have quickly become a benchmark for addressing cardiac disorders, particularly, in the successful reconstruction of vascular grafts and heart valves	However, these biomaterial-based implants exhibit a relatively short degradation time, which can vary between batches. Additionally, the purification process poses challenges making it difficult to consistently maintain the quality of the biomaterials	[196–198]
Heart valves	Heart valves provide a permanent solution for patients reducing the likelihood of rejection and eliminating the need for organ transplantation	These implants can occasionally trigger inflammation of lymphocytes, pose risks of complications such as calcification and may at times malfunction necessitating replacement	[199, 200]
Stents	Stents are characterized by their biodegradability, biocompatibility, nontoxicity and importantly low thrombogenicity, thereby reducing the risk of stent restenosis and improving the overall health condition of the patient	Challenges associated with stent implants include the potential for stent occlusion due to thrombus formation and the occurrence of restenosis. In some cases, there is also a serious concern about the risk of bleeding in patients experiencing adverse cardiovascular complications	[201, 202]

(continued)

Table 1 (continued)

Name of the implant	Advantages	Disadvantages	References
Cochlear replacements	Cochlear replacements stand as the optimal choice for hearing aids, especially for individuals with malformations of the middle and outer ear. These replacements effectively enhance hearing loss, contributing to the well-being of the patients with efficiency and cost effectiveness	Challenges associated with cochlear replacements include potential toxicity, a lack of mechanical stability, complex processing, and the potential to induce chronic or acute inflammation in the patient's body	[203, 204]
Dental implants	Dental implants are effective solutions for replacing missing teeth offering essential characteristics such as tensile strength and stiffness while preventing inflammation around the implanted area	The selection of materials for implant development is crucial. Inappropriate choices may result in failure, causing inflammation around the implant or leading to corrosion. Careful consideration in material selection is essential for the long-term success of dental implants	[205–207]
Artificial ligaments and tendons	They play a vital role in the generation of collagen fibers contributing to the repair and regeneration of essential ligament and tendon tissues	A significant drawback is the potential for tendon and ligament ruptures, resulting in a loss of strength over time. Additionally, the deposition of the graft debris can occur, posing challenges to the long-term effectiveness of these implants	[208]
Bone cement	Bone cement is considered highly safe, easy to operate with minimal trauma, and demonstrates a positive prognosis effect	It may encounter limitations such as a lack of bone formation, reduced bioactivity, and inferior mechanical properties compared to natural bone, potentially leading to failure in certain cases	[209, 177]

(continued)

Table 1 (continued)

Name of the implant	Advantages	Disadvantages	References
Joint replacements	Joint replacements provide the advantages of preserving the nearby joint, allowing for the continued growth of the remaining epiphysis, and ensuring superb limb functionality	Potential issues associated with joint replacements include the risk of loosening, deformity, breakage, and a discrepancy in limb length which can impact the patient's growth. Careful consideration and monitoring are essential to address these challenges	[210]
Intraocular lenses	Intra-ocular lenses ensure the continuous delivery of drugs eliminating the need for further implantations	Continuous drug delivery may impact the quality of images processed through the retina. Balancing the therapeutic benefits with potential visual effects is crucial in the evaluation of intraocular lenses	[211, 212]
Skin tissue substitutes	Skin tissue substitutes contribute to the ability to create skin structures that closely mimic the native skin's architecture and diverse nature leading to clinically applicable skin constructs	There are concerns regarding the safety of allogenic cells and the utilization of skin tissue substitutes is often associated with high expenses. Balancing safety considerations with cost-effectiveness is essential for the widespread adoption of these substitutes	[213]

and PET [153]. The concept of using synthetic grafts was initially introduced by Alexis Carrel and was later advanced by Voorhees et. al. who through further research, developed their use as implants to facilitate proper blood circulation, preventing complications associated with vascular diseases [154].

5.2 *Cardiovascular Biomaterials*

Cardiovascular biomaterials are specifically designed for application in cardiovascular diseases [155]. Biomedical devices tailored for cardiovascular applications have firmly established their presence in modern medicine, encompassing a diverse array of critical products such as artificial vascular grafts, heart valves, peacemakers, and more all aimed at saving lives [156]. Biomaterials play a pivotal role in advancing cardiovascular devices and implants significantly contributing to enhanced treatments, better patient outcomes, and improved cardiac care [157].

5.3 Heart Valves

To address heart valve disorders, scientists have developed innovative solutions, particularly, in the form of heart valves [158]. Valvular heart diseases are a major concern in the present world, and they could potentially be addressed through tissue-engineered heart valves made of cells or scaffolds that mimic the natural form of native valves promoting tissue growth [159]. With a focus on durability and longevity, these biomaterial-based heart valves have introduced a new perspective in cardiac care. They play a vital role in cardiac regeneration by providing a habitat with a supportive environment that facilitates tissue repair and promotes heart recovery rates [160].

5.4 Stents

Biomaterial stents, a component of Cardiac Vascular Devices (CVD), represent an innovative approach used to treat blocked or narrowed arteries. These stents provide structural support while aiding in vessel healing, and they are crafted from biocompatible materials that interact well with the body, thereby lowering associated risks [161]. Stents made from bioresorbable materials are designed to gradually dissolve over time, ensuring that as the healing process concludes, the arteries can regain their natural function without the need for a permanent implant to remain in place [162]. Biomaterial stents offer several advantages, including the reduction of inflammation and enabling better healing, eliminating the need for long-term artificial stents [163]. Research in this field has progressed, and biomaterial stents have become an evolving concept in interventional cardiology, aiming to improve patient outcomes and reduce complications [164].

5.5 Cochlear Implants

Cochlear implants are innovative devices specifically designed to restore hearing in individuals suffering from severe to profound hearing loss or deafness [165]. These implants aid damaged parts of the inner ear and auditory nerve, enabling users to perceive sound signals [166]. Cochlear implants comprise internal and external components, featuring an external microphone and a speech processor typically worn behind the ear. These elements capture and process the incoming sound [167]. While cochlear implants do not fully replicate the same as natural hearing, they significantly improve communication abilities and the perception of speech, enhancing the quality of life for individuals suffering from hearing loss [168].

5.6 *Dental Implants*

Dental implants act as substitutes for missing teeth, aiming to restore both function and aesthetics for individuals [33]. The dental implants consist of a titanium post, an abutment, and a crown [169]. The initial step involves the implanting titanium post into the jawbone, serving as the foundation for the artificial tooth. Through a process known as Osseointegration, the post fuses with the bone ensuring stability. The abutment acts as the connection between the crown and the post, resembling the characteristics of natural teeth [170]. Dental implants offer several advantages including the ability to chew and speak, as well as preventing bone loss in the jaw, ultimately contributing to the achievement of a genuine smile [171].

5.7 *Artificial Ligaments and Tendons*

The human body relies on non-fibrous matrices, fibrous matrices, and cells that bind with adjacent tissues through connective tissue. In cases where these structures are weakened or injured, support can be provided by artificial ligaments and tendons [172]. Recent research has shifted its focus toward developing implants for bone and cartilage, utilizing polymeric composite materials and tissue engineering approaches due to their biocompatibility, elasticity, resistance, and strength [173]. These implants are designed to replace damaged tissues in affected areas [174]. The ongoing progress in biomedical implant research contributes to improving compatibility, durability, and advancements in artificial ligaments and tendons, ultimately aiming to enhance their effectiveness in improving the quality of life for patients after injuries [175].

5.8 *Bone Cement*

Bone cement plays a vital role in orthopedic surgeries, particularly, in joint replacement procedures such as knee or hip replacements [176]. Typically, bone cement consists of Polymethyl Methacrylate (PMMA), a polymer that establishes a robust bond between the artificial joint components and the natural bone [177]. This biomaterial-based bone cement aims to enhance the bond strength between the implant and the bone, ensuring compatibility with the body and minimizing the risk of complications [178].

Bioactive additives like hydroxyapatite primarily composed of calcium phosphate can be incorporated into the bone cement to stimulate bone growth, promoting integration with the surrounding natural bone tissue [179]. This type of bone cement presents a promising avenue in orthopedic surgeries, reducing infection rates and associated risks related to artificial implants [180].

5.9 Joint Replacements

Joint replacements, employing biomaterials have emerged as a transformative approach in orthopedic surgeries especially for severe arthritis [181]. The biomaterials utilized in joint replacements encompass metals such as titanium, cobalt-chromium alloys, ceramics, and certain polymers [182]. Advances in biomaterial technology have resulted in improved designs, enhanced wear resistance, and prolonged shelf life for joint replacements [183]. Biomaterial-based joint replacements play a pivotal role in substantially enhancing the quality of patients' lives by alleviating discomfort and improving overall joint functionality [184].

5.10 Intra-Ocular Lenses

Intra-ocular lenses crafted from biomaterial implants, play a crucial role in eye surgery and vision correction [185]. These lenses are typically composed of biomaterials like acrylic serving as replacements for the natural lenses of the eyes [186]. Foldable intraocular lenses are inserted into the eye through a small incision, effectively replacing the natural lens [187]. Multifocal intraocular lenses contribute to enhanced vision, reducing the dependence on conventional eyeglasses [188]. The development of intraocular lenses using biomaterials has significantly improved the effectiveness of cataract surgery, thereby enhancing the quality of life for individuals worldwide [189].

5.11 Skin Tissue Substitutes

Skin tissue substitutes developed using biocompatible materials serve as implants for healing chronic wounds [190]. These substitutes closely mimic natural skin and contribute to the treatment of various skin wounds, incorporating stem cells in the process [191]. By facilitating wound healing and reducing the risk of infection, these implants act either as temporary replacements for natural tissue or as permanent solutions for permanently lost skin [192]. Researchers are actively exploring bioengineered skin substitutes as promising alternatives for treating injuries such as skin burns and skin loss with a focus on restoring both dermal and epidermal tissues making it an emerging field [193].

6 Challenges and Future Prospects

The progress in implantable medical textile materials has revolutionized healthcare, notably impacting vascular grafts, tissue engineering scaffolds, and various other medical applications. The selection of a vascular graft is contingent upon the patient's unique medical condition, the specific blood vessel involved, and other individualized factors. Challenges in this context encompass the potential formation of blood clots within the graft (thrombosis), the risk of bacterial or fungal infections, the imperative of maintaining graft openness and functionality (graft patency), and the need to minimize the risk of rejection or adverse reactions (ensuring biocompatibility). Healthcare professionals typically make surgical decisions by evaluating the patients' medical history and the characteristics of the vascular disease aiming to address these challenges effectively and optimize the overall success of the vascular graft procedure.

Ongoing research in vascular grafts is focused on refining materials to enhance biocompatibility, reduce infection risks, and improve mechanical properties with a future outlook on optimizing hemocompatibility and addressing challenges related to long-term degradation and mechanical stability. Concurrently, the field of tissue engineering scaffolds is witnessing innovative developments with an emphasis on mimicking native extracellular matrix (ECM) functions more closely, aiming to create scaffolds that efficiently transport nutrients, ensure controlled degradation and minimize inflammation and immune responses, thereby holding immense promise for tissue regeneration. Beyond these applications, implantable medical textiles are anticipated to find use in wound closure, orthopedic implants, and neurological devices, contributing to the evolution of medical interventions.

Looking ahead, several trends are expected to shape the landscape of implantable medical textile materials, these include the integration of smart textiles equipped with sensors and actuators for real-time monitoring, personalized medicine driving customization efforts based on individual patient characteristics, and the use of bioactive coatings to enhance tissue integration and mitigate inflammation. Advancements in 3D printing techniques will enable precise fabrication of complex textile structures, expanding possibilities in the field. However, challenges persist, necessitating standardized testing methods and regulations to ensure safety and efficacy, long-term durability concerns, associated with degradation and potential side effects, must be addressed for sustained success. Achieving optimal biocompatibility remains a critical objective to minimize immune responses and maximize patient well-being. In navigating these challenges, implantable medical textile materials are poised to bring transformative innovations in healthcare, promising improved therapeutic outcomes and enhanced patient care.

7 Conclusions

In conclusion, the continuous advancements in implantable medical textile materials represent a paradigm shift in healthcare applications, notably within vascular grafts, tissue engineering scaffolds, and diverse medical realms. These innovations hold significant promise for improving biocompatibility, reducing infection risks, and enhancing mechanical properties in vascular grafts. Similarly, developments in tissue engineering scaffolds aim to replicate native extracellular matrix functions more closely, fostering efficient nutrient transport and controlled degradation. Beyond these applications, implantable medical textiles are expected to find versatile uses, from wound closure to orthopedic implants and neurological devices, contributing to the evolution of medical interventions. As research progresses, integrating smart textiles, personalized medicine, and bioactive coatings will likely shape the future of this field. While challenges persist, particularly in standardized testing and long-term durability, the transformative potential of implantable medical textile materials underscores their role in advancing therapeutic outcomes and elevating patient care standards.

Acknowledgements The authors would like to acknowledge to Department of Biotechnology, Vignan's Foundation for Science Technology and Research, Guntur for providing a platform to carry out the research work.

References

1. Akbari M, Tamayol A, Bagherifard S et al (2016) Textile technologies and tissue engineering: a path toward organ weaving. Adv Healthc Mater 5(7):751–766
2. OatleyG CT, Buckman P et al (2021) Smart textiles for improved quality of life and cognitive assessment. Sensors 21(23):8008
3. Petros S, Tesfaye T, Ayele M et al (2020) A review on gelatin-based hydrogels for medical textile applications. J Eng 2020:1–12
4. Betz UA, Arora L, Assal RA et al (2023) Game changers in science and technology-now and beyond. Technol Forecast Soc Chang 193:122588
5. Junaid SB, Imam AA, Balogun AO et al (2022) Recent advancements in emerging technologies for healthcare management systems: a survey. Healthcare 10(10):1940
6. Hasan M, Mridha J, Alam AM et al (2022) Medical textiles: application of implantable medical textiles. Global J Med Res 19(K4):17–24
7. Al-Zyoud W, Haddadin D, Hasan SA et al (2023) Biocompatibility testing for implants: a novel tool for selection and characterization. Materials 16(21):6881
8. Das P, Roy MD, Ghosh S et al (2023) Implantable medical devices by using textile materials. In: Functional and technical textiles. Woodhead Publishing, London, pp 521–542
9. Mariani E, Lisignoli G, Borzì RM et al (2019) Biomaterials: foreign bodies or tuners for the immune response. Int J Mol Sci 20(3):636
10. Carnicer-Lombarte A, Chen ST, Malliaras GG et al (2021) Foreign body reaction to implanted biomaterials and its impact in nerve neuroprosthetics. Front Bioeng Biotechnol 2021:271
11. Eltom A, Zhong G, Muhammad A et al (2019) Scaffold techniques and designs in tissue engineering functions and purposes: a review. Adv Mater Sci Eng 2019:Article ID 3429527

12. Nikolova MP, Chavali MS (2019) Recent advances in biomaterials for 3D scaffolds: a review. Bioact Mater 4:271–292
13. Suamte L, Tirkey A, Barman J et al (2023) Various manufacturing methods and ideal properties of scaffolds for tissue engineering applications. Smart Mater Manuf 1:100011
14. Abyzova E, Dogadina E, Rodriguez RD et al (2023) Beyond tissue replacement: the emerging role of smart implants in healthcare. Mater Today Bio 22:100784
15. Agarwal KM, Singh P, Mohan U et al (2020) Comprehensive study related to advancement in biomaterials for medical applications. Sens Int 1:100055
16. Todros S, Todesco M, Bagno A et al (2021) Biomaterials and their biomedical applications: from replacement to regeneration. Processes 9(11):1949
17. Kantaros A, Ganetsos T (2023) From static to dynamic: smart materials pioneering additive manufacturing in regenerative medicine. Int J Mol Sci 24(21):15748
18. Festas AJ, Ramos A, Davim JP et al (2020) Medical devices biomaterials—a review. Proc Inst Mech Eng Part L J Mater Des Appl 234(1):218–228
19. Chen FM, Liu X (2016) Advancing biomaterials of human origin for tissue engineering. Prog Polym Sci 53:86–168
20. Pesode P, Barve S (2023) Additive manufacturing of metallic biomaterials: sustainability aspect, opportunity, and challenges. J Ind Prod Eng 40(6):464–505
21. Gobbi SJ, Reinke G, Gobbi VJ et al (2020) Biomaterial: concepts and basics properties. EurInt J SciTechnol 9:23–42
22. Pandey A, Awasthi A, Saxena KK et al (2020) Metallic implants with properties and latest production techniques: a review. Adv Mater Process Technol 6(2):405–440
23. Al-Shalawi FD, Mohamed Ariff AH, Jung DW et al (2023) Biomaterials as implants in the orthopedic field for regenerative medicine: metal versus synthetic polymers. Polymers 15(12):2601
24. Morsiya C (2022) A review on parameters affecting properties of biomaterial SS 316L. Aust J Mech Eng 20(3):803–813
25. Dwivedi SP, Sharma S, Srivastava AP et al (2023) Homogeneity, metallurgical, mechanical, wear, and corrosion behavior of Ni and B4C coatings deposited on 304 stainless steels developed by microwave cladding technique. J Market Res 27:5854–5867
26. Unune DR, Brown GR, Reilly GC et al (2022) Thermal based surface modification techniques for enhancing the corrosion and wear resistance of metallic implants: a review. Vacuum 2022:111298
27. Jin W, Chu PK (2019) Orthopedic implants. Encyclop Biomed Eng 1(3):425–439
28. Al Mahmud MZ, Mobarak MH, Hossain N et al (2023) Emerging breakthroughs in biomaterials for orthopedic applications: a comprehensive review. Bioprinting 36(1):e00323
29. Gautam S, Bhatnagar D, Bansal D et al (2022) Recent advancements in nanomaterials for biomedical implants. Biomed Eng Adv 3:100029
30. Yan L, Lim JL, Lee JW et al (2020) Finite element analysis of bone and implant stresses for customized 3D-printed orthopaedic implants in fracture fixation. Med Biol Eng Compu 58:921–931
31. Allizond V, Comini S, Cuffini AM et al (2022) Current knowledge on biomaterials for orthopedic applications modified to reduce bacterial adhesive ability. Antibiotics 11(4):529
32. Nicholson WJ (2020) Titanium alloys for dental implants: A review. Prosthesis 2(2):11
33. Saini M, Singh Y, Arora P et al (2015) Implant biomaterials: a comprehensive review. World J Clin Cases WJCC 3(1):52
34. Hoque ME, Showva NN, Ahmed M et al (2022) Titanium and titanium alloys in dentistry: current trends, recent developments, and future prospects. Heliyon 8(11):e11300
35. Sarraf M, RezvaniGhomi E, Alipour S et al (2021) A state-of-the-art review of the fabrication and characteristics of titanium and its alloys for biomedical applications. Bio Des Manuf 5:371–395
36. Prestat M, Thierry D (2021) Corrosion of titanium under simulated inflammation conditions: clinical context and in vitro investigations. Actabiomaterialia 136:72–87

37. Kang CW, Fang FZ (2018) State of the art of bioimplants manufacturing: part I. Adv Manufa 6:20–40
38. Anene FA, AizaJaafar CN, Zainol I et al (2021) Biomedical materials: a review of titanium-based alloys. Proc Inst Mech Eng C J Mech Eng Sci 235(19):3792–3805
39. Vaicelyte A, Janssen C, Le Borgne M et al (2020) Cobalt-chromium dental alloys: metal exposures, toxicological risks, CMR classification, and EU regulatory framework. Crystals 10(12):1151
40. Minnath MA (2018) Metals and alloys for biomedical applications. In: Fundamental biomaterials metals. Woodhead Publishing, London, pp 167–174
41. Hallab NJ, Samelko L, Caicedo M et al (2021) Implant material bio-compatibility, sensitivity, and allergic reactions. In: Handbook of spine technology. Springer, Cham, pp 127–149
42. Ran Z, Dai W, Xie K et al (2022) Advances of biodegradable magnesium based implants for orthopedics. Life Res 5(7):1–12
43. Roy M, Bandyopadhyay A, Bose S et al (2017) Ceramics in bone grafts and coated implants. In: Materials for bone disorders. Academic Press, pp 265–314
44. Grossin D, Monton A, Navarrete-Segado P et al (2021) A review of additive manufacturing of ceramics by powder bed selective laser processing (sintering/melting): calcium phosphate, silicon carbide, zirconia, alumina, and their composites. Open Ceramics 5:100073
45. Saxena V, Shukla I, Pandey LM et al (2019) Hydroxyapatite: an inorganic ceramic for biomedical applications. In: Materials for biomedical engineering. Elsevier, London, pp 205–249
46. Hossain N, Mobarak MH, Islam MA et al (2023) Recent development of dental implant materials, synthesis process, and failure—a review. Results Chem 6:101136
47. Jiang X, Yao Y, Tang W et al (2020) Design of dental implants at materials level: an overview. J Biomed Mater Res Part A 108(8):1634–1661
48. Abd El-Ghany OS, Sherief AH (2016) Zirconia based ceramics, some clinical and biological aspects. Fut Dent J 2(2):55–64
49. Zadeh PN, Lümkemann N, Sener B et al (2018) Flexural strength, fracture toughness, and translucency of cubic/tetragonal zirconia materials. J Prosthet Dent 120(6):948–954
50. Sequeira S, Fernandes MH, Neves N et al (2017) Development and characterization of zirconia–alumina composites for orthopedic implants. Ceram Int 43(1):693–703
51. Islam MT, Felfel RM, Abou Neel EA et al (2017) Bioactive calcium phosphate-based glasses and ceramics and their biomedical applications: a review. J Tissue Eng 8:2041731417719170
52. Arcos D, Vallet-Regi M (2020) Substituted hydroxyapatite coatings of bone implants. J Mater Chem B 8(9):1781–1800
53. Chellan P, Sadler PJ (2015) The elements of life and medicines. Philos Trans R Soc A Math Phys Eng Sci 373(2037):20140182
54. Song R, Murphy M, Li C et al (2018) Current development of biodegradable polymeric materials for biomedical applications. Drug Des Develop Ther 12:3117–3145
55. Garcia-Estrada P, Garcia-Bon MA, Lopez-Naranjo EJ et al (2021) Polymeric implants for the treatment of intraocular eye diseases: trends in biodegradable and non-biodegradable materials. Pharmaceutics 13(5):701
56. Chan DS, Fnais N, Ibrahim I et al (2019) Exploring polycaprolactone in tracheal surgery: a scoping review of in-vivo studies. Int J Pediatr Otorhinolaryngol 123:38–42
57. Li C, Guo C, Fitzpatrick V et al (2020) Design of biodegradable, implantable devices towards clinical translation. Nat Rev Mater 5(1):61–81
58. Idumah CI, Nwabanne JT, Tanjung FA et al (2021) Novel trends in poly(lactic) acid hybrid bionanocomposites. Cleaner Mater 2:100022
59. Joseph TM, Kallingal A, Suresh AM et al (2023) 3D printing of polylactic acid: recent advances and opportunities. Int J Adv Manuf Technol 125(3–4):1015–1035
60. Ozdil D, Wimpenny I, Aydin HM et al (2017) Biocompatibility of biodegradable medical polymers. In: Science and principles of biodegradable and bioresorbable medical polymers, pp 379–414

61. Prabhu B, Karau A, Wood A et al (2018) Bioresorbable materials for orthopedic applications (lactide and glycolide based). In: Orthopedic biomaterials: progress in biology, manufacturing, and industry perspectives, pp 287–344
62. Genina N, Hollander J, Jukarainen H et al (2016) Ethylene vinyl acetate (EVA) as a new drug carrier for 3D printed medical drug delivery devices. Eur J Pharm Sci 90:53–63
63. Feng X, Zhang F (2018) Twin-screw extrusion of sustained-release oral dosage forms and medical implants. Drug Deliv Transl Res 8:1694–1713
64. Ashammakhi N, Hernandez AL, Unluturk BD et al (2021) Biodegradable implantable sensors: materials design, fabrication, and applications. Adv Func Mater 31(49):2104149
65. Mendolia I, Zanca C, Ganci F et al (2021) Calcium phosphate/polyvinyl acetate coatings on SS304 via galvanic co-deposition for orthopedic implant applications. Surf Coat Technol 408:126771
66. Griffin M, Castro N, Bas O et al (2020) The current versatility of polyurethane three-dimensional printing for biomedical applications. Tissue Eng Part B Rev 26(3):272–283
67. Naureen B, Haseeb ASMA, Basirun WJ et al (2021) Recent advances in tissue engineering scaffolds based on polyurethane and modified polyurethane. Mater Sci Eng, C 118:111228
68. Cozma V, Rosca I, Radulescu L et al (2021) Antibacterial polysiloxane polymers and coatings for cochlear implants. Molecules 26(16):4892
69. Alshehri M, Albaqiah H (2017) Antimicrobial efficacy of materials used for sealing the implant abutment screw hole: an in-vitro evaluation. Implant Dent 26(6):911–914
70. Chen N, Li M, Wu H et al (2023) An extracellular matrix (ECM)-mimetic coating with dual bionics for cardiovascular stents. Regen Biomater 10:rbad055
71. Lepicka M, Barros-Silva S, Licciardello N et al (2021) Silane-based coating charged with TiO_2 NPs for dental implant applications. Surf Coat Technol 413:127066
72. Tortolano L, Yen-Nicolay S, Rogliano PF et al (2018) Permeability of expander breast implants: in vitro and in vivo analyses. J Mech Behav Biomed Mater 78:427–432
73. Karki S, Gohain MB, Yadav D et al (2021) Nanocomposite and bio-nanocomposite polymeric materials/membranes development in energy and medical sector: a review. Int J Biol Macromol 193:2121–2139
74. Verma D, Dong Y, Sharma M et al (2022) Advanced processing of 3D printed biocomposite materials using artificial intelligence. Mater Manuf Processes 37(5):518–538
75. Alizadeh-Osgouei M, Li Y, Wen C et al (2019) A comprehensive review of biodegradable synthetic polymer–ceramic composites and their manufacture for biomedical applications. Bioact Mater 4:22–36
76. Piitulainen JM, Mattila R, Moritz N et al (2017) Load-bearing capacity and fracture behavior of glass fiber-reinforced composite cranioplasty implants. J Appl Biomater Funct Mater 15(4):356–361
77. Akter M, Uddin MH, Tania IS et al (2022) Biocomposites based on natural fibers and polymers: a review on properties and potential applications. J Reinf Plast Compos 41(17–18):705–742
78. Bechir F, Bataga SM, Tohati A et al (2021) Evaluation of the behavior of two CAD/CAM fiber-reinforced composite dental materials by immersion tests. Materials 14(23):7185
79. Chua CYX, Liu HC, Di Trani N et al (2021) Carbon fiber reinforced polymers for implantable medical devices. Biomaterials 271:120719
80. Moldovan MA, Bosca AB, Roman CR et al (2020) Bone reaction to a newly developed fiber-reinforced composite material for craniofacial implants. Mater Plast 57(2):131–139
81. Vallittu PK (2018) An overview of development and status of fiber-reinforced composites as dental and medical biomaterials. Acta Biomater Odontolog Scand 4(1):44–55
82. Socci MC, Rodríguez G, Oliva E et al (2023) Polymeric materials, advances and applications in tissue engineering: a review. Bioengineering 10(2):218
83. Guru R, Kumar A, Grewal D et al (2022) Study of the implantable and non-implantable application in medical textile. In: Next-generation textiles. IntechOpen, London, pp 1–16
84. Ratner B (2023) Vascular grafts: technology success/technology failure. BME Front 4:0003
85. Fang S, Ellman DG, Andersen DC et al (2021) Tissue engineering of small-diameter vascular grafts and their in-vivo evaluation in large animals and humans. Cells 10(3):713

86. Jouda H, Larrea Murillo L, Wang T et al (2022) Current progress in vascular engineering and its clinical applications. Cells 11(3):493
87. Li MX, Wei QQ, Mo HL et al (2023) Challenges and advances in materials and fabrication technologies of small-diameter vascular grafts. Biomater Res 27(1):58
88. Vander Slegt J, Steunenberg SL, Donker JM et al (2014) The current position of precuffed expanded polytetrafluoroethylene bypass grafts in peripheral vascular surgery. J Vasc Surg 60(1):120–128
89. Wystrychowski W, McAllister TN, Zagalski K et al (2014) First human use of an allogeneic tissue-engineered vascular graft for hemodialysis access. J Vasc Surg 60(5):1353–1357
90. Mallis P, Kostakis A, Stavropoulos-Giokas C et al (2020) Future perspectives in small-diameter vascular graft engineering. Bioengineering 7(4):160
91. Pashneh-Tala S, MacNeil S, Claeyssens F et al (2016) The tissue-engineered vascular graft-past, present, and future. Tissue Eng Part B Rev 22(1):68–100
92. Oli AN, Rowaiye AB, Adejumo SA et al (2022) Classic and current opinions in human organ and tissue transplantation. Cureus 14(11):e30982
93. Shakeel A, Corridon PR (2023) Mitigating challenges and expanding the future of vascular tissue engineering—are we there yet. Front Physiol 13:1079421
94. Soares JS, Saunders SK, Potere F et al (2022) Engineered tissue vascular grafts: are we there yet. Appl Eng Sci 2022:100114
95. Choi J, Lee EJ, Jang WB et al (2023) Development of biocompatible 3D-printed artificial blood vessels through multidimensional approaches. J Funct Biomater 14(10):497
96. Leal BB, Wakabayashi N, Oyama K et al (2021) Vascular tissue engineering: polymers and methodologies for small caliber vascular grafts. Front Cardiovasc Med 7:592361
97. Hu X, Wang T, Li F et al (2023) Surface modifications of biomaterials in different applied fields. RSC Adv 13(30):20495–20511
98. Tan W, Boodagh P, Selvakumar PP et al (2023) Strategies to counteract adverse remodelling of vascular graft: a 3D view of current graft innovations. Front Bioeng Biotechnol 10:1097334
99. Deb P, Deoghare AB, Borah A et al (2018) Scaffold development using biomaterials: a review. Mater Today Proc 5(5):12909–12919
100. Echeverria Molina MI, Malollari KG, Komvopoulos K et al (2021) Design challenges in polymeric scaffolds for tissue engineering. Front Bioeng Biotechnol 9:617141
101. Krishani M, Shin WY, Suhaimi H et al (2023) Development of scaffolds from bio-based natural materials for tissue regeneration applications: a review. Gels 9(2):100
102. Mao AS, Mooney DJ (2015) Regenerative medicine: current therapies and future directions. Proc Natl Acad Sci 112(47):14452–14459
103. Negrini NC, Volponi AA, Higgins CA et al (2021) Scaffold-based developmental tissue engineering strategies for ectodermal organ regeneration. Materials Today Bio 10:100107
104. Osorio LA, Silva E, Mackay RE et al (2021) A review of biomaterials and scaffold fabrication for organ-on-a-chip (OOAC) systems. Bioengineering 8(8):113
105. Ruiz-Alonso S, Lafuente-Merchan M, Ciriza J et al (2021) Tendon tissue engineering: cells, growth factors, scaffolds and production techniques. J Control Rel 333:448–486
106. Pedde RD, Mirani B, Navaei A et al (2017) Emerging bio fabrication strategies for engineering complex tissue constructs. Adv Mater 29(19):1606061
107. Farag MM (2023) Recent trends on biomaterials for tissue regeneration applications. J Mater Sci 58(2):527–558
108. Oleksy M, Dynarowicz K, Aebisher D et al (2023) Advances in biodegradable polymers and biomaterials for medical applications—a review. Molecules 28(17):6213
109. Joyce K, Fabra GT, Bozkurt Y et al (2021) Bioactive potential of natural biomaterials: Identification, retention and assessment of biological properties. Signal Transduct Target Ther 6(1):122
110. Flores-Rojas GG, Gómez-Lazaro B, Lopez-Saucedo F et al (2023) Electrospun scaffolds for tissue engineering: a review. Macromol 3(3):524–553
111. Huzum B, Puha B, Necoara RM et al (2021) Biocompatibility assessment of biomaterials used in orthopedic devices: an overview. Exp Ther Med 22(5):1–9

112. Moghadasi K, Isa MSM, Ariffin MA et al (2022) A review on biomedical implant materials and the effect of friction stir based techniques on their mechanical and tribological properties. J Market Res 17:1054–1121
113. Masaeli R, Zandsalimi K, Tayebi L et al (2019) Biomaterials evaluation: conceptual refinements and practical reforms. Therap Innov Regul Sci 53(1):120–127
114. BapatRA PA, Chaubal T et al (2021) Recent update on potential cytotoxicity, biocompatibility and preventive measures of biomaterials used in dentistry. Biomater Sci 9(9):3244–3283
115. Salandova M, van Hengel IA, Apachitei I et al (2021) Inorganic agents for enhanced angiogenesis of orthopedic biomaterials. Adv Healthc Mater 10(12):2002254
116. Al-Maawi S, Orlowska A, Sader R et al (2017) In-vivo cellular reactions to different biomaterials—physiological and pathological aspects and their consequences. In: Seminars in immunology. Academic Press, pp 49–61
117. Karami P, Stampoultzis T, Guo Y et al (2023) A guide to preclinical evaluation of hydrogel-based devices for treatment of cartilage lesions. Acta Biomater 158:12–31
118. Bernard M, Jubeli E, Pungente MD et al (2018) Biocompatibility of polymer-based biomaterials and medical devices–regulations, in-vitro screening and risk-management. Biomater Sci 6(8):2025–2053
119. Komeri R, Kasoju N, Kumar PA et al (2022) In-vitro cytotoxicity and cytocompatibility assays for biomaterial testing under regulatory platform. In: Biomedical product and materials evaluation. Woodhead Publishing, London, pp 329–353
120. Liu X, Rodeheaver DP, White JC et al (2018) A comparison of in-vitro cytotoxicity assays in medical device regulatory studies. Regul Toxicol Pharmacol 97:24–32
121. Ammann KR, Hossainy SF, Hossainy S et al (2021) Hemocompatibility of polymers for use in vascular endoluminal implants. J Appl Polym Sci 138(43):51277
122. Thanigaivel S, Priya AK, Balakrishnan D et al (2022) Insight on recent development in metallic biomaterials: strategies involving synthesis, types and surface modification for advanced therapeutic and biomedical applications. Biochem Eng J 187:108522
123. Weber M, Steinle H, Golombek S et al (2018) Blood-contacting biomaterials: in-vitro evaluation of the hemocompatibility. Front Bioeng Biotechnol 6:99
124. Labarrere CA, Dabiri AE, Kassab GS et al (2020) Thrombogenic and inflammatory reactions to biomaterials in medical devices. Front Bioeng Biotechnol 8:123
125. Cvetkovic VJ, Takic Miladinov D, Stojanovic S et al (2018)Genotoxicity and mutagenicity testing of biomaterials. In: Biomaterials in clinical practice: advances in clinical research and medical devices. Springer, London, pp 501–527
126. Wang N, Fuh JYH, Dheen ST et al (2021) Functions and applications of metallic and metallic oxide nanoparticles in orthopedic implants and scaffolds. J Biomed Mater Res B Appl Biomater 109(2):160–179
127. De Jong WH, Carraway JW, Geertsma RE et al (2020) In-vivo and in vitro testing for the biological safety evaluation of biomaterials and medical devices. In: Biocompatibility and performance of medical devices. Woodhead Publishing, London, pp 123–166
128. Kovochich M, Monnot A, Kougias DG et al (2021) Carcinogenic hazard assessment of cobalt-containing alloys in medical devices: review of in-vivo studies. Regul Toxicol Pharmacol 122:104910
129. Melnick RL (2019) Commentary on the utility of the national toxicology program study on cell phone radiofrequency radiation data for assessing human health risks despite unfounded criticisms aimed at minimizing the findings of adverse health effects. Environ Res 168:1–6
130. Wang CH, Hsieh DJ, Periasamy S et al (2020) Regenerative porcine dermal collagen matrix developed by supercritical carbon dioxide extraction technology: role in accelerated wound healing. Materialia 9:100576
131. Stoppelkamp S, Würschum N, Stang K et al (2017) Speeding up pyrogenicity testing: identification of suitable cell components and readout parameters for an accelerated monocyte activation test (MAT). Drug Test Anal 9(2):260–273
132. Yesilkir-Baydar S, Oztel ON, Cakir-Koc R et al (2017) Evaluation techniques. In: Nano biomaterials science, development and evaluation. Woodhead Publishing, London, pp 211–232

133. Barchowsky A (2020) Systemic and immune toxicity of implanted materials. In: Biomaterials science. Academic Press, pp 791–799
134. DeSesso JM, Scialli AR (2018) Bone development in laboratory mammals used in developmental toxicity studies. Birth Def Res 110(15):1157–1187
135. Prakasam M, Locs J, Salma-Ancane KL et al (2017) Biodegradable materials and metallic implants—a review. J Funct Biomater 8(4):44
136. Revathi A, Borras AD, Muoz AI et al (2017) Degradation mechanisms and future challenges of titanium and its alloys for dental implant applications in oral environment. Mater Sci Eng, C 76:1354–1368
137. Tsakiris V, Tardei C, Clicinschi FM et al (2021) Biodegradable Mg alloys for orthopedic implants—a review. J Magn Alloys 9(6):1884–1905
138. Basu B, Basu B (2017) Probing toxicity of biomaterials and biocompatibility assessment. Biomater Musculoskel Regen Concepts 2017:291–351
139. Whitaker R, Hernaez-Estrada B, Hernandez RM et al (2021) Immunomodulatory biomaterials for tissue repair. Chem Rev 121(18):11305–11335
140. Mukherjee P, Roy S, Ghosh D et al (2022) Role of animal models in biomedical research: a review. Lab Anim Res 38(1):18
141. Li Y, Chen SK, Li L et al (2015) Bone defect animal models for testing efficacy of bone substitute biomaterials. J Orthop Transl 3(3):95–104
142. Mohan S, Huneke R (2019) The role of IACUCs in responsible animal research. ILAR J 60(1):43–49
143. de LacerdaSchickert S, van den Beucken JJ, Leeuwenburgh SC et al (2020) Pre-clinical evaluation of biological bone substitute materials for application in highly loaded skeletal sites. Biomolecules 10(6):883
144. Pero M, Casani L, Castells-Sala C et al (2021) Rabbit as an animal model for the study of biological grafts in pelvic floor dysfunctions. Sci Rep 11(1):10545
145. Smith D, Anderson D, Degryse AD et al (2018) Classification and reporting of severity experienced by animals used in scientific procedures: FELASA/ECLAM/ESLAV Working Group report. Lab Anim 52(1):5–57
146. Pippi R (2017) Post-surgical clinical monitoring of soft tissue wound healing in periodontal and implant surgery. Int J Med Sci 14(8):721
147. Othman Z, Pastor BC, van Rijt S et al (2018) Understanding interactions between biomaterials and biological systems using proteomics. Biomaterials 167:191–204
148. O'Brien MT, Schuh JC, Wancket LM et al (2022) Scientific and regulatory policy committee points to consider for medical device implant site evaluation in nonclinical studies. Toxicol Pathol 50(4):512–530
149. Koydemir HC, Ozcan A (2018) Wearable and implantable sensors for biomedical applications. Annu Rev Anal Chem 11:127–146
150. Haleem A, Javaid M (2019) Polyether ether ketone (PEEK) and its 3D printed implants applications in medical field: an overview. Clin Epidemiol Global Health 7(4):571–577
151. Hu K, Li Y, Ke Z et al (2022) History, progress and future challenges of artificial blood vessels: a narrative review. Biomater Transl 3(1):81
152. Gutowski P, Gage SM, Guziewicz M et al (2020) Arterial reconstruction with human bioengineered acellular blood vessels in patients with peripheral arterial disease. J Vasc Surg 72(4):1247–1258
153. Obiweluozor FO, Emechebe GA, Kim DW et al (2020) Considerations in the development of small-diameter vascular graft as an alternative for bypass and reconstructive surgeries: a review. Cardiovasc Eng Technol 11:495–521
154. Brewster L, Brey EM, Greisler HP et al (2020) Blood vessels. In: Principles of tissue engineering. Academic Press, London, pp 617–634
155. ScafaUdrişteA NAG, Grumezescu AM et al (2021) Cardiovascular stents: a review of past, current, and emerging devices. Materials 14(10):2498
156. Heath DE (2017) Promoting endothelialization of polymeric cardiovascular biomaterials. Macromol Chem Phys 218(8):1600574

157. Di Franco S, Amarelli C, Montalto A et al (2018) Biomaterials and heart recovery: cardiac repair, regeneration and healing in the MCS era: a state of the "heart." J Thorac Dis 10(20):S2346
158. Motta SE, Lintas V, Fioretta ES et al (2018) Off-the-shelf tissue engineered heart valves for in situ regeneration: current state, challenges and future directions. Expert Rev Med Dev 15(1):35–45
159. Xue Y, Sant V, Phillippi J et al (2017) Biodegradable and biomimetic elastomeric scaffolds for tissue-engineered heart valves. Acta Biomater 48:2–19
160. Sacco AM, Castaldo C, Di Meglio F et al (2023) The long and winding road to cardiac regeneration. Appl Sci 13(16):9432
161. Raikar AS, Priya S, Bhilegaonkar SP et al (2023) Surface engineering of bioactive coatings for improved stent hemocompatibility: a comprehensive review. Materials 16(21):6940
162. Ang HY, Ng J, Bulluck H et al (2018) Fundamentals of bioresorbable stents. In: Functionalised cardiovascular stents, pp 75–97
163. Borhani S, Hassanajili S, Ahmadi Tafti SH et al (2018) Cardiovascular stents: overview, evolution, and next generation. Prog Biomater 7:175–205
164. Peng X, Qu W, Jia Y et al (2020) Bioresorbable scaffolds: contemporary status and future directions. Front Cardiovasc Med 7:589571
165. Baumann U, Stover T, Weigerber T et al (2020) Device profile of the MED-EL cochlear implant system for hearing loss: overview of its safety and efficacy. Expert Rev Med Devices 17(7):599–614
166. Chen F, Ni W, Li W et al (2019) Cochlear implantation and rehabilitation. Hearing Loss: Mechanisms, Prevention and Cure, pp 129–144
167. Ghosh R, Ali H, Hansen JH et al (2021) CCi-MOBILE: a portable real time speech processing platform for cochlear implant and hearing research. IEEE Trans Biomed Eng 69(3):1251–1263
168. Nogueira W, Nagathil A, Martin R et al (2018) Making music more accessible for cochlear implant listeners: recent developments. IEEE Signal Process Mag 36(1):115–127
169. Aldhayan I, Qassadi W, AlShehri T et al (2018) Review on dental implantology. Egypt J Hosp Med 71(1):2217–2225
170. Saha S, Roy S (2022) Metallic dental implants wear mechanisms, materials, and manufacturing processes: a literature review. Materials 16(1):161
171. Geraets W, Zhang L, Liu Y et al (2014) Annual bone loss and success rates of dental implants based on radiographic measurements. Dentomaxillofac Radiol 43(7):20140007
172. Freedman BR, Mooney DJ (2019) Biomaterials to mimic and heal connective tissues. Adv Mater 31(19):1806695
173. Elhadad AA, Alcudia A, Begines B et al (2022) A multidisciplinary perspective on the latest trends in artificial cartilage fabrication to mimic real tissue. Appl Mater Today 29:101603
174. Ma Z, Bao G, Li J et al (2021) Multifaceted design and emerging applications of tissue adhesives. Adv Mater 33(24):2007663
175. Bedian L, Villalba-Rodríguez AM, Hernández-Vargas G et al (2017) Bio-based materials with novel characteristics for tissue engineering applications—a review. Int J Biol Macromol 98:837–846
176. Gallo J, Gibon E, Goodman SB et al (2017) Implants for joint replacement of the hip and knee. In: Materials for bone disorders. Academic Press, London, pp 119–196
177. Soleymani Eil Bakhtiari S, Bakhsheshi-Rad HR, Karbasi S et al (2020) Polymethyl methacrylate-based bone cements containing carbon nanotubes and graphene oxide: an overview of physical, mechanical, and biological properties. Polymers 12(7):1469
178. He W, Benson R (2017) Polymeric biomaterials. In: Applied plastics engineering handbook. William Andrew Publishing, London, pp 145–164
179. Yousefi AM (2019) A review of calcium phosphate cements and acrylic bone cements as injectable materials for bone repair and implant fixation. J Appl Biomater Funct Mater 17(4):2280800019872594
180. Schwarz EM, McLaren AC, Sculco TP et al (2021) Adjuvant antibiotic-loaded bone cement: concerns with current use and research to make it work. J Orthop Res 39(2):227–239

181. Kuncicka L, Kocich R, Lowe TC et al (2017) Advances in metals and alloys for joint replacement. Prog Mater Sci 88:232–280
182. Szczęsny G, Kopec M, Politis DJ et al (2022) A review on biomaterials for orthopaedic surgery and traumatology: from past to present. Materials 15(10):3622
183. Poliakov A, Pakhaliuk V, Popov VL et al (2020) Current trends in improving of artificial joints design and technologies for their arthroplasty. Front Mech Eng 6:4
184. ZyliNska B, Silmanowicz P, Sobczynska-Rak A et al (2018) Treatment of articular cartilage defects: focus on tissue engineering. In-Vivo 32(6):1289–1300
185. Zvornicanin J, Zvornicanin E (2018) Premium intraocular lenses: the past, present and future. J Curr Ophthalmol 30(4):287–296
186. Tsui JK, Bell S, da Cruz L et al (2022) Applications of three-dimensional printing in ophthalmology. Surv Ophthalmol 67(4):1287–1310
187. Karayilan M, Clamen L, Becker ML et al (2021) Polymeric materials for eye surface and intraocular applications. Biomacromol 22(2):223–261
188. Breyer DR, Kaymak H, Ax T et al (2017) Multifocal intraocular lenses and extended depth of focus intraocular lenses. Asia Pac J Ophthalmol 6(4):339–349
189. Williams DF (2021) Assessing the triad of biocompatibility, medical device functionality and biological safety. Med Dev Sens 4(1):e10150
190. Ruggeri M, Bianchi E, Rossi S et al (2020) Nanotechnology-based medical devices for the treatment of chronic skin lesions: from research to the clinic. Pharmaceutics 12(9):815
191. Wang L, Liu SK et al (2022) New prospects in skin tissue engineering and fabrication. Bio Fabric Orthop Methods Tech Appl 2:423–459
192. Kim HS, Sun X, Lee JH et al (2019) Advanced drug delivery systems and artificial skin grafts for skin wound healing. Adv Drug Deliv Rev 146:209–239
193. Dearman BL, Boyce ST, Greenwood JE et al (2021) Advances in skin tissue bioengineering and the challenges of clinical translation. Front Surg 8:640879
194. Chandra P, Atala A (2019) Engineering blood vessels and vascularized tissues: technology trends and potential clinical applications. Clin Sci 133(9):1115–1135
195. Geelhoed WJ, Moroni L, Rotmans JI et al (2017) Utilizing the foreign body response to grow tissue engineered blood vessels in vivo. J Cardiovasc Transl Res 10:167–179
196. Bharadwaj A (2021) An overview on biomaterials and its applications in medical science. In: IOP conference series: materials science and engineering, vol 1116 (1). IOP Publishing, London, p 12178
197. Hashemzadeh MR, Yazdi MET, Amiri MS et al (2021) Stem cell therapy in the heart: biomaterials as a key route. Tissue Cell 71:101504
198. Theus AS, Tomov ML, Cetnar A et al (2019) Biomaterial approaches for cardiovascular tissue engineering. Emerg Mater 2:193–207
199. Priya CH, Divya M, Ramachandran B et al (2020) Recent investigation on biomaterial-based tissue engineered heart valve (TEHV). Mater Today Proc 33:4467–4478
200. Williams DF, Bezuidenhout D, De Villiers J et al (2021) Long-term stability and biocompatibility of pericardial bioprosthetic heart valves. Front Cardiovasc Med 8:728577
201. Majewska P, Oledzka E, Sobczak M et al (2020) Overview of the latest developments in the field of drug-eluting stent technology. Biomater Sci 8(2):544–551
202. Tetali SSV, Fricker ATR, van Domburg YA et al (2023) Intelligent biomaterials for cardiovascular applications. Curr Opin Biomed Eng 28(6605):100474
203. Goycoolea M, Ribalta G, Tocornal F et al (2020) Clinical performance of the Osia™ system, a new active osseointegrated implant system. Results from a prospective clinical investigation. Acta Oto Laryngol 140(3):212–219
204. Shivgotra R, Soni B, Kaur M et al (2023) Advancement in biomaterials in the form of implants. In: Engineered biomaterials: synthesis and applications. Springer, London, pp 281–322
205. Haugen HJ, Chen H (2022) Is there a better biomaterial for dental implants than titanium? A review and meta-study analysis. J Funct Biomater 13(2):46
206. Kumar ST, Devi SP, Krithika C et al (2020) Review of metallic biomaterials in dental applications. J Pharm Bioallied Sci 12(1):S14

207. Najeeb S, Mali M, Syed AUY et al (2019) Dental implants materials and surface treatments. In: Advanced dental biomaterials. Woodhead Publishing, London, pp 581–598
208. Rinoldi C, Kijenska-Gawrońska E, Khademhosseini A et al (2021) Fibrous systems as potential solutions for tendon and ligament repair, healing, and regeneration. Adv Healthc Mater 10(7):2001305
209. Mu S, Wang J, Gong S et al (2021) Application of medical imaging based on deep learning in the treatment of lumbar degenerative diseases and osteoporosis with bone cement screws. Comput Math Methods Med 2021:2638495
210. Takeuchi A, Yamamoto N, Hayashi K et al (2019) Joint-preservation surgery for pediatric osteosarcoma of the knee joint. Cancer Metastasis Rev 38(4):709–722
211. Kanclerz P, Toto F, Grzybowski A et al (2020) Extended depth-of-field intraocular lenses: an update. Asia Pac J Ophthalmol (Philadelphia, Pa) 9(3):194
212. Topete A, Saramago B, Serro AP et al (2021) Intraocular lenses as drug delivery devices. Int J Pharm 602:120613
213. Tavakoli S, Klar AS (2021) Bioengineered skin substitutes: advances and future trends. Appl Sci 11(4):1493

Extracorporeal Devices: Artificial Kidney

T. Hemamalini and V. R. Giridev

Abstract The field of artificial kidneys has the interest among researchers in recent years due to efficiency of the portable extracorporeal devices in addressing renal failure. Biocompatibility is the major concern as the interaction between blood components and artificial surfaces can trigger immune responses and clotting issues. Hollow fibers are used for membrane application which offers pores that allows the transport of uremic toxins and purification of the blood. This chapter reviews the design, materials used, and evaluation of membrane for dialyzer application.

Keywords Artificial kidney · Dialyzer · Membrane · Textile substrate

1 Introduction

Failure of organs is heart threatening which requires either replacement of organs or additional aid namely extracorporeal devices to increase the lifespan of the particular human. The term extracorporeal devices can be defined as devices designed to remain outside the body during treatment of the patients in intensive care management. The multidisciplinary option combines the textile along with medicine for the failure of organs such as kidneys, lungs, and liver by means of the evolution of artificial kidneys, artificial lungs, and artificial liver as the organs failed to perform the intended function [1]. The primary component of the urinary system is the kidney which is a bean-shaped organ located on either side of the spine. The primary function of the kidney is to eliminate waste from blood and reintroduce healthy blood into the body. The blood enters the body through the renal arteries whereas renal veins allow the blood to return to the body after it has been cleansed. The purification of blood takes place as a kidney composed of a glomerulus that is joined to a tubule, to form each nephron. Fluid and waste products are filtered out of the blood as flow through the nephron and are excreted as urine. Urine travels through a tube called the ureter to enter the bladder which drains the fluids from the body through the urethra [2]. Membrane

T. Hemamalini · V. R. Giridev (✉)
Department of Textile Technology, A.C.Tech., Anna University, Chennai, India
e-mail: vrgiridev@yahoo.com

S. Sundaramoorthy et al. (eds.), *Textile Materials for Good Health and Wellbeing*, SDGs and Textiles, https://doi.org/10.1007/978-981-97-6496-9_14

technology is used in an artificial kidney, commonly referred to as a dialyzer, to substitute the kidney in the elimination of pollutants. The membrane serves as a core to the dialyzer which is the major component for the extracorporeal device used for the treatment of renal replacement therapy or artificial kidney. The major requirement of the membrane is to thin separation layer with the enhanced trans-membrane flow, hydrophilic surface for spontaneous wetting, reduced and narrow pore size to remove uremic toxins by preventing the loss of unnecessary proteins, mechanical stability to withstand the treatment process based on single or reusable type membrane, offering rough surface to reduce the interaction with the blood and activate the blood coagulation components, to avoid the transfer of cytokines from the dialyzer to the bloodstream and suitable structure for the transport of high molecule proteins [3]. The dialyzers are commercially available in various categories based on membrane type, operational type, and sterilizing technique. This chapter summarizes the polymers used for the membrane preparation, design consideration, sterilization, and evaluation methods used for dialyzer.

2 Membranes in Artificial Kidney

A healthy human being requires normal functioning of organs in the body which involves complex phenomena such as filtration of the waste, detoxification, metabolism, and transfer/exchange of mass/nutrients to/ within the other parts of the body. Failure of any of the mechanisms results in the collapse of the body and is fatal. Thus, extracorporeal devices have been developed to offer patients whose organs are failing by aiding temporary assistance or a replacement for their failing organs [4].

The human kidney performs two essential functions such as hemodialysis and ultrafiltration to maintain the body fluids by removing waste from the blood. Blood filters namely hemodialyzer offer artificial renal support to kidney failure patients over several decades which was composed of two compartments. A semi-permeable membrane typically separates two compartments in a hemodialyzer as one holds the patient's blood, and the other is filled with a dialysate bath (a water solution containing electrolytes, the concentration of which should be close to normal levels found in human blood serum) [5]. Dialysate is a physiological solution made up of glucose and inorganic ions from the body. It is an isotonic solution made of 6 electrolytes such as sodium (Na^+), potassium (K^+), calcium (Ca^{2+}), magnesium (Mg^{2+}), chloride (Cl^-), and bicarbonate (HCO^{3-}) and nonelectrolyte solution such as glucose or dextrose [6]. The mechanism of blood purification involves the separation of larger molecules by means of a fibrous textile membrane as small molecules like water and electrolytes can easily flow through the semi-permeable membrane separating the two compartments, whereas larger molecules like fat, protein, and red and white blood cells will not be able to pass through and will remain in the blood compartment. Therefore, waste from the blood in the body will be removed from performing the functions of the human kidney. Based on the operation of the artificial kidney, it can be classified into three categories such as hemodialysis, hemofiltration, and hemodiafiltration

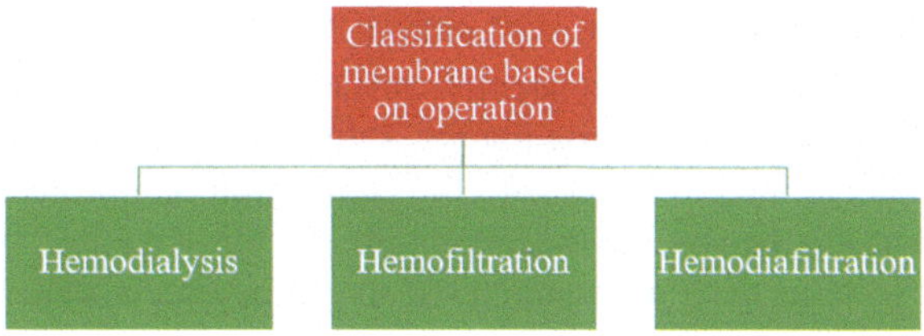

Fig. 1 Classification of membrane based on operation/function

which aids in the removal of waste through the semipermeable membrane as shown in Fig. 1 [7]. The pore diameter of a semipermeable membrane is 3 nm which enables the removal of small and medium molecules while intercepting proteins and blood cells. Through the convection effect, hemofiltration eliminates medium molecules. The medium molecules pass through the membrane from the dialysate side to the blood side at the specified trans-membrane pressure. The concept of hemodiafiltration integrates the ideas of hemodialysis and hemofiltration, providing a useful method for eliminating macromolecules [8].

3 Mechanism of Blood Purification in Artificial Kidney

The blood components are purified in the bioreactor or dialyzer unit by exchanging the water and solute transport across the semipermeable membrane. Osmosis and ultrafiltration enhance fluid (water) transfer. Adsorption affects the removal of solutes by causing the solute to adhere to the membrane [9]. Ultrafiltration is the process of moving fluid through a filter's porous membrane while it is being used for the dialysis process. The plasma oncotic pressure and hydrostatic pressure gradient between the blood and effluent/ultrafiltrate compartments control the water movement. When fluid compartment cells are divided by semipermeable membranes, a biophysical phenomenon known as osmosis frequently occurs in biological systems. The process of a solvent diffusing through a semipermeable membrane is called osmosis. The concentration difference of the solutes in the solutions separated by the semipermeable barrier thereby drives the solvent shift. Solutes cannot pass through the barrier, compared to the solvents. In biological systems, water, a common solvent, moves from the compartment with a lower solute concentration to a higher solute concentration. When the concentration of osmotic active molecules in these compartments is equal, the movement is stopped (net fluid flux). Osmosis is not only the movement of solutes but also the movement of water which is caused by the two basic processes such as diffusion and convection [10]. Convection is the movement of solutes with larger dimensions to pass through the membrane pores along with water flow over a semipermeable membrane which is caused by either an osmotic or a hydrostatic pressure gradient which is also called solvent drag. Solute have an unrestricted passage through the membrane as the concentration between the filtrate and original solution whereas larger molecules have a restricted movement as it is sieved causing the decrease in concentration in filtrate compared to the original solution. The process of diffusion involves molecules traveling randomly through a semipermeable barrier. Within the two compartments, solutes migrate from a more concentrated to a less concentrated location until the equilibrium is reached. Small particles move easily through the pores in the membrane, not allowing the larger particles to pass through. The latter gets accumulated at the filtrate/membrane interface and tends to move in two ways, returning to the source or tending to move in the crossflow direction which depends on the shear force acting at the interface. The movement of solute can be restricted for diffusion depending on the molecular size [11, 12]. Solute removal was done using three mechanisms such as diffusion, convection, and adsorption. The three main categories of uremic solutes removed by extracorporeal purification include small water-soluble compounds namely urea with a molecular weight not more than 500 Da can be removed by diffusion mechanism through any type of dialysis membrane whereas larger middle-molecular-weight molecules between 500 and 15,000 Da that can be removed by dialyzer membranes with large pores. An enhanced transport capacity can be achieved using high-flux dialyzers and finally molecules that are bound to proteins that are of molecular weight larger than 500 Da are removed by adsorption mechanism [13]. Examples of small molecular weight compounds such creatine, creatinine, asymmetric dimethylarginine, guanidine, hyaluronic acid, guanidinoacetate, guanidinosuccinate, oxalate, symmetric dimethylarginine, urea, and uric acid whereas for medium weight compounds such as β2-Microglobulin, adiponectin, α1-acid glycoprotein, cystatin C, prolactin, osteocalcin, vascular endothelial growth factor and protein-bound molecules such as p-Cresylsulfate, indoxyl sulfate, phenol, indol-3-acetic acid, hippuric acid, homocysteine, carboxymethyllysine, acrolein [14]. The mass transfer coefficient and the surface area of the dialyzer affect the solute removal efficiency. High-flux dialyzers enable the convective separation of solutes and low-molecular-weight proteins from big serum proteins and blood components by improving porosity and effective mass transfer. Proteins and macromolecules can attach to the membrane surface by the process of adsorption, which prevents the molecules from penetrating the membrane structure. The hydrophobicity and internal pore structure of the membrane are the main factors influencing this process [15]. Removal of urea from the blood is an essential characteristic required for an ideal dialyzer. Increasing the size of the membrane pore results in higher clearance of β2-microglobulin. Owing to its molecular weight of 11,000 Da, β2-microglobulin clearance serves as a surrogate marker for membrane flux and a marker of intermediate molecule elimination. Dialyzer can be called a high-flux dialysis device when it offers an ultrafiltration coefficient higher than 15 mL/h/mmHg by removing 20 mL/min of β2-microglobulin whereas it is called as a super high flux if the β2-microglobulin clearance is greater than 50 mL/min with the components containing molecular weight of 65,000 Da as it can clear inflammatory cytokines and middle- and large-sized uremic poisons [16].

4 Components of Blood Purification in Artificial Kidney

The ideal renal replacement extracorporeal unit is responsible for the removal of impure blood from the patient's body, purifying the blood using the filter, and returning the purified blood to the body. The task must be performed without causing harm to blood components, leaking blood into the surroundings, or exposing the patient to potentially hazardous substances from the extracorporeal circuit or surroundings as shown in Fig. 2 [17]. The process involves the removal of blood from the patients by placing the catheter in the blood vessel to and from the extracorporeal devices through blood tubing. Two tubes are used, one is the inflow (arterial) segment to remove the impure blood and the second is the outflow (venous) segment to resend the purified blood to the body. Blood carrying tubes are made of plasticized polyvinylchloride and may be sterilized with ethylene oxide or gamma irradiation and can be coated with an anticoagulant. To uniformly supply the blood from the patient to the filtering unit, a blood pump is used as the central venous pressure is not sufficient for the blood flow. Peristaltic pumps are used, as it possesses no direct contact with blood and accurate control of blood flow is possible.

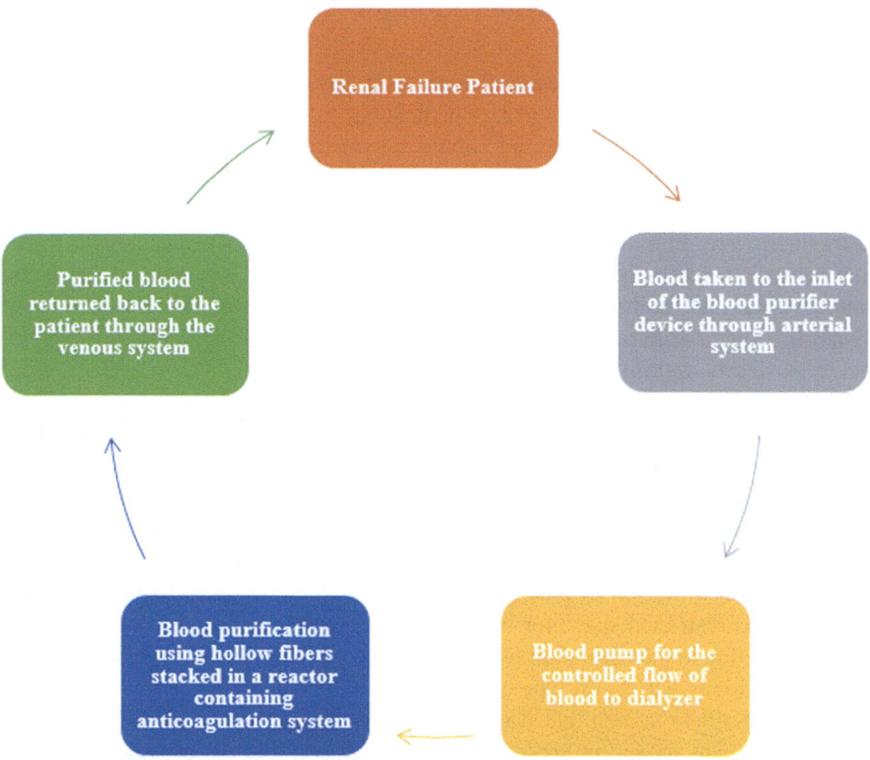

Fig. 2 Process involved in the purification of blood in dialyzer

Blood purification is carried out using a hollow-fiber membrane which allows the exchange of solutes either by diffusion or convection and the removal of water by ultrafiltration. While hemofiltration, which is accomplished by convection, enables the removal of greater molecular weight solutes, hemodialysis mostly relies on diffusion, especially for small waste products, electrolytes, and non-protein-bound medications and hemodiafiltration, which depend on both convection and diffusion [18]. As the blood tends to coagulate in the presence of foreign materials and air, hence anticoagulant surfaces are to be infused in extracorporeal devices. The anticoagulant can be infused in two ways involving the patients and devices or at devices alone. Heparin can be injected into patients that prevent blood coagulation wherein case of local region, Citrate can be injected along with the blood in entry to the dialyzing unit and calcium solution can be passed through the outlet tube from the devices to patients thereby preventing the blood coagulation within the body. The dialysis is carried out by monitoring the inflow and outflow pressure, without the loss of blood. The removal of fluids from the blood plays a crucial role as excessive removal of fluid leads to hypovolemic condition and less removal leads to hypervolemia [19].

5 History of Design Consideration for Artificial Kidney

The ideal dialyzer should possess solute removal and excellent performance during usage with low blood volume for the purification process. The major requirement of the device is biocompatible and should be sterilized with steam to avoid the residues of chemicals (ethylene oxide). Different types of dialyzers were studied as an alternative for damaged kidneys to support renal failure patients. Rotating drum-based filters were initially used for the purification of blood, the mechanism works by wrapping the cellophane tube around the cylinder which was rotated inside the stationary dialysate bath. The flood flows inside the diameter of the cellophane tube toward the lowest side of the tube in a helical screw structure which was driven by gravity powered pumping mechanism.

The blood flow was governed by the arterial cannula, not by an inlet blood pump. However the limitation of the rotating drum kidney was very low blood compartment and trans-membrane pressure resulting in minimal ultrafiltration of plasma that was overcome by the application of osmotic pressure. Upon modification, the extracorporeal device required a larger volume of blood 500–700 mL as it was not possible to filter the low volume of blood. Rotating drum kidney was modified to increase the ultrafiltration by placing the cellophane tube between the two cylinders which created restricted space for blood flow thereby reducing the pressure within the dialysate due to the formation of higher blood compartments [20]. Another type of artificial kidney was a coil type of dialyzer on which a cellophane tube was kept inside the fiberglass screen that was circumferentially placed around the cylinder. The dialysate was passed in cross flow toward the coil and the pressure built was reduced by spacing the screens whereas the pressure difference was used for ultrafiltration of plasma. Twin coil kidneys were used as two coils were wound onto the cylinder offering the advantages of large-scale manufacture and used for the treatment of acute and chronic renal failure [21]. To narrow the blood flow channel, a support surface was placed between the cellophane in the case of multiple-use parallel flow dialyzers which creates a shorter blood transfer path thereby reducing the blood resistance. The parallel flow dialyzers were constructed by placing four parallel cellophane channels which were constructed by placing the plastic board in between, the dialysate flow in the counter direction to the blood flow on the grooves of the plastic board. In order to decrease the path length and narrow the blood channel, a membrane made of cuprophan was placed. This arrangement resulted in multiple uses of the filters without causing damage upon sterilization. The drawback of the system was large-scale manufacturing of plastic boards which were replaced with polypropylene [22, 23].

Single-flow parallel dialyzers were developed, as the multiple-use parallel flow dialyzers possess the limitation of multiple sterilization using formaldehyde that affects the blood surface. The disposable blood compartments were prepared by two parallel sheets of cellulose membrane between grooved polystyrene plates which were pressed into a single unit. The advantages of the dialyzer were equal distribution of blood flow to different compartments and prevented the collapse of individual blood channels [24]. The limitation of these devices was overcome by the hollow fiber dialyzer as the coil dialyzer offers high blood resistance and compliance which creates hemodynamic instability whereas parallel coil dialyzer offers lower blood compartment resistance and less complaint than the coil and hollow fiber-based dialyzer. A hollow fiber artificial kidney was developed by encapsulating the ends of regenerated cellulose membrane with silicone rubber that was connected to the arterial and venous of the dialysate without requiring the support surface. The system enabled smooth running by maintaining the blood flow compartment without altering the blood rheology [25]. Thus, hollow fiber bioreactor was used widely for the blood purification process in which a bioreactor is composed of a semi-permeable membrane made of a large number of fibers within a cylindrical jacket or shell. Two compartments for blood flow were created within the fiber such as the intra-tubular space of the fiber called lumen or capillary and extra-capillary space between the fibers.

6 Materials and Manufacturing of Membrane for Artificial Kidney

The polymeric material used for the membrane media in the dialyzer should offer requirements such as the removal of toxins from the blood, balancing the electrolyte in the body without activating the blood components, and possessing excellent mechanical, chemical, and thermal stability during usage and sterilization process [26]. The most crucial requirement for any polymer material to be utilized in close proximity to living things is biocompatibility. Blood compatibility is crucial for the blood purification membrane used in hemodialysis because it prevents thrombus or blood coagulation as it activates the leucocytes which cause allergic reactions during dialysis resulting in inflammation, and immune dysregulation. Further, the platelet and fibrinogen in the bloodstream interact with extracorporeal device components resulting in clot formation which hinders the purification process during dialysis. The material should possess anti-fouling properties as there is a possibility of interaction between small-sized toxins and albumin in the blood and gets clogged in the pores which is not possible to remove by a diffusion process. The polymer should possess good physical strength, transport of blood across the structure by means of diffusion and convection. Additionally, the polymeric membrane should hold the toxins outside the dialyzer. Based on the type of membrane, the material utilized for preparing renal replacement therapy is classified into two types such as natural and synthetic as shown in Fig. 3. Cellulose is a complex polysaccharide, and is a major structural component in the cell walls of plants. Its molecular structure is a linear chain of glucose molecules linked together by β-1, 4-glycosidic bonds. Cuprophan is the first cellulose membrane produced using the cuprammonium process in which cellulose is dissolved in an ammonium solution of cupric oxide and is then precipitated in acid (dilute sulfuric acid) and sterilized. Cuprophan membranes used for dialysis have a pore radius in the range of 1.72 nm and offer diffusive transport for low-weight molecules in the size between 500 and 1000 Da [27]. The regenerated cellulose is hydrophilic due to the presence of hydroxyl groups which would activate the complement system which causes further harmful biological processes, such as leucopenia, by releasing enzymes from granulocytes and monocytes, accumulating granulocytes in lung capillaries, or inhibiting granulocyte metabolism. The drawback of Cuprophan can be overcome by substituting the hydroxyl group with either acetylation, diethylamino groups, or benzyl groups which increases the hemo-compatibility. Upon acetylation, adsorption of protein molecules was higher compared to cuprophan due to the hydrophobic surface resulting in lower water uptake. This can be related to the degree of substitution of acetyl groups based on which it can be classified as cellulose acetate, cellulose diacetate, and cellulose triacetate. Diethylaminoethanol and etherified cellulose are blended to create Hemophans membranes which offer higher biocompatibility with no complement activation Cellulose is etherified with benzene groups to create synthetically modified membranes, or SMCs. Surface coating with polyethylene glycol, polyacrylonitrile or vitamin E improves the biocompatibility of cellulosic membranes compared to untreated cellulose membranes, due to the presence of Vitamin E resulting in reduced activation and migration of monocytes and granulocytes offering higher filtration efficiency [28]. Ethylenevinylalcohol membrane is composed of hydrophilic and hydrophobic parts in which hydrophilic segments interact with water creating dynamic movement that reduces blood–membrane interaction. This type of membrane prevents the adsorption of proteins and platelets along with the anticoagulant used during the process thereby causing less cell activation and possessing excellent biocompatibility compared to regenerated cellulose [29].

Synthetic polymers are widely used for membrane preparation by blending with hydrophilic polymers for dialysis applications among which polysulfone or

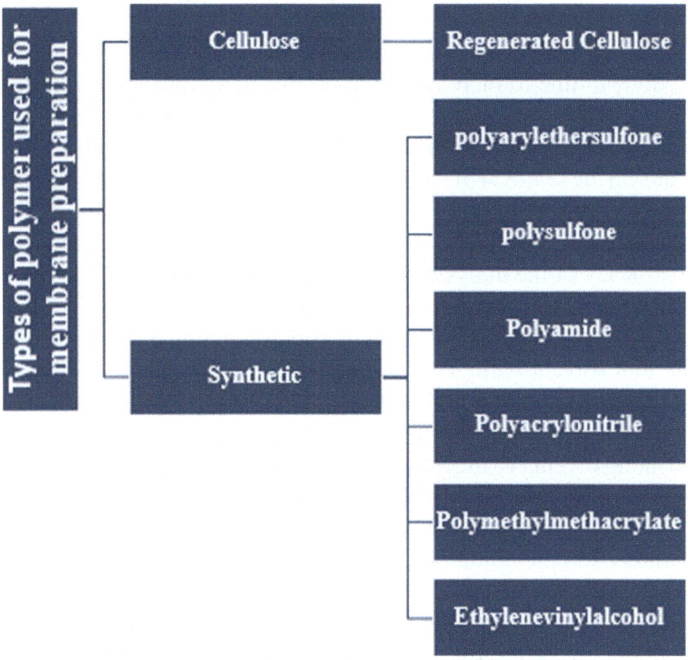

Fig. 3 Polymer used for membrane preparation for dialysis application

polyethersulfone (polyarylethersulfone) are widely used for their hydrophobic nature as they offer higher transport properties due to excellent thermal, chemical, and physical stability, and Polyvinylpyrrolidone (PVP) utilized for hydrophilicity in the membrane. Polysulfone and polyarylethersulfone differ in the chemical structure by means of having alkyl or aryl groups along with sulfone groups whereas polysulfone contains isopropylidene. The hydrophilicity in the membrane can be achieved by either blending with copolymers like methallylsulfonate or additives like polyvinylpyrrolidone (PVP) or polyethylene glycol (PEG). PVP is produced by vinylation of pyrrolidone followed by polymerization and blending with synthetic polymer, upon production of the membrane, PVP polymer can be washed out partially to create a gradient structure with higher hydrophilicity in the core. The membrane prepared using those polymers offered minimal competent activation by reducing the release of leucocyteelastase. The drawback of synthetic polymers such as polysulfone contains water whereas cuprophan is free from stabilizers [30].

Polyamide such as nylon is employed for membrane application due to its hydrophobic nature by blending with PVP. The nylon-based membrane offered low complement activation and cell activation with reduced oxidative stress [31]. Polyacrylonitrile membranes are produced by acrylonitrile polymer for dialysis application but offer limitation of formation of larger pores. To overcome the limitation, the polymer is blended with methallyl sulfonate to produce finer pores to facilitate diffusive and convective transport properties. But the membrane activates competent particularly the activation of kallikrein (KK) and secession of bradykinin due to a negatively charged surface. The surface charge was reduced by coating with polycationic polyethylenimine to increase blood purification efficiency [32]. Polymethylmethacrylate (PMMA) polymer is used for the preparation of a high-flux dialyzer for hemofiltration and hemodiafiltration due to its higher permeability to liquid. The membrane aids in the removal of microglobulin by adsorption mechanism and retains the albumin after stabilizing the pores with glycerol. The limitation of the PMMA membrane is the undesired platelet adsorption resulting in fibrin production which reduces the efficacy of the dialysis process during treatment [33].

Manufacturing of hollow fibers utilizes a wet spinning technique in which the polymer is dissolved in a suitable solvent and upon spinning into filament it is precipitated in the non-solvent bath. The total polymer content is 20 wt% depending on the polymer type. To create a hollow fiber, the polymer solution is forced through an annular die, also known as a spinneret. A bore liquid, which is a mixture of solvent and non-solvent, is introduced into the inner region of the spinneret and forms the inner void of the hollow fiber. The hollow fiber is directed through a non-solvent liquid called a coagulation bath. Through diffusive solvent and non-solvent exchange (immersion precipitation), the homogenous liquid-polymer solution must be transformed into a two-phase system with the aid of the non-solvent bath and bore liquid. Hollow fiber is formed due to vitrification and pores are formed due to the removal of liquids from the fibrous structure. Concentration, molecular weight of the polymer, viscosity and temperature of the polymer solution, the use of additives, nature of the fiber such as crystallized or amorphous, spinneret design, composition of the coagulation bath, and the conditions between the spinneret and coagulation bath decide the spin-ability of hollow fibers for membrane application. In the case of a coagulation bath, the internal structure of the hollow filament is formed which is influenced by the composition and temperature of the coagulation bath. The resulting filament is dried and wound into cross wound packages [34].

7 Sterilization of Membrane Dialyzer

Sterilization of the dialyzer is crucial in order to kill the pathogens and allergic reactions to the users in dialyzer manufacturing. Common sterilizing techniques often involve the use of heat, irradiation (beta or gamma), or ethylene oxide gas (ETO). The process used to sterilize a dialyzer has clinical implications when patient-specific sensitivities exist [35]. ETO sterilization is widely carried out for medical filters as it is affordable and kills the bacteria by alkylation of sulfur-containing proteins and the technique possesses low physical and thermal stress without affecting the dialysis membrane. However the drawback of the sterilizing agent is the residue that creates an allergenic reaction in the patients due to the adsorption and slow release of the agent by a polymeric membrane and capping material used for the production [36].

The dialyzers are exposed to either beta or gamma radiation to achieve radiation sterilization. This leads to the high-energy ionization of the atoms and the production of free radicals, which dimerize the bases of DNA and scissor the sugar-phosphate backbone, preventing the growth of bacteria. This process is easy to follow and safe. The penetrating strength of beta radiation accelerators typically is comparable to that of gamma radiation. The ability to dose and target sterilization more precisely is the only benefit of beta radiation over gamma radiation. For materials with a higher density and sensitivity to heat, gamma irradiation is frequently used, whereas beta irradiation has a shorter exposure time and causes less material damage. The quantity of radiation adsorbed, expressed in gray, is correlated with the degree of sterilization. For the optimal amount of sterilization, the materials are positioned in close proximity to the radiation source for seconds in beta radiation and minutes or hours in gamma radiation as the radiation tends to degrade the polymeric membrane utilized for the development of dialyzer [37].

Heat denaturation of the proteins and cell walls kills the bacteria during thermal sterilization. This process prevents any gamma and beta-ray-induced alterations to the dialyzer components and leaves no chemical residue behind. It is regarded as the most effective sterilization technique. To ensure that the killed germs are removed, dialyzers must be rinsed with saline before use after this kind of sterilization. Steam or heat sterilization seems to be more effective compared to radiation and ethylene oxide but not all the materials are thermally strong [38]. Based on the specification, the dialyzer can be sterilized either using ethylene oxide, radiation (beta or gamma), or steam.

8 Evaluation of Membrane for Artificial Kidney

Diffusion of hollow fiber membranes has a major impact as the purification of blood takes place through pores in the fibers. Diffusion of the saline solution can be studied by stacking the hollow fibers and diffusive permeability of the membrane can be calculated based on the inlet and outlet concentration but the technique is suitable for the removal of small particles not for medium or high molecules found in the blood [39]. Another technique utilizes radioactive isotope-based solute and passes through the hollow fiber, which allows quantitative assessment of residual solute. However this technique is limited due to the high precision of handling and overexposure of radioactive compounds [40]. Diffusion through the membrane can be measured by capping the optical glass at both ends and light was passed on the one end, transmittance was calculated from the other end using a silicon photodiode. This technique is suitable as it works independent diffusion through convection and osmotic flow through the membranes [41].

Porosity and permeability are the essential characteristics required for mass transfer across the membrane in a dialyzer. In the case of porosity, direct and indirect tests can be carried out. Microscopic, bubble point, and mercury porosimetry are the direct test methods to study the pore size and pore size distribution of the hollow fibers. The surface morphology of the hollow fibers can be studied using a scanning electron microscope (SEM), transmission electron microscope (TEM), field emission scanning electron (FESEM), and atomic force microscopy (AFM). SEM and TEM are not widely employed for analyzing the surface morphology as pass a high beam energy which damages the structure and sample preparation is complex as the samples were coated with an external agent which deteriorates the membrane structure. In the case of FESEM, low beam energy can be used whereas the AFM, no electron beam energy is used and samples are tested in contact with air. AFM is widely used to study the surface morphology of membranes. The indirect method includes water permeation, gas permeation, and solute transport across the membranes which measure the mean radius of the pores based on permeation and rejection performance in relation to reference standard [42, 43]. Surface property of the membrane plays a major role in offering cell adhesion and blood computability. The hemo-compatibility parameters such as plasmatic immune systems, coagulation activation, cell activation, platelet count, and platelet factor are not essential for a successful dialyzer in extracorporeal devices. The surface characteristics of the membrane can be measured using a goniometer or a sessile drop with a telescope or microscope as it measures the contact angle of the membrane. The membrane is considered to be hydrophobic if the contact angle is greater than 90° whereas less than 90° indicates the hydrophilic nature of the membrane [44]. The wettability of the membrane can be measured using the Wilhelm method, in which advancing and receding contact angles (dynamic) can be measured [45]. Fourier transform infrared attenuated total reflection (FTIR-ATR) and X-ray photoelectron spectroscopy (XPS) techniques can be used to study the functional groups pertaining to the surface nature of the membrane [46]. Protein adsorption on the membrane can be studied using radioisotope counting or autoradiography whereas cell adhesion was studied using laser confocal scanning microscopy as cells play a major role in the function of living cells during blood purification within the dialysis [47, 48].

9 Future Trends

The future of artificial kidneys for extracorporeal devices holds promise for further improvements in performance, miniaturization, and portability. Nanotechnology and bio-mimetic design principles offer potential solutions to enhance biocompatibility and reduce the risk of thrombosis. Furthermore, ongoing research in regenerative medicine and tissue engineering may lead to the development of bio-artificial kidneys with improved functionality and longevity.

10 Conclusions

The development of artificial kidneys for extracorporeal devices represents a dynamic and evolving field with significant potential to revolutionize the treatment of renal failure. While advancements have been made, addressing challenges such as biocompatibility and long-term durability is crucial for the successful translation of the technologies into clinical practice. The quality of the renal failure patient's life can be improved by utilizing the fibrous media for filtering the uremic toxins.

References

1. Shekar K, Mullany DV, Thomson B, Ziegenfuss M, Platts DG, Fraser JF (2014) Extracorporeal life support devices and strategies for management of acute cardiorespiratory failure in adult patients: a comprehensive review. Crit Care 18(3):219. https://doi.org/10.1186/cc13865
2. Josipovic V (1959) Physiology of the kidney. Med Glas 13:402–406. https://doi.org/10.1093/milmed/147.12.1024
3. Bloom M, Mourttsen OG (1995) The evolution of membranes. Handb Biol Phys 1(C):65–95. https://doi.org/10.1016/S1383-8121(06)80019-9
4. Kumar A, Tripathi A, Jain S (2011) Extracorporeal bioartificial liver for treating acute liver diseases. J Extra Corpor Technol 43(4):195–206
5. Clark WR, Gao D (2002) Properties of membranes used for hemodialysis therapy. Semin Dial 15(3):191–195. https://doi.org/10.1046/j.1525-139X.2002.00019.x
6. Pittard JD (2017) Safety monitors in hemodialysis, 5th edn. Elsevier, London
7. Nguyen BTD, Thi HYN, Thi BPN, Kang DK, Kim JF (2021) The roles of membrane technology in artificial organs: current challenges and perspectives. Membranes (Basel). https://doi.org/10.3390/membranes11040239
8. Bello G, Di Muzio F, Maviglia R, Antonelli M (2012) New membranes for extracorporeal blood purification in septic conditions. Minerva Anestesiol 78(11):1265–1281
9. Clark WR, Macias WL, Molitoris BA, Wang NH (1995) Plasma protein adsorption to highly permeable hemodialysis membranes. Kidney Int 48(2):481–488. https://doi.org/10.1038/ki.1995.317
10. Misra M (2008) Basic mechanisms governing solute and fluid transport in hemodialysis. Hemodial Int 12(Suppl. 2):25–28. https://doi.org/10.1111/j.1542-4758.2008.00320.x
11. Ronco C, Ghezzi PM, Brendolan A, Crepaldi C, La Greca G (1998) The haemodialysis system: basic mechanisms of water and solute transport in extracorporeal renal replacement therapies. Nephrol Dial Transplant off Publ Eur Dial Transpl Assoc Eur Ren Assoc 13(Suppl 6):3–9. https://doi.org/10.1093/ndt/13.suppl_6.3
12. Tang YS, Tsai YC, Chen TW, Li SY (2022) Artificial kidney engineering: the development of dialysis membranes for blood purification. Membranes (Basel) 12(2):1–16. https://doi.org/10.3390/membranes12020177
13. Venkataraman R, Subramanian S, Kellum JA (2003) Clinical review: extracorporeal blood purification in severe sepsis. Crit Care 7(2):139–145. https://doi.org/10.1186/cc1889
14. Neirynck N, Vanholder R, Schepers E, Eloot S, Pletinck A, Glorieux G (2013) An update on uremic toxins. Int Urol Nephrol 45(1):139–150. https://doi.org/10.1007/s11255-012-0258-1
15. Magnani S, Atti M (2021) Uremic toxins and blood purification: a review of current evidence and future perspectives. Toxins (Basel). https://doi.org/10.3390/toxins13040246
16. Abe M, Masakane I, Wada A, Nakai S, Nitta K, Nakamoto H (2021) Dialyzer classification and mortality in hemodialysis patients: a 3-year nationwide cohort study. Front Med 8(August):1–14. https://doi.org/10.3389/fmed.2021.740461

17. Alwall N (1947) On the artificial kidney; apparatus for dialysis of the blood in vivo. Acta Med Scand 128(4):317–325
18. Grandi F, Bolasco P, Palladino G, Sereni L, Caiazzo M, Atti M, Maria P (2013) Adsorption in extracorporeal blood purification: how to enhance solutes removal beyond diffusion and convection. Hemodialysis. https://doi.org/10.5772/52272
19. Zaccaria M, Neri M, Garzotto F, Ronco C (2019) Principles of extracorporeal circulation and transport phenomena, 3rd edn. Elsevier, London
20. Merrill JP, Thorn GW, Walter CW, Callahan EJ, Hollingsworth Smith LJ (1950) The use of an artificial kidney I. Technique. J Clin Invest 29(4):412–424. https://doi.org/10.1172/JCI102274
21. Palmer RA, Price JD, English ET, Thomas N (1957) Clinical trials with the Koiff twin coil artificial kidney. Can Med Assoc J 77(9):850–855
22. Ae M, Je D, Anthone R, Anthone S (1959) Technic with parallel flow, straight tube blood dialyzer. N Y State J Med 59:4137–4149
23. Skeggs LTJ, Leonards JR (1948) Studies on an artificial kidney: I. Preliminary results with a new type of continuous dialyzer. Science 108(2800):212–213. https://doi.org/10.1126/science.108.2800.212
24. Lavender AR, Markley FW, Forland M (1968) A new pumpless, parallel-flow hemodialyzer. Preliminary report. Trans Am Soc Artif Intern Organs 14:92–98
25. Gotch F, Lipps B, Weaver JJ, Brandes J, Rosin J, Sargent J, Oja P (1969) Chronic hemodialysis with the hollow fiber artificial kidney (HFAK). Trans Am Soc Artif Intern Organs 15:87–96
26. Sheng D, Li X, Sun C, Zhou J, Feng X (2023) The separation membranes in artificial organs. Mater Chem Front 7(17):3455–3474. https://doi.org/10.1039/D3QM00255A
27. Klein E, Holland FF, Eberle K (1979) Comparison of experimental and calculated permeability and rejection coefficients for hemodialysis membranes. J Membr Sci 5(C):173–188. https://doi.org/10.1016/S0376-7388(00)80446-9
28. Zaluska WT, Ksiazek A, Rolíski J (2001) Effect of vitamin E modified cellulose membrane on human lymphocyte, monocyte, and granulocyte CD11b/CD18 adhesion molecule expression during hemodialysis. ASAIO J 47(6):619–622. https://doi.org/10.1097/00002480-200111000-00010
29. Su Y-L, Beltsios KG, Su JF, Cheng L-P (2023) Preparation of poly(vinyl alcohol-co-ethylene) hollow fiber membranes for high-flux ultrafiltration applications. Polymer (Guildf) 266:125607. https://doi.org/10.1016/j.polymer.2022.125607
30. Schaefer RM, Heidland A, Hörl WH (1985) Release of leukocyte elastase during hemodialysis. Effect of different dialysis membranes. Contrib Nephrol 46:109–117
31. Kostenbauder HB, Boxenbaum HG, Deluca PP (1969) Nylon as a dialysis membrane. J Pharm Sci 58(6):753–757. https://doi.org/10.1002/jps.2600580624
32. Floege J, Bartsch A, Schulze M, Shaldon S, Koch KM, Smeby LC (1991) Clearance and synthesis rates of beta 2-microglobulin in patients undergoing hemodialysis and in normal subjects. J Lab Clin Med 118(2):153–165
33. Aucella F, Vigilante M, Gesuete A (2010) Review: the effect of polymethylmethacrylate dialysis membranes on uraemic pruritus. NDT Plus 3(Suppl 1):i8–i11. https://doi.org/10.1093/ndtplus/sfq031
34. Zweigart C, Boschetti-de-fierro A, Neubauer M, Storr M, Böhler T, Krause B (2017) 4.11 progress in the development of membranes for kidney-replacement therapy. Elsevier, London
35. Takesawa S, Ohmi S, Konno Y, Sekiguchi M, Shitaokoshi S, Takahashi T, Hidai H, Sakai K (1987) Varying methods of sterilisation, and their effects on the structure and permeability of dialysis membranes. Nephrol Dial Transplant off Publ Eur Dial Transpl Assoc Eur Ren Assoc 1(4):254–257
36. Mendes GCC, Brandão TRS, Silva CLM (2007) Ethylene oxide sterilization of medical devices: a review. Am J Infect Control 35(9):574–581. https://doi.org/10.1016/j.ajic.2006.10.014
37. Hanrahan CT, Himmele R, Diaz-Buxo JA (2012) The challenges of heat sterilization of peritoneal dialysis solutions: Is there an alternative? Adv Perit Dial 28:126–130
38. Allard B, Begri R, Potier J, Coupel S (2013) Dialyzers biocompatibility and efficiency determinants of sterilization method choice. Pharm Hosp Clin. https://doi.org/10.1016/j.phclin.2013.10.071

39. To N, Sanada I, Ito H, Prihandana GS, Morita S, Kanno Y, Miki N (2015) Water-permeable dialysis membranes for multi-layered microdialysis system. Front Bioeng Biotechnol 3:70. https://doi.org/10.3389/fbioe.2015.00070
40. Kanamori T, Sakai K, Awaka T, Fukuda M (1994) An improvement on the method of determining the solute permeability of hollow-fiber dialysis membranes photometrically using optical fibers and comparison of the method with ordinary techniques. J Membr Sci 88(2):159–165. https://doi.org/10.1016/0376-7388(94)87003-9
41. Ohmura T, Tatsuguchi T, Nishikido J, Yamamoto T, Fushimi F, Nishida O, Sakai K (1989) A new method of determining the solute permeability of hollow-fiber dialysis membranes by means of laser lights traveling along optic fibers. ASAIO Trans 35(3):601–603. https://doi.org/10.1097/00002480-198907000-00140
42. Nakao S (1994) Determination of pore size and pore size distribution. J Membr Sci 96(94):131–165
43. Zhao C, Zhou X, Yue Y (2000) Determination of pore size and pore size distribution on the surface of hollow-fiber filtration membranes: a review of methods. Desalination 129:107–123
44. Zawada AM, Lang T, Ottillinger B, Kircelli F, Stauss-Grabo M, Kennedy JP (2022) Impact of hydrophilic modification of synthetic dialysis membranes on hemocompatibility and performance. Membranes (Basel) 12(10):1–16. https://doi.org/10.3390/membranes12100932
45. Lin Y-C, Brayfield CA, Gerlach JC, Rubin JP, Marra KG (2009) Peptide modification of polyethersulfone surfaces to improve adipose-derived stem cell adhesion. Acta Biomater 5(5):1416–1424. https://doi.org/10.1016/j.actbio.2008.11.031
46. Ulbricht M, Belfort G (1996) Surface modification of ultrafiltration membranes by low temperature plasma II. Graft polymerization onto polyacrylonitrile and polysulfone. J Membr Sci 111(2):193–215. https://doi.org/10.1016/0376-7388(95)00207-3
47. Hlady V, Buijs J, Jennissen HP (1999) Methods for studying protein adsorption. Methods Enzymol 309:402–429. https://doi.org/10.1016/s0076-6879(99)09028-x
48. Zhou L, Tan GX, Ning CY (2015) Modi fi cation of biomaterials surface by mimetic cell membrane to improve biocompatibility. Front Mater Sci 8:325–331

Weaving the Future: Exploring Electromagnetic Shielding Textiles

Hariharasudan Natarajan and Periyasamy Siddhan

Abstract This chapter focuses on the newly emerging threat to mankind, the Electromagnetic Radiation, abbreviated as EMR, and Electromagnetic Interference, abbreviated as EMI, that is instigated as a result of electric transmission, various electric and electronic appliances, which are day-to-day essentials for the present and unavoidable for the future as the world has become completely dependent on them. Here we focus on harnessing the textile potential toward prevention of the user and the environment from these radiations which are some times very harmful and often create degradation at a slower rate. This part deliberates the different types of materials that can shield EMR and the techniques and technology behind and further how these materials can be incorporated or turn out to be a wearable piece that can withstand and protect.

1 Introduction—Textile Metamorphosis: From Aesthetic to Functional

Electronic devices are increasingly becoming an integral part of our daily lives, providing us with a range of communication, entertainment, work and educational opportunities. This is due to a variety of factors, such as convenience, affordability, portability and functionality. Electronic devices enable us to stay in touch with our loved ones, access information rapidly and complete tasks with ease. Also, they are becoming increasingly affordable, making them more accessible to a broader range of individuals. Moreover, most electronic devices are typically hand-held, allowing users to carry them with them on the go. The electronic industry is continually innovating, resulting in the development of more powerful and versatile devices that can be used for a wide range of tasks. Additionally, these devices are now being used for a range of domestic tasks, as well as for more complex applications and research where the scale and efficiency of the equipment are unrivaled. Despite these

H. Natarajan (✉) · P. Siddhan
Central Silk Technological Research Institute, Central Silk Board, Bangalore, India
e-mail: hari.csb@nic.in

S. Sundaramoorthy et al. (eds.), *Textile Materials for Good Health and Wellbeing*, SDGs and Textiles, https://doi.org/10.1007/978-981-97-6496-9_15

advantages, there are also some potential drawbacks to the usage of electronic devices [1].

The accelerating pace of technology has brought humanity into an era of complex electronic devices and wireless communications. In the midst of this web of connectivity, worries about electromagnetic interference and radiation exposure have become increasingly prominent in technological discourse. Traditionally, protective measures against these unseen forces have been pragmatic and aesthetically pleasing. However, as textiles and technology converge, a new frontier of innovation opens up, leading to the emergence of electromagnetic shielding textiles. This chapter dives into the emerging world of these textiles and reveals their potential to redefine protection against electromagnetic intrusion.

The innate versatility of textiles lies in their ability to merge form and function. At the crossroads of this versatility, electromagnetic shielding textiles stand as a testament to human ingenuity. Through a synthesis of advanced materials, innovative designs and cutting-edge manufacturing techniques, textiles cease to be mere coverings and emerge as guardians against a silent but pervasive threat. This chapter examines the conceptual shift from viewing textiles as passive adornments to active participants in safeguarding human environments from the ever-growing presence of electromagnetic disruptions.

Electromagnetic radiations (EMRs) are a type of energy that exists naturally and can be generated and travels through the universe in the form of waves. EMRs are composed of oscillating electrical and magnetic fields and can range in wavelength from very long, such as radio waves, to very short, such as gamma rays. Depending on the intensity, EMRs can be either strong or weak. Electromagnetic Radiations (EMR) are classified according to the wavelength, with the wavelength representing the distance between successive peaks of the wave. These wavelengths of an EMR are measured in meters and the frequency is measured in Hertz (Hz). It is important to note that EMR wavelengths and frequencies are inversely related, meaning that the longer the wavelength, the lower the frequency. Enumerated below are the approximate wavelength and frequency of the electromagnetic spectrum (Table 1).

Electromagnetic radiation has a wide range of modes of interaction with matter, including absorption, reflection, scattering and transmission. The manner in which electromagnetic radiation interacts with matter is contingent upon its wavelength.

Table 1 Classification of Electromagnetic Radiation

Rays	Wavelength (m)	Frequency (Hz)
Gamma-ray	Less than 1×10^{-11}	Greater than 3×10^{19}
X-ray	1×10^{-11} to 1×10^{-8}	3×10^{16} to 3×10^{19}
UV	1×10^{-8} to 4×10^{-7}	7.5×10^{14} to 3×10^{16}
Optical	4×10^{-7} to 7×10^{-7}	4×10^{14} to 7.5×10^{14}
Infrared	7×10^{-7} to 1×10^{-3}	3×10^{11} to 4×10^{14}
Microwave	1×10^{-3} to 1×10^{-1}	3×10^{9} to 3×10^{11}
Radio	Greater than 1×10^{-1}	Less than3×10^{9}

This is similar to the absorption and refraction of visible light by various objects; some objects exhibit reflection, while others do not. EMR is a fundamental component of our environment and is utilized for communication, medical research, science and technology. Electromagnetic radiation is an energy source that can have various applications from communication to more advanced medical diagnostic application, however, some parts lower wavelength spectrum compared to visible light that has adverse effects on living organisms. It is essential to comprehend the various types of electromagnetic radiation and the manner in which they interact with matter in order to ensure the safe and effective utilization of electromagnetic radiation.

Electromagnetic Radiation (EMR) can be divided into four main categories: radio waves, microwave, infrared and gamma rays. Radio waves are the longest and most common type of EMR and are used in communications, radar and broadcasting. Microwaves have a shorter wavelength and higher frequency than radio waves which are used for cooking and heating applications. Infrared radiation is emitted from warm objects and is felt as heat with shorter wavelength and higher frequency than micro waves, while visible light which is present next has a shorter wavelength range and higher frequency than infrared radiation. Ultraviolet radiation, on the other hand, has lower frequency than infrared radiation and visible light, which can cause skin cancer and damage cells. X-rays and Gamma rays are emitted from radioactive materials that are used in medical imaging and have the highest frequency and lowest wavelength.

The utilization of electromagnetic radiation is a complex and rapidly evolving field. As new technologies are developed, electromagnetic radiation is being used for a variety of applications, including medical treatments such as cancer therapy and new communication technologies such as quantum computing.

Electromagnetic radiation (EMR) creates Electromagnetic Interference (EMI), that can disrupt the functioning of electronic devices and living space. This type of radiation is composed of ionizing waves, which has sufficient energy to ionize atoms, resulting in the formation of ions, which can cause damage to cells and DNA that lead to cancer, as well as other health and environmental risks. Other sources of electromagnetic radiation, such as radio waves and microwaves, as well as ultraviolet radiation, including X-rays and gamma rays, can also cause EMI. The risks associated with EMR and EMI that are also caused by EMR vary depending on the type, environment and level of exposure.

The degree of exposure to electromagnetic radiation is a critical factor in its health effects. Exposure to high levels of EMF over a short period of time can lead to immediate health issues, such as skin burns and cancer. Conversely, lower levels of EMR exposure over a longer period of time may increase the likelihood of developing cancer and other health issues.

The International Agency for Research and Cancer (IARC) has classified electromagnetic fields as potentially carcinogenic to humans in Group 2B, meaning there is some evidence to suggest that exposure to electromagnetic fields can lead to cancer, but further research is necessary to confirm the extent of exposure. Additionally, the IARC has classified ultraviolet radiation in Group 1, meaning there is strong evidence to suggest that ultraviolet radiation exposure can lead to cancer.

Besides the various domestic, industrial, medical and scientific applications, EMR is associated with a range of potential risks, including but not limited to the effects of ionizing radiation, ultraviolet radiation and microwave radiation on living organisms and some electrical and electronic instruments also. Ionizing radiation has been linked to a variety of health issues, such as cancer and cataracts. Ultraviolet radiation, on the other hand, has been linked to skin cancer, premature aging, suppression of the immune system and damage to the body's electrical signals. Radiofrequency has been associated with increased cancer risk, although further research is necessary to determine the exact nature of this risk. Government agencies have established guidelines to reduce the risk of EMR exposure as much as possible. It is essential to adhere to government agencies' guidelines to reduce one's exposure to EMR. These guidelines can include limiting cell phone usage, maintaining a safe distance from power lines and implementing safe operating procedure with other personal care for daily exposure.

EMI (Electromagnetic Interference) is a disturbance that affects an electrical circuit due to the presence of an electromagnetic field. EMI can be caused by a variety of sources, including radio waves, microwaves, and X-rays. It can also be caused by man-made sources, such as power lines, cell phones and computers. EMI can cause a variety of problems in electronic equipment, including interference with signal reception, data corruption and equipment failure. It can also be a health hazard, as exposure to high levels of EMI can cause headaches, dizziness and nausea. Textiles which are not only limited to aesthetics and clothing but also can be used to protect from EMI by blocking or absorbing EMR and EMI.

In the present scenario, electromagnetic radiation and waves become increasingly detrimental to industries, health and well-being. Textile materials find their application to protect human beings, domestic, industrial or special area entraps like laboratories, etc., from EMR and EMI by providing a shielding effect by absorbing and conducting or reflecting the radiation and the magnetic interference caused. To make textiles more conductive, metal particles or fibers can be incorporated into the fabric and coatings of metal or conductive polymer can be applied. Specialized textiles for electromagnetic shielding offer a range of advantages over traditional materials for EMI shielding, such as their lightweight and flexibility, which can be easily incorporated into clothing or other textiles, as well as their low cost. This chapter delves into the evolution of manufacturing methods, from the rhythm of looms to the accuracy of additive manufacturing and delves into the intricate weaving of conductive fibers/fabrics, as well as the emergence of innovative methods such as printing and coating.

2 Threads of Defense: Materials in Electromagnetic Shielding Textiles

Textiles for electromagnetic shielding are the result of a new paradigm that combines textiles with technology. These textiles are designed to provide functionality and versatility through the use of carefully selected materials. In a world characterized by the proliferation of electronic devices, wireless communication and the need to protect against electromagnetic interference, the need for shielding has become increasingly important. In the past, the solution to this problem was to use rigid metal enclosures or protective shields, which could impede the mobility and ease of use that modern society requires.

EMI shielding materials are designed to protect individuals, devices and other objects from electromagnetic radiation by either absorbing or reflecting it, often a combination of both. This type of electromagnetic shielding is primarily used to reduce exposure to electromagnetic radiation, particularly in the cyber security context, where it is necessary to protect electronic equipment from external electromagnetic disruptions. Textile fabrics are often chosen for their flexibility, drapability, lightweight and relative thickness.

The secondary electromagnetic irradiation caused by the reflection of conductive materials is often seen as the optimal mechanism for measurement of electromagnetic shielding [2]. This is becoming increasingly important due to the proliferation of emitters and new standards, such as those for medical electrical equipment. Generally, testing of shielding materials is conducted at frequencies ranging from 104 to 1012 Hz, depending on the type of equipment being tested, such as power lines, motors or computers [3]. Other experiments focus on low frequency or quasi-quantitative measurements, such as those used for magnetic resonance tomography. Different physical properties of shielding materials are required for different frequency ranges, such as magnetic properties for static magnetic fields, grounded conductive properties for static electric fields and conductive properties for high-frequency electromagnetic fields.

Electromagnetic shielding textiles depend on the selection of materials that possess unique electrical properties. Conductive materials are of particular importance, including metallic fibers (such as silver, copper and other metals), conductive polymer materials and even certain types of graphene, which act as conductive paths that effectively redirect and repel electromagnetic waves, thus preventing them from entering sensitive areas. In addition, the material selection has a direct impact on the shielding effectiveness.

The shielding efficiency SE represents the amount of shielding due to reflection, absorption and multiple-reflection within the shielding material [2].

The Shielding Efficiency (SE) can be expressed as

$$\mathrm{SE} = 20 \log T^{-1}$$

In other terms Shielding Efficiency can be defined as

$$SE = 10\log(P_0/P_t)$$

where T is the transmission coefficient of the radiation with P_0 and P_t as power of the radiation without shielding and with shielding respectively.

The transmission coefficient of an electromagnetic wave is defined by

$$T = (E_t/E_0) \quad \text{which is equivalent to} \quad T = (H_t/H_0) \tag{1}$$

where E_0—represents the electric field intensity without shielding, E_t—represents the electric field intensity with shielding, H_0—represents the magnetic field intensity without shielding, H_t—represents the magnetic field intensity with shielding.

As previously discussed, EMI shielding relies on reflection, absorption and multiple reflections within the shielding fabric. Therefore, materials with high electrical conductivity and magnetic permeability are advantageous for EMI shielding, while those with high absorption losses necessitate the presence of electric or magnetic dipoles. The thickness, porosity and number of conductive, magnetic and dielectric material components of a textile will naturally influence its EMI shielding properties. Additionally, multiple reflections must be taken into consideration for thin shielding layers with a thickness equivalent to skin depth.

The Shielding Efficiency is typically expressed as

$$\text{SE} = \left[20\log(1/T)\right] = \left[20\log(E_0/E_t)\right] = \left[20\log(H_0/H_t)\right] = \left[10\log(P_0/P_t)\right] \tag{2}$$

where P_0 and P_t being the power of the measurement without and with shielding [4].

The Shielding Efficiency SE is a combination of Shielding caused by Reflection (R), Absorption (A) and Multiple Reflection (M).

$$\text{SE} = \text{SE}_R + \text{SE}_A + \text{SE}_M \tag{3}$$

The high-frequency characteristics of shielding materials are determined by the electrical conductivity of the material and its thickness. To measure the conductivity of textile fabrics, a more complex setup is required, as the appropriate contact between the fabric and the measurement equipment is challenging. This can be overcome, for example, by using four-probe measurement, well-fitting contact clamping, standard pressure and additional solder lines. The conductive properties of textiles are of great importance in many smart textile applications and thus are frequently measured and improved. Magnetic properties are also more common, as they are correlated to fewer applications. Superconducting Quantum Interference Devices (SQUIDs), Alternating Gradient Magnetometers (AGMs), Vibrating Sample Magnetometers (VSMs) and the like are commonly used to measure magnetic properties [5]. However, magnetic measurements, such as the Magneto-Optically induced Kerr Effect (MOKE) are difficult on rough surfaces and have become unavailable for textile materials [2].

The EMI shielding of various materials can be investigated using a variety of methods, including open field (free-space), shielded-box and shielded-room methods. These methods can be applied in a variety of frequency ranges [6]. This may require varying levels of time and equipment, as well as the use of coaxial transmission-lines, for example, according to *ASTM D4935* standard.

In order to provide EMI shielding, there are various types of textiles that can be specially treated. The most common types of textiles are metal-coated, carbon-based and polymer-coated. Metal-coated fabrics are composed of a thin layer of copper or aluminum, which reflects electromagnetic waves, thus preventing them from entering the fabric [7]. Carbon-based textiles are composed of carbon fibers, carbon nanotubes and other materials that are highly conductive. Polyaniline and polypyrrole are two of the most common conductive polymers used in this process.

Electromagnetic shielding textiles are composed of conductive fibers, which are often combined with traditional textile fibers such as cotton or polyester. The conductivity of metals such as silver and copper allows them to effectively absorb or reflect electromagnetic waves, thus providing a barrier against EMI [8]. Conductive fibers can be incorporated into textiles using established textile manufacturing techniques, allowing them to retain their existing tactile and aesthetic characteristics while also providing additional functionality. The combination of conductive materials and textiles is not only about functionality, but also about creating a fabric that is both comfortable and practical. Conductive fibers are flexible and malleable, allowing them to conform to textile materials in a smooth and pliable manner, which is especially beneficial for textiles that are intended for wearables, medical garments or interior furnishings.

Material science has advanced to the point where conductive polymers can be incorporated into electromagnetic and electromechanical shielding textiles [1]. Not only do these polymers provide electrical conductivity, but they also offer increased flexibility and decreased weight. Furthermore, these polymers can be applied to fabrics or combined with existing textile materials, providing designers and engineers with a variety of options for creating textiles that are efficient and adaptable.

As nanotechnology advances, so do the opportunities for materials in electromagnetically shielding textiles. Nano-based materials, such as Carbon nano-tubes and Graphene, have sparked interest in their use in shielding textiles [9]. They possess remarkable conductivity and mechanical characteristics and their incorporation into textiles could lead to the emergence of lightweight, highly efficient shielding fabrics.

2.1 Conductive Fibers and Conductive Textiles: The Weave of Innovation

Conductive fibers are revolutionizing material science by providing an essential building block for conductive textiles. These fibers are designed to have extraordinary electrical conductivity, allowing them to transmit and spread electrical energy or signals efficiently [10]. The most notable types of conductive fibers include those infused with metals such as silver, copper and nickel. These metals' inherent conductivity transforms them from decorative items to essential components that can be used to conduct electricity along the fiber's length. When these conductive fibers are woven into textiles, the process of weaving is characterized by a delicate balance between the electrical characteristics of the conductive material and the mechanical properties of textiles. The result is a fabric that is flexible and powerful, capable of transmitting electricity while still being comfortable to the touch [11].

The potential of conductive fibers and textiles is yet to be fully realized, as researchers and engineers struggle to ensure consistent conductivity across fabrics, maintain durability over multiple wash cycles and address cost and scalability issues. To address these challenges, advances in textile manufacturing such as additive manufacturing and nano-coatings have been developed, pushing the limits of what can be achieved with conductive fabrics. As the technology landscape continues to evolve, advances in nanotechnology have the potential to further refine conductive fibers, allowing them to become more conductive, flexible and lightweight. This integration with other emerging technologies, such as flexible electronics and the IoT (Internet of Things), is expected to revolutionize the way we experience the world.

Conductive fibers are a broad range of materials that are capable of conducting electricity. These fibers are used in a variety of applications, such as electronics and wearable technology. To illustrate the range of uses for conductive fibers, a list is provided:

Metal-Coated Fibers such as Silver-Coated Nylon (SCN) is a type of nylon fiber that has been coated with a thin layer of silver. This coating provides a high degree of electrical conductivity, as well as flexibility, making it a popular choice for smart textile and EMI shielding applications. Copper-coated fibers offer similar electrical conductivity to SCN, but are typically used in applications that are concerned with weight. Fiber coated with nickel offers a combination of the conductive properties of metallic coatings with the durability and elasticity of nickel. Nickel-coated fibers are widely used in electronics and sensors, as well as in EMI shielding applications. Like other types of metal-coated fibers, zinc-coated fibers offer electrical conductivity and resistance to corrosion. They are employed in applications where oxidation resistance is of paramount importance. Metal fiber blends are a combination of traditional textile fibers and metal fibers (for example, stainless steel), resulting in textiles with integral conductivity. Such textiles are commonly used for Electromagnetic shielding and intelligent textiles.

Carbon-based fibers are composed of carbon nano-tubes, which are characterized by their high conductivity and light weight, making them suitable for a variety of applications, such as flexible electronics and textiles that are conductive [12]. Graphene is a single layer composed of carbon atoms that has remarkable electrical properties, making it an ideal material for the production of carbon-based electromagnetic interference shielding materials [13]. Carbon fiber composites are composed of carbon fibers, which are conductive in themselves, and can be made partially or fully conductive through the use of non-conductive and conductive resins or coatings respectively. Carbon fiber composites are commonly used in lightweight conductive materials. The polymer-carbon nano-composite (PCN) fibers are composed of polymer matrices that have been embedded with carbon nano-structures, such as carbon nano-tubes and graphene [14]. These fibers provide a combination of polymer flexibility and the conductivity of nano-materials. Silver-coated carbon fiber is a combination of silver's electrical conductivity and carbon fiber's mechanical properties. These hybrid fibers are highly versatile and can be used in a variety of applications, including aerospace, automotive and medical devices. Graphene-based fibers can be used to improve the conductibility and mechanical strength of textiles.

Certain polymers are capable of becoming intrinsically conductive when dosed with conductive compounds [15]. Examples of such polymers are polyaniline and polythiophene, which can be spun to form moderate conductivity fibers. Polymer-metal hybrid fibers are composed of materials that are encapsulated in polymer sheaths, allowing for the encapsulation of conductive metals. This combines metal conductivity with polymer flexibility, resulting in fibers with distinct properties, which can be used in a variety of applications, such as flexible electronics. Intrinsically Conductive Polymer (ICP) fibers, on the other hand, do not require any additional doping to maintain conductivity, unlike traditional polymers. ICPs are composed of polymers that are woven into fabrics that possess intrinsic electrical conductivity.

Traditional fibers can be coated with a variety of conductive nano-materials, including nanoparticles and nanowires, in order to impart conductivity without compromising textile properties. Additionally, polymers or polymer composites can be applied to the fibers to further enhance their conductivity while maintaining flexibility. Silver nanowires, for example, can be incorporated into fibers to create flexible and clear conductive materials for use in touch screens or flexible displays. Besides these, some natural fibers, including cotton and silk, can be coated or impregnated with conductive materials to become conductive. These fibers offer a combination of eco-friendliness and conductivity, making them suitable for a variety of applications.

Polymer-embedded fibers are composed of polymers that have been embedded with conductive filler, such as a carbon black or metallic particle. These fibers can be spun to provide a combination of elasticity and conductivity, while hybrid fibers can be formed from a combination of various conductive materials. Concrete polymers can be processed to produce conductive fibers, while conjugated polymers, such as polyacetylene, have a unique electronic structure that allows them to exhibit considerable electrical conductivity.

This special type of conductive fiber illustrates the wide variety of conductive textile materials available. As technology advances, the incorporation of these conductive fibers into a variety of applications promises to revolutionize the manner in which we interact and benefit from electronic devices and intelligent textiles. Furthermore, these special conductive fibers illustrate the broad range of innovation that is taking place within the conductive materials domain. As the technology advances, these conductive fibers are expected to remain a key factor in the transformation of industries from electronics to health care and beyond. The shielding efficiency of specially treated textiles is contingent upon a variety of variables, including the composition of the textile, the viscosity of the metal or conductive coating and the electromagnetic wave frequency. Generally, thicker metal coating and conductive polymer viscosity are associated with higher shielding efficiencies. The shielding efficiency also increases with frequency.

Liquid metal-filled fibers are composed of liquid metal alloys which retain their conductive properties even when they are in a flexible or deformable state. These fibers are suitable for applications that necessitate shape-altering circuits and wearable sensors, as well as for applications that require both elasticity and electrical conductivity. Elastomeric conductive fibers, which are filled with conductive materials, are utilized in electronics that require flexibility and comfort. The range of conductive fibers available demonstrates the range and versatility of the materials that can be incorporated into textiles to provide conductivity. The choice of the most appropriate conductive fiber is contingent upon the specific application's needs in terms of conductivity, elasticity, durability and other performance criteria. Textile fibers are the foundation upon which the effectiveness, versatility and comfort of the electromagnetic shielding textiles are based. The choice of conductive materials (e.g. metallic fibers or conductive polymers) or novel nano-materials will determine the effectiveness of the textiles in reducing electromagnetic interference. As technology advances, materials science will be essential in refining and advancing the electromagnetic shielding textile, ensuring that they serve as both guardians of security and agents of innovation in a rapidly evolving world.

3 Designing for Disruption: Creative Patterns in Electromagnetic Shielding Textiles

The integration of aesthetics and function into the design of textiles has enabled the emergence of new paradigms for design innovation. These textiles, which are capable of deflecting and absorbing electromagnetic waves, are no longer limited to utilitarian designs, but instead offer a range of creative patterns that enhance both visual appeal and shielding efficacy. The content examines how these intricate patterns are not only aesthetically pleasing, but also contribute to the efficacy of the textiles.

At the core of this development lies the identification of patterns in electromagnetic shielding textiles that can be used to improve their performance. The days of shielding effectiveness being limited to materials and manufacturing methods are long gone. The advent of complex patterns has opened up a new realm of possibilities, enabling engineers and designers to create a protective symphony that is both practical and aesthetically pleasing.

In the past, the effectiveness of electromagnetic shielding textiles was largely determined by the materials and manufacturing methods used. However, with the identification of creative patterns that can significantly affect their performance, a new field of investigation has opened up. Patterns are not merely decorative elements, but are the means by which electromagnetic waves pass through conductive materials and influence their behavior in fundamental ways.

Electromagnetic shielding textiles are the result of a combination of scientific knowledge and artistic creativity. The arrangement of the conductive elements in the textile design process has a major influence on the effectiveness of the shielding. Frequency Selective Surface (FSS) pattern is a type of electromagnetic shielding that utilizes the principles of resonance to filter out particular frequencies of electromagnetic wave. The pattern is composed of conductive elements arranged in specific geometries. This pattern is capable of blocking undesired frequencies while allowing for the transmission of desired signals without interference. FSS pattern, for example, utilizes resonance to permit or block certain frequencies of electromagnetic radiation, while allowing signals to pass through unobstructed. This approach has sophisticated combination of pattern design and science which is used in a variety of concepts and applications, ranging from telecommunications to wireless communication systems, aeronautics and military applications where customized electromagnetic shielding is essential that requires frequency filtering with tailored shielding.

These textiles have been embraced by interior designers and architects due to their ability to shape spaces. The textiles with built-in electromagnetic shielding capabilities are used as a versatile tool for architectural decoration. These textiles can be used as curtains, wall dividers and wall coverings. Not only do they provide protection, but they also blend in with interior motifs. The patterns used can range from delicate abstractions that are perfect for contemporary interiors to complex designs that convey a timeless elegance, thus combining the function and the aesthetics.

Creative patterns in shielding textiles not only provide protection, but also provide a visual experience that evokes curiosity and amazement. Digital design tools have enabled designers to explore complex weaves and motifs that not only captivate the eye, but also serve functional purposes. Electromagnetic shielding textiles have also been adopted by architects and interior designers. Building designs often feature textiles with built-in shielding capabilities as curtains, dividing walls or wall covering. This discreet incorporation of protection into the architectural design allows for creative patterns to be incorporated into the design. Patterns that imitate traditional designs or contemporary abstractions complement interior motifs while providing the additional benefit of reducing electromagnetic interference.

The pursuit of novel patterns for electromagnetic shielding textiles is accompanied by challenges, as is the case with any creative endeavor. Researchers and designers must confront a range of challenges in order to incorporate creative patterns into electromagnetic shielding textile designs. These challenges include ensuring consistent shielding performance over a wide range of frequencies, maintaining durability during production and addressing scalability and budget concerns. Research and development efforts are focused on ensuring consistent shielding performance over a wide range of frequencies, preserving durability through production processes and addressing scalability and cost considerations. Looking ahead, the potential of creative patterns for shielding textiles is limitless. As technology progresses, computational tools may be able to optimize patterns for particular shielding requirements. By combining pattern design with cutting-edge materials such as meta-materials and nano-composites, shielding textiles may be able to achieve unprecedented shielding efficiency and visual appeal.

The story of the creative patterns found within electromagnetic shielding textiles tells a story of transformation, in which functional requirements and artistic expression merge seamlessly. As we move forward into a technologically complex world, these patterns serve as a testament to humanity's ingenuity. They give textiles a dual purpose—to protect against electromagnetic interference and to create works of art that evoke awe. The development of these patterns contributes to the enrichment of both textiles and technology, while also demonstrating the capacity of creativity to create a harmonious combination of function and beauty.

To sum up, the emergence of innovative patterns within electromagnetic shielding textiles has blurred the boundaries of traditional textile design, creating a new era in which functionality and aesthetics are inextricably linked. These patterns have gone beyond mere decoration, becoming portals of both aesthetic pleasure and electromagnetic security. As technology advances, the combination of innovative design and functional innovation is likely to create a future in which textiles are not merely custodians of electromagnetic peace, but also marvels of human ingenuity.

4 Manufacturing Techniques for Electromagnetic Textiles

Electromagnetic shielding textiles can be realized through the use of manufacturing techniques that bridge the conceptual design and practical application of the textiles. The combination of traditional textile techniques and modern technologies has resulted in a wide range of manufacturing methods for the production of these specialized textiles. This part outlines various manufacturing techniques that can be used to produce electromagnetic textiles.

The spinning process allows for the direct incorporation of conductive fibers into yarns. This process results in yarns that are integrated with conductivity, allowing for the weaving or knitting of these into fabrics. Additionally, electro-spinning allows for the generation of ultrafine fibers through the electrically charging of a polymer solution, followed by the drawing of the polymer into fine threads, allowing for the production of conductive nano-mats that offer a large surface area and improved conductivity. Further, melt spinning involves the melting of conductive polymer compounds or polymer mixtures and the extrusion of these materials through spinnerets, resulting in the formation of conductive fibers that can be integrated into textiles.

The majority of electromagnetic shielding textiles are derived from traditional weaving techniques, which involve the weaving of conductive and non-conductive fibers, such as silver-coated and copper-coated, through looms. This method allows for the production of fabrics that are both flexible and durable, while also incorporating integrated conductivity into them. Knitting is a particularly versatile and adaptable technique for producing electromagnetic shielding fabrics, as it allows for the integration of conductive fibers into knitted garments, resulting in garments that not only protect against electromagnetic interference, but also offer comfort and durability [16]. Additionally, embroidery involves the stitching of conductive threads onto fabrics, allowing for the creation of intricate patterns or even whole circuits, which is suitable for applications that require precision, including flexible circuits and smart textiles.

Textiles can be incorporated with electromagnetic shielding properties through the use of dyeing, printing and additive manufacturing techniques. Dyeing involves the application of metal particles containing conductive dyes to textiles, which preserves the look and feel of the fabric while providing functionality. Printing involves the use of conductive inks to precisely place conductive elements on fabrics, while inkjet coating involves the use of inkjet printing technology to precisely deposit conductive inks on textile surfaces. Additive manufacturing, on the other hand, involves the use of 3D printing to create complex three-dimensional structures with precise electromagnetic properties.

Coating is a distinct method of applying a textile substrate with conductive material, such as polymers or metals. This method can be applied in a variety of ways, such as dip coating, spray coating and hot melt coating. Dip coating involves immersing the textile substrate into a solution with conductive particles, or materials and then drying or curing the fabric. Spray coating, on the other hand, involves the application of a fine-sprayed conductive material to the textile surface. Lamination is a method of bonding conductive layer/films or fabrics onto traditional textiles, allowing for the addition of shielding capabilities or the creation of multilayered fabrics with improved performance [17] (Figs. 1, 2).

Plasma treatment is a process of altering the surface characteristics of fibers or fabrics by exposing them to vacuum or atmospheric plasma where textile surfaces are exposed to a plating environment in which conducting molecules are deposited on the textile surface [18]. This process has the potential to enhance adhesion and conductibility, thereby improving the electromagnetic shielding properties retention of textile properties. Sol–gel coating is a chemical process in which conductive

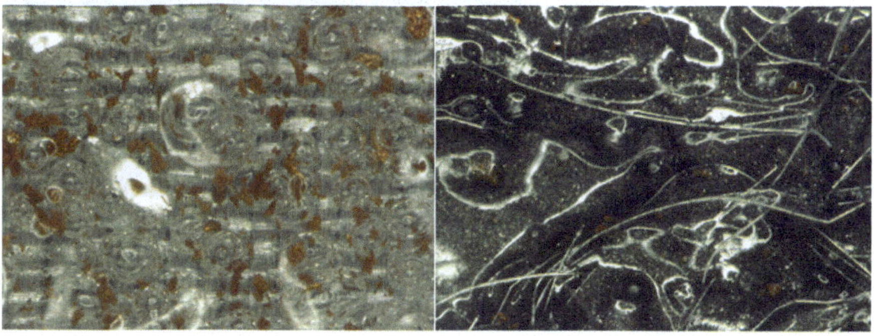

(a) (b)

Fig. 1 Optical microscope image of the silicone coating with 10% copper content on polyamide fabric (**a**) and with 10% graphite, copper and steel fibers on polyamide fabric (**b**)

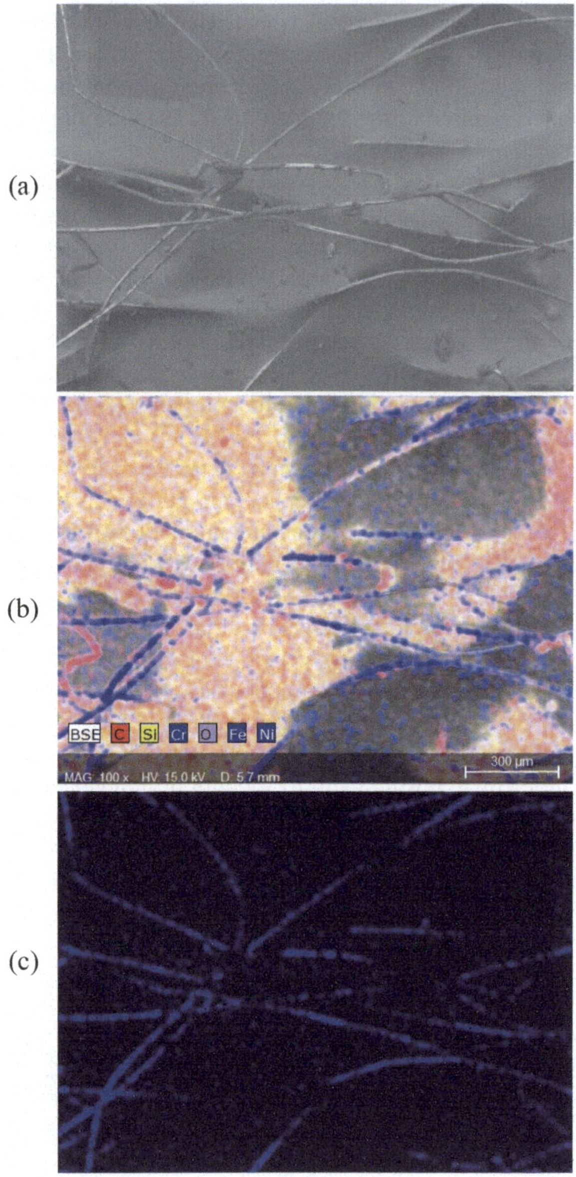

Figure 2 **a** Electron microscopic image of a silicone coating with embedded steel fibers, **b**, **c** the elemental composition on the sample surface determined with EDX

material is deposited on textile substrates. This process allows for precise control of the coating thickness and can provide uniform shielding performance.

Needle punching is a method of mechanically connecting conductive fibers to non-conductive fibers using barbed needles, resulting in three-dimensional structures that are more conductive and durable. Thermal bonding, on the other hand, involves the use of heat and pressure to join conductive fibers together, resulting in fabrics that have integrated shielding properties [19]. This process is often employed in the manufacture of non-woven electromagnetic shielding fabrics [13]. Ultrasonic welding involves the use of high-frequency vibration and pressure to attach conductive components to textiles, making it suitable for the formation of reliable and long-lasting connections in intelligent textiles.

Electro-spraying is a process that is similar to electro-spinning in which an electric field is applied to the conductive solutions, resulting in the formation of fine droplets that are deposited onto textile substrates. This method is suitable for the production of fine, controlled-thickness conductive coatings. Electrochemical deposition also called as electroplating is the chemical process of coating a material under a solvent using electric potential in which conductive materials, preferably metals are deposited onto textiles, resulting in conformal, durable and uniform coatings with a controlled thickness (Fig. 3).

Hybrid integration is the combination of manufacturing processes to produce high-quality electromagnetic shielding textile. For instance, the weaving of conductive fibers into a fabric, followed by the application of conductive coating, can lead to textiles with improved shielding properties (Fig. 4).

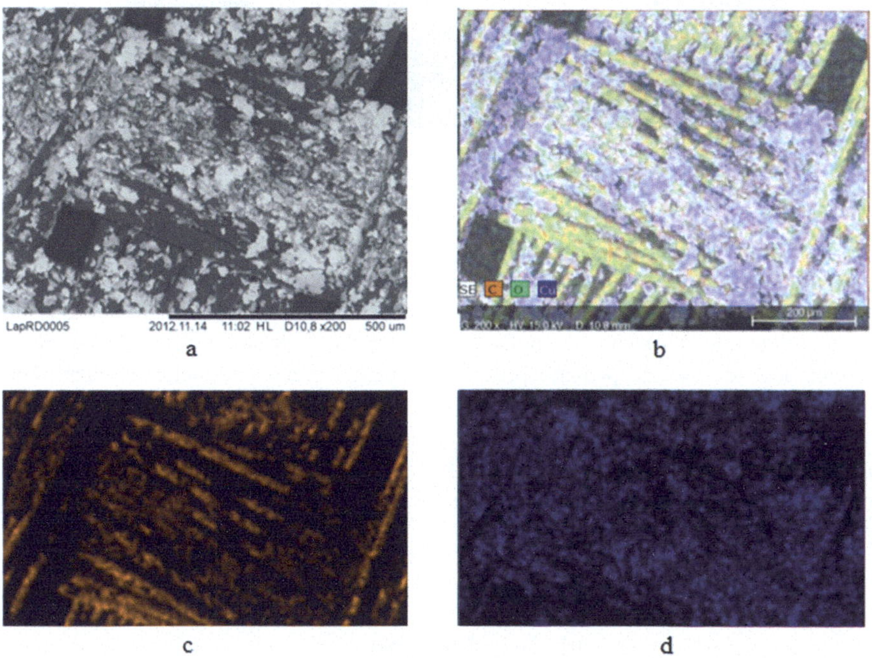

Figure 3 **a** Electron microscopic image of copper and graphite platelet pigments on polyester fabric, **b** EDX analysis of the pigment plated polyester fabric, **c** the elemental composition on the sample surface determined for graphite with EDX, and **d** the elemental composition on the sample surface determined for copper with EDX

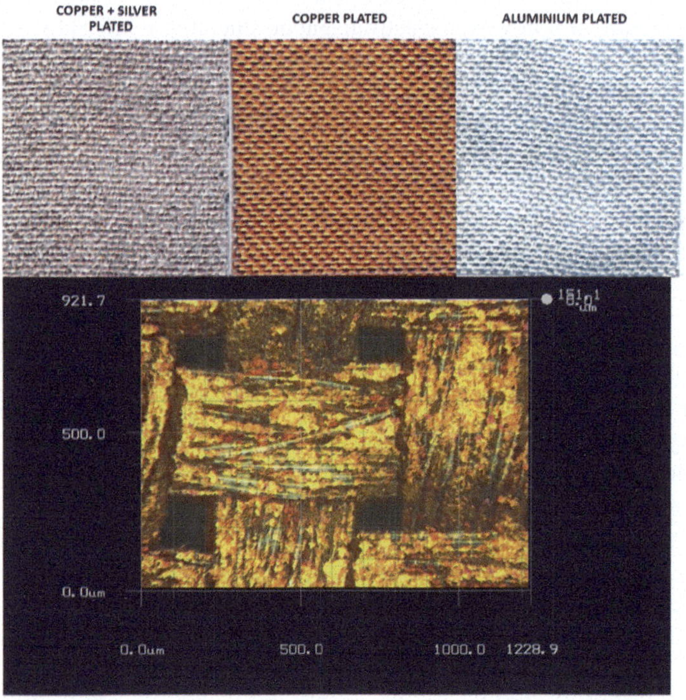

Fig. 4 Polyester fabric with different types of platelet pigments, along with an optical micrograph of the sample with copper pigment

All these manufacturing techniques illustrate the versatility and contribute to the development and production of electromagnetic shielding textiles, which can be incorporated into a variety of applications, ranging from wearable technology to architectural textiles, to provide protection against electromagnetic damage. The choice of a particular technique is based on considerations such as desired conductivity, elasticity, scalability and intended application. Taken together, they showcase the ingenuity and versatility of electromagnetic shielding textiles, which can be tailored to meet the specific shielding needs and design preferences of a variety of applications.

5 Incorporating Shields into Fashion

Electromagnetic shielding textiles have a wide range of applications in a variety of industries due to their capacity to reduce Electromagnetic Interference (EMI). This topic outlines the most prominent applications of EMS in clothing industry and fashion. Electromagnetic shielding textiles have a broad range of applications and can be used to protect sensitive systems and enhance user experiences in a variety of industries. As technology advances, these textiles play an essential role in reducing electromagnetic interference, from protecting human health to guaranteeing the dependability of essential infrastructure and advanced technology. The application of wearable EMI shielding textiles is expansive and has a wide range of applications. Medical attire with integrated shielding provides protection for those in the healthcare sector, while military and defense sectors where electromagnetic security is paramount.

In today's world, where electronic devices are ubiquitous in our lives, it is essential to protect ourselves from EMI. Wearable electromagnetic shielding textiles are revolutionary solutions that combine technology and fashion. Not only do they provide protection from EMI exposure, but they also integrate seamlessly into our daily outfits. This combination of protection and fashion is a major step forward in the development of both textile and wearable technologies. Wearable technology has evolved beyond simply wearing a smart watch or fitness tracker to include textiles that are actively contributing to our health. Wearing a wearable EMI shielding fabric not only protects us from potential health risks, but also highlights the potential of fashion as a means of promoting a healthier lifestyle.

In wearable technology, Electromagnetic shielding textiles are incorporated into garments to shield the wearer from potentially hazardous electromagnetic radiation emitted by electronic devices such as smart phones, tablets, smart watches and more. Electromagnetic shielding textiles are incorporated into smart clothing, wearable sensors and other electronic devices to protect them from external electromagnetic interference. This allows for the accurate collection and communication of data in a variety of applications, such as fitness and health tracking. Additionally, Electromagnetic shielding textiles enabled accessories, such as laptop sleeves and camera bags, as well as RFID blocking wallets, use Electromagnetic shielding textiles-enabled textiles to prevent unauthorized access to or interference with sensitive electronic devices. In aerospace, Electromagnetic shielding textiles are employed to protect essential electronic systems from external EMIs, thus ensuring the safe functioning of aircraft avionic and communication systems.

Textiles with an electromagnetic shielding property are incorporated into building materials in order to create areas free of electromagnetic interference. Examples of this type of material include hospitals, research laboratories and other settings where exact electronic equipment is being operated. In consumer electronics, these textiles help to protect the device and user from the harmful effects of electromagnetic radiation. For this purpose, they are employed in power plants, electrical substations and other areas where electromagnetic emissions are to be contained. These textiles are also used as enclosures in electromagnetic compatibility testing chambers, in which the electromagnetic emissions of electronic devices are tested. In a hospital and diagnostic facilities, Electromagnetic shielding textiles are used to protect medical equipment from external EMIs. Healthcare professionals can work safely in an environment full of electronic devices by wearing Electromagnetic shielding textiles such as gowns, caps and aprons.

Electromagnetic shielding textiles are used in automotive applications to reduce the amount of electromagnetic interference in the electronic systems of vehicles, thereby enhancing the performance of communications, navigation systems and entertainment. These textiles are also employed in the design of communication facilities and server rooms, as well as data centers, to reduce the risk of electromagnetic interference which could interfere with signals and data transmission. In military applications, these textiles are utilized to shield sensitive equipment and to prevent the interference of communication systems from eavesdropping.

Additionally, military personnel use electromagnetic shielding textiles enriched in uniforms, equipment and other garments to protect against external sources of electromagnetic interference and to maintain the effectiveness of communication and navigational systems. Finally, electromagnetic shielding fabrics can be used to create Gaskets and Seals, which prevent the entry or escape of electromagnetic interference into or out of enclosures.

Electromagnetic shielding textiles are utilized in research environments to create a controlled electromagnetic environment for precise and reproducible experimental results. Electromagnetic shielding textiles are used as a test material in EMI testing facilities to assess the sensitivity of electronic equipment to an external electromagnetic field. Wearable electromagnetic shielding textiles can help reduce electromagnetic interference that may affect the performance of Virtual Reality (VR) and Augmented Reality (AR) devices, thereby improving user experience in immersive environments.

The remarkable feature of wearable EMI shielding textiles is their flawlessness, allowing for a seamless transition from traditional textiles to practical protection. This is achieved through the incorporation of conductive fibers, metals or coatings in the fabric structure, utilizing cutting-edge manufacturing techniques. This allows for clothing to appear and feel like any other while discreetly shielding us from the electromagnetic noise that permeates the environment (Fig. 5).

Design and function go hand in hand and wearable EMI shielding textiles offer a variety of innovative patterns, designs and textures to suit a wide range of tastes. From delicate patterns that are perfect for formal wear to bold motifs that are perfect for casual wear, these textiles provide a personalized statement that resonates with individual fashion tastes. In the consumer world, these textiles have found their place in everyday apparel, allowing individuals to remain connected without sacrificing their health. Smart garments equipped with sensors that monitor vital signs, stress and posture are ushering in a new age of proactive health management that combines technology and textiles.

Despite the immense potential of wearable EMI shielding textiles, obstacles remain. These obstacles include consistent shielding performance between different fabrics, durability in the face of washing and daily wear and comfort and breathing considerations. To overcome these obstacles, advancements in conductive materials, textile engineering and manufacturing techniques are necessary. The emergence of wearable electromagnetic shielding textiles is revolutionizing the fashion and technology worlds. As the world becomes more interconnected, the combination of protection and fashion in these textiles is a testimony to human creativity. These

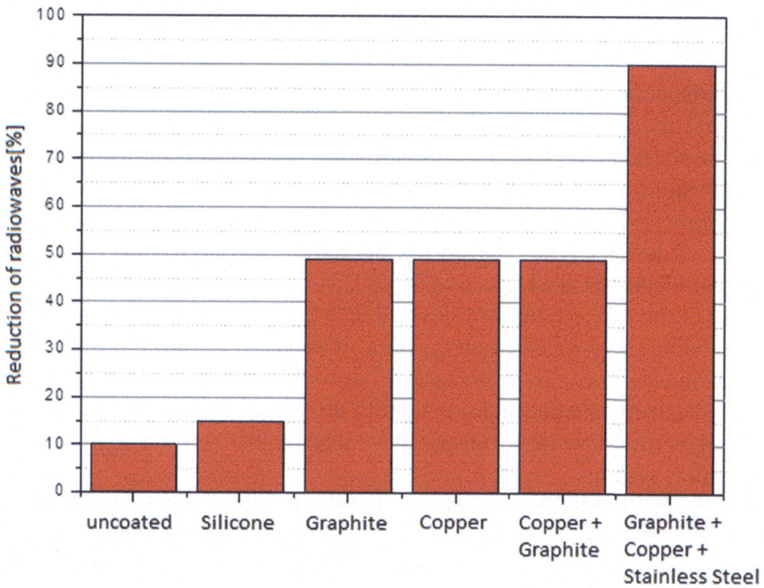

Fig. 5 Reduction of electromagnetic radiation in the radio wave range through uncoated polyamide fabric and a comparison of polyamide fabric with various modified silicone coatings. Loading level with the various additives amounts to a total of 10%

textiles reject the notion that protection detracts from appearance, instead providing a harmonious amalgamation of both. By wearing wearable EMI shielding textile, individuals are embarking on a journey of fashion and function, creating a world in which personal style and well-being are interwoven into one.

6 Beyond Apparel: Electromagnetic Shielding Textiles in Industries

The emergence of electromagnetic shielding textiles has been instrumental in addressing the widespread problem of electromagnetic interference across a wide range of industries. As the prevalence of electronic devices and their interconnectedness has increased significantly, EMI has become a major threat to operations and the security of sensitive equipment. Electromagnetic shielding textiles are a versatile defense mechanism that can be used in a variety of industries where dependability and protection from electromagnetic interference are essential.

The use of electromagnetic shielding textiles in the healthcare sector is essential for the protection of medical equipment. EMI can cause inaccurate readings or equipment malfunctions when used on sensitive devices such as Magnetic Resonance Imaging (MRI) machines, Clinical Targeting scanners (CT scanners) and

Patient Monitoring Systems (vitals monitoring devices). Protective attire made of shielding textiles allows healthcare professionals to operate these devices in a non-electromagnetic environment.

Electromagnetic shielding textiles are essential for the aerospace industry, as they are used to protect avionics and essential electronic components from outside interference. Furthermore, in industrial applications, appliances, such as microwaves, fridges and TVs, employ electromagnetic shielding textiles enhanced fabrics to contain electromagnetic emissions generated during operation, thereby reducing interference with adjacent electronic devices.

The core of the telecommunications and data center infrastructure, where a large number of electronic devices come together, is prone to disruption caused by electromagnetic interference. To protect sensitive equipment from outside electromagnetic noise, electromagnetic shielding textiles are employed to construct enclosures, curtains and partition walls to ensure the integrity of the data transmission and communication system.

As vehicle technology advances, the need for Electromagnetic Compatibility (EMC) has become increasingly important. EMC utilizes the advantage of electromagnetic shielding textiles that are incorporated into automotive interior design to eliminate electromagnetic interference that occurs from outside and reduce the Electromagnetic interference within the vehicle's electronic components. This application improves the functionality of communications, entertainment and safety features while preserving the comfort of passengers.

The defense sector depends on reliable and secure communications. Military uniforms, equipment and enclosures use electromagnetic shielding textiles to protect against electromagnetic interference that could disrupt communications, navigation and monitoring systems, ensuring operational performance in demanding environments. Furthermore, in transportation systems such as trains and subways, electromagnetic shields can be used to protect communications and control systems, thus ensuring the smooth operation of these systems and the safety of passengers.

As the consumer electronics sector continues to expand, the need for protection against electromagnetic interference and measurement of it has become increasingly important. In order to ensure the accuracy of experimentation in laboratory settings, it is essential to create conditions that are interference-free. This is achieved through the use of electromagnetic shielding textiles, which are used to enclose and enclose scientific equipment in chambers that are impervious to external electromagnetic noise. This helps to ensure the accuracy and reliability of research results.

Power stations, electrical substations and utility infrastructure are vulnerable to EMIs that could interfere with control systems and equipment, thus necessitating the use of electromagnetic shielding textiles to construct enclosures and barriers to contain electromagnetic emissions and ensure reliable operation and the protection of critical monitoring systems.

Additionally, the architectural industry can benefit from the incorporation of electromagnetic shields in building materials, as they can be used to create electromagnetic interference-free zones in various facilities and laboratories, as well as in research centers, to guarantee the functioning of sensitive electronic equipment without disruption.

Electromagnetic shielding textiles play an essential role in maintaining the performance, safety and dependability of essential electronic systems across a wide range of industries. As technology advances, shielding textiles will continue to play a critical role in achieving smooth operations and uncompromising performance.

7 Future of Electromagnetic Shielding Textiles: Pioneering Protection in a Connected World

Electromagnetic shielding textiles play an essential role in maintaining the performance, safety and dependability of essential electronic systems across a wide range of industries. As technology advances, shielding textiles will continue to play a critical role in achieving smooth operations and uncompromising performance.

Advances in manufacturing methods, such as additive production and digital printing, are expected to lead to the development of tailored electromagnetic shielding textiles that can be tailored to the individual's preferences. Through the incorporation of custom patterns, designs and shielding levels, users will be able to combine protection with individual style. Additionally, architectural textiles will be incorporated into building materials to create smart spaces that protect against unwanted electromagnetic interference, creating secure environments for sensitive devices and human health. Furthermore, electromagnetic shielding textiles are expected to play a major role in healthcare innovations, from protecting devices and equipment, to improving human safety during various processes and procedures [20]. Ultimately, the future of electromagnetic shielding will be focused on human well-being, with wearable shielding garments designed to not only protect against EMI, but also contribute to general health, comfort and quality of life.

As the Internet of Things continues to expand, electromagnetic shielding textiles will become an essential part of the IoT infrastructure [21]. Electromagnetic shielding textiles will provide dependable communication and data transfer across a network of connected devices. Through cross-disciplinary collaboration between engineers in the textile, materials, electrical engineering and design fields, Electromagnetic shielding textiles will be able to create textiles that combine protection, aesthetic and functionality. As the significance of electromagnetic interference protection increases, educational initiatives will be implemented to increase awareness among industries, practitioners and the public. Through training programs and resources, individuals will be able to make informed decisions regarding EMI protection.

The future of electromagnetic shielding textiles lies in a combination of innovation, security and integration into everyday life. From fashion, healthcare, smart cities and space exploration, these technologies will pave the way for a connected world in which technology and human well-being co-exist in harmony.

References

1. Chung DDL (2000) Materials for electromagnetic interference shielding. J Mater Eng Perform 9:350–354
2. Blachowicz T, Wójcik D, Surma M, Magnuski M, Ehrmann G, Ehrmann A (2023) Textile fabrics as electromagnetic shielding materials—a review of preparation and performance. Fibers 11(3):29. https://doi.org/10.3390/fib11030029
3. Singh R, Surwase P (2014) Electromagnetic shielding efficiency tester for textiles. Man Made Text India 42(12):449–453
4. Akman F, Ozkan I, Ozel F, Aydemir H, Kaçal MR, Agar O (2022) Production of barite-doped yarns and testing their usability for ionizing and electromagnetic radiation shielding. Radiat Phys Chem 201:110464. https://doi.org/10.1016/j.radphyschem.2022.110464
5. Liao Z, Zi Y, Zhou C, Zeng W, Luo W, Zeng H, Xia M, Luo Z (2022) Recent advances in the synthesis, characterization, and application of carbon nanomaterials for the removal of endocrine-disrupting chemicals: a review. Int J Mol Sci 23(21):13148. https://doi.org/10.3390/ijms232113148
6. Liu Z, Duan J, Wang X (2023) Shielding performance of electromagnetic shielding fabric implanted with "split-ring resonator." Polymers 15(6):1366. https://doi.org/10.3390/polym15061366
7. Amberg M, Grieder K, Barbadoro P, Heuberger M, Hegemann D (2008) Electromechanical behavior of nanoscale silver coatings on PET fibers. Plasma Processes Polym 5(9):874–880. https://doi.org/10.1002/ppap.200800059
8. Pan Y et al (2023) Efficient electromagnetic interference shielding of three-dimensional hydrophobic Cu/wood/Cu porous composites. J Wood Chem Technol 43(4):206–220. https://doi.org/10.1080/02773813.2023.2213691
9. Rosu G, Velicu V, Boitan A, Mihai G, Tuta L, Baltag O (2022) On the electromagnetic shielding properties of carbon fiber materials. Electr Eng Electromech 1:38–43. https://doi.org/10.20998/2074-272X.2022.1.05
10. Tong H, Wan J, Wang S, Yu Y, Zhang W, Liu R, Huang J (2023) Preparation of woven copper tube and its application in electromagnetic shielding. J Mater Sci Mater Electron. https://doi.org/10.1007/s10854-023-10705-2
11. Quercio M et al (2023) Electromagnetic shielding properties if LPBF produced Fe 2.9 wt% Si alloy. J Phys Energy 5(4):5003. https://doi.org/10.1088/2515-7655/ace92f
12. Xu Z, Dou T, Wang Y, Zuo H, Chen X, Zhang M, Zou L (2023) Three-dimensional-printed carbon nanotube/polylactic acid composite for efficient electromagnetic interference shielding. Polymers 15(14):3080. https://doi.org/10.3390/polym15143080
13. Zhang P et al (2023) Ultra-thin freestanding graphene films for efficient thermal insulation and electromagnetic interference shielding. RSC Adv 13(28):19388–19402. https://doi.org/10.1039/D3RA00638G
14. Nan Z, Wei W, Lin Z, Chang J, Hao Y (2023) Flexible nano composite conductors for electromagnetic interference shielding. Nano Micro Lett. https://doi.org/10.1007/s40820-023-01122-5
15. Kausar A, Ahmad I (2023) Conducting polymer nanocomposites for electromagnetic interference shielding—radical developments. J Compos Sci 7(6):240. https://doi.org/10.3390/jcs7060240
16. Wang D, Hu S, Kremenakova D, Militky J (2023) Evaluation of the wearing comfort properties for winter used electromagnetic interference shielding sandwich materials. J Ind Text. https://doi.org/10.1177/15280837231159869
17. Natarajan H, Alagundi PM (2023) Metal-containing coatings on textiles—concepts and properties. Int J Adv Res Sci Eng Technol 10(10):21156–21163
18. Hegemann D, Amberg M, Ritter A, Heuberger M (2009) Recent developments in Ag metallised textiles using plasma sputtering. Mater Technol 24(1):41–45. https://doi.org/10.1179/175355509X417981

19. Wang D, Hu S, Kremenakova D, Militky J (2022) The electromagnetic shielding effectiveness of the copper plated nonwoven fabric and its related comfort properties. J Eng Fibers Fabrics 17:1–12. https://doi.org/10.1177/15589250221133462
20. Protz K (2005) Moderne Wundversorgung. Elsevier, München
21. Zhang H, Lin S (2023) Research progress with membrane shielding materials for electromagnetic/radiation contamination. Membranes 13(3):315. https://doi.org/10.3390/membranes13030315

Thermal Protective Wear

Steeve Roy and Mahesh Pratap Dubey

1 Introduction

Not only is clothing one of every human being's basic needs but it's also seen as a 'second skin,' an extension of the body. While clothing can offer psychological benefits like self-expression and social identification, its fundamental purpose is to shield wearers from their environment, providing thermal and weather protection, as well as physical comfort [1, 2]. Fourt et al. [3] stated that, beyond these functions, clothing can even offer physiological advantages. By carefully designing features like fabric type, garment construction, and fit, clothing helps regulate the balance between the body's heat production and its release to the surroundings (through dry heat loss or sweat evaporation), keeping wearers within a comfortable temperature range.

Traditional attire has long been recognized for offering psychological, physical, and physiological ease to individuals in various natural surroundings (chilly, moderate, or hot) [2]. However, such clothing is inadequate for individuals working in hazardous ambient environments [4]. Bajaj et al. [5] stated that technological advancements have heightened the risk of human exposure to various occupational hazards, including chemical, biological, nuclear, mechanical, and thermal dangers. Workers in sectors like pharmaceuticals, petrochemicals, fertilizers, and nuclear energy may encounter chemical, biological, and nuclear hazards. Government personnel, such as defense, police, and military personnel, may face mechanical hazards like bullets or knives. Fire-fighters, dealing with flame, radiant heat, hot surfaces, hot liquids, and steam, are exposed to thermal hazards. In order to offer extra

S. Roy (✉)
Department of Textile Technology, Anna University, Chennai, India
e-mail: steeveroy5@gmail.com

M. P. Dubey
Department of Handloom and Textile Technology, Indian Institute of Handloom Technology, Varanasi, India

S. Sundaramoorthy et al. (eds.), *Textile Materials for Good Health and Wellbeing*, SDGs and Textiles, https://doi.org/10.1007/978-981-97-6496-9_16

safeguarding against these dangers, numerous specialized protective garment varieties have been devised and are extensively employed in industrial and governmental fields. These encompass chemical-resistant attire, bio protective gear, radiation shielding clothing, robust protective wear or garments fortified with puncture or cut resilience, and flame-retardant apparel or heat-resistant garments. Thermal protective clothing holds particular significance, given the persistent occurrence of fire hazards in daily life. The likelihood of fire risks requires proficient fire control by fire-fighters to reduce human and financial damages. Consequently, reliable thermal shielding attire is essential for ensuring fire-fighter safety [6].data-gpts-theme="light"/>

2 Factors Influencing the Performance of Thermal Protective Clothing.

Protective clothing, specifically thermal protective clothing, serves as a crucial safeguard for individuals, including fire-fighters and workers in the oil and gas industry, protecting them from diverse thermal exposures such as flame, heat, hot surfaces, molten metals, steam, and hot liquids [7]. Typically, fire-retardant or resistant fiber-based fabrics are engineered into thermal protective clothing to conform to the intricate geometry of the wearers' bodies. Subsequently, numerous aspects within textile engineering, encompassing both fabric and garment considerations, exert a substantial influence on the thermal insulation features of protective clothing. The porous and hygroscopic nature of a fabric reveals a structure comprising two distinct phases: (a) a gaseous phase, encompassing water vapor and/or dry air, and (b) a solid phase, comprising fibers/yarn, and potentially bound water if hygroscopic [8]. These phases alone contribute to the transference of thermal energy through the fabrics. The properties of fibers, yarns, coupled with fabric properties, play a pivotal role in trapping dead air in the woven or nonwoven fabrics. This, in turn, governs the thermal energy transfer through the gaseous phase, thereby influencing the thermal insulation characteristics of the fabrics. These properties similarly dictate the thermal energy transfer through the solid phase of the fabrics, thereby impacting the overall thermal insulation characteristics [9]. Essentially, diverse components within textile engineering, encompassing fiber attributes, yarn qualities, fabric characteristics, and garment attributes, along with the formed microenvironment, collaboratively play a role in shaping the thermal insulation properties of fabric and attire. These characteristics bear a strong correlation with the protective and comfort performances of the clothing, highlighting the intricate interplay between these factors in the realm of thermal protective textiles.

Gilligan [1] says comfort can be characterized as a state of pleasant harmony encompassing physiological, psychological, and physical well-being between a human being and their environment. Physiological comfort is associated with the body's capability to sustain life, while psychological comfort pertains to the mind's resilience to the impact of the external environment on the body. The concept of comfort encompasses both thermal and non-thermal components, intricately linked to various wear situations, including working, non-critical, and critical conditions [3]. Achieving comfort involves navigating the intricate interactions among fabric properties, climatic conditions, physiological responses, and psychological factors.

2.1 *Properties of Fiber, Yarn, and Fabric*

The properties of fibers, yarns, and fabrics play a crucial role in determining the thermal insulation qualities of textiles. These fiber attributes encompass fineness, cross-sectional shape, and crimp. Yarn characteristics include varieties and twist, while fabric properties can be classified into surface features, structural elements (weave/knit pattern, design, porosity), physical attributes (weight, thickness, and density), as well as thermal and moisture retention capacities [10].

2.1.1 Fiber Properties

The attributes of fibers play a crucial role in determining the thermal insulating qualities of the fabric. These fiber attributes include thickness, cross-sectional shape, length, and waviness [11, 12].

Fineness or Linear Density

The capacity to accommodate more fine fibers in a specific volume arises from the smaller diameter of fine fibers compared to coarse fibers. This results in a higher fiber surface-to-volume ratio for fine fibers. Since fiber surfaces trap dead air (boundary air layer), it becomes evident that fine fibers can trap more dead air than coarse fibers. Consequently, fabrics composed of fine fibers exhibit superior thermal insulation characteristics compared to those made from coarse fibers [13–15].

Cross-Section

The cross-sectional shape of a fiber plays a crucial role in determining the thermal insulation characteristics of fabrics. The presence of more trapped dead air in a fiber's cross-section tends to enhance thermal insulation. For instance, a hollow fiber, compared to a commonly available solid circular fiber, traps more dead air within its structure. This explains why a fabric made from hollow cotton fibers can offer superior thermal insulation compared to a fabric made from solid circular synthetic fibers [14]. Murukami et al. [10] states that similarly, noncircular or profiled fibers, such as trilobal or scalloped oval shapes, are more effective in trapping dead air than circular fibers. The increased presence of trapped dead air in fabrics made from these noncircular fibers ultimately enhances their thermal insulation characteristics [12].

Crimp

Crimp is defined as the natural wave formation in a fiber and is identified as another crucial property influencing thermal insulation [16, 17]. When a fiber has a highly crimped structure, it is characterized by numerous loops on its surface, effectively trapping dead air on the fibre's surface [18]. Conversely, a non- or semi-crimped fiber lacks or has few loops in its structure. Consequently, a fabric comprising non- or semi-crimped fibers traps less dead air compared to a fabric with highly crimped fibers. As a result, a fabric based on crimped fibers exhibits superior thermal insulation characteristics compared to a fabric relying on non- or semi-crimped fibers [19].

2.1.2 Yarn Properties

The type of yarn, and twist imparted to it have an impact on the thermal insulation properties of fabrics [20–22].

Types

Various types of yarn are utilized in the fabrication of fabrics, including texturized yarn, spun yarn, rotor yarn, and filament yarn. Notably, texturized, spun, and rotor yarns are observed to have protruding fibers on their surface. These protruded fibers effectively trap the air, enhancing the thermal insulation of the yarn. In contrast, filament yarn features a smooth surface, limiting its ability to trap dead air. Consequently, fabric made from filament yarn exhibits lower thermal insulation characteristics compared to fabrics made from texturized, spun, or rotor yarns [21, 23].

Twist

In the case of spun yarns, a group of fibers is twisted along its axis to form the yarn. The structure and property of yarn are significantly influenced by the number of twists (turns per inch) applied to the yarn. It has been noted that a highly twisted yarn exhibits a more compact structure compared to a less twisted yarn. Due to this compact structure, a highly twisted yarn traps less dead air than a loosely twisted yarn. Consequently, fabric made from a loosely twisted yarn provides greater thermal insulation than the fabric made from a highly twisted yarn. However, in a very loose yarn structure, the movement of trapped air may occur freely through natural convection, potentially reducing the fabric's ability to retain heat [23].

2.1.3 Fabric Properties

Structural Properties

Fabrics are typically classified into three categories: woven, knitted, and nonwoven. Nonwoven fabrics, despite having the same weight as other fabrics, hold more dead air within their structure [19]. Consequently, nonwoven fabrics are extensively employed in the production of thermal insulators for use in thermal protective clothing [24]. Nonwoven materials are good thermal insulators because they reduce conductive and convective thermal energy transmission through the cloth. However, these fabrics may not function well as thermal insulators due to their regular fiber surface area, particularly when radiative thermal energy transfer prevails [25]. Numerous studies have shown that increasing fiber volume in nonwoven fabrics can diminish the transient rate of radiative thermal energy transfer for a fixed fiber diameter. Yet, fiber diameter has a negligible impact on the unsteady transfer of radiative thermal energy through these fabrics for a fixed fiber volume [26], Qashou et al. [25, 27]. A notable consideration is that nonwoven fabrics hold lower amounts of dead air during compression, potentially reducing their thermal insulation. Additionally, these fabrics exhibit poor compression recovery, which may permanently lower thermal insulation after prolonged compression [28].

Knitted fabrics outperform woven fabrics in trapping dead air within their structure due to the presence of loops. In contrast, woven fabric structures, developed mainly through the interlacement of warp and weft, lack loops and consequently trap a lesser amount of dead air than knitted fabrics. Therefore, the thermal insulation characteristic of knitted fabric is typically much higher than that of woven fabric [8, 29]. Fabric porosity is another factor influencing thermal insulation characteristics [13, 30]. Depending on the weave or knit design, fabric porosity varies, for instance, knitted fabrics generally exhibit higher porosity than woven fabrics, and plain weave fabrics have much higher porosity than twill weave fabrics [29]. In general, a fabric with a moderately porous structure can trap more dead air than a less porous structure, resulting in better thermal insulation. However, a highly porous fabric structure may allow thermal energy penetration during radiant heat and/or flame exposures, thus decreasing the fabric's thermal insulation properties [27].

Physical Properties

Fabric thickness stands out as the paramount property, given that a thicker fabric consistently captures more dead air compared to a thinner counterpart. Notably, a thicker fabric, especially a nonwoven structured batting commonly used in thermal liners, has the ability to absorb and retain thermal energy within its structure [8, 25]. These attributes significantly enhance the thermal insulation characteristics of thicker fabrics, establishing a positive correlation between fabric thickness and thermal protective performance under radiant heat and flame exposures [29]. It's essential to highlight that a multilayered thicker fabric can offer superior insulation compared to a single-layered fabric with the same thickness. This is attributed to the layered structure's capacity to create dead air layers and provide more space for storing absorbed thermal energy within its structure. However, it's crucial to acknowledge that the stored energy inside a multilayered thicker fabric may naturally dissipate over time and/or be forcefully discharged through compression during wearers' activities or movements. This discharge of energy can potentially lead to additional burn injuries during and after the exposure [25].

Research indicates that a fabric with a higher weight has the potential to trap more dead air compared to a lower weight fabric, assuming their weave structures and thicknesses remain constant. When weave structures and thicknesses are comparable, increasing fabric weight entails using more threads per unit length (warp and weft/inch). This causes greater interlacing between the warp and weft, which traps dead air and improves heat insulation. However, a key disadvantage is that a cloth with a heavier weight exerts more heat stress on wearers' bodies and inhibits their movement during work [31].

Fabric density, measured in g/m^3, is another critical property that influences a fabric's performance. Lower-density fabrics are observed to contain fewer solid fibers/yarn compared to high-density fabrics. This characteristic allows low-density fabrics to trap more dead air, resulting in significantly higher thermal insulation compared to high-density fabrics. It appears that fabric density plays a substantial role in determining the thermal insulation characteristics of a fabric.

Moisture Accumulation Properties

Fabrics have the capacity to absorb internally generated moisture from the wearer's body, leading to its accumulation within the fabric's structure. In conditions of elevated ambient temperature or intense physical activity, individuals, particularly fire-fighters, can perspire profusely (with a perspiration rate reaching 2010 g/h). Consequently, fabrics in direct contact with the skin can become saturated with moisture. External causes, such as sprayed water from a hose or water on the floor when a fire-fighter crawls to avoid smoke and severe radiant heat, can further damp the garments [25]. Research indicates that the presence of both internal and external moisture significantly influences the thermal insulation characteristics of a fabric. This impact varies based on the fabric types, the quantity of moisture, and the types and intensity of thermal exposures. Generally, nonwoven fabrics used in fire-fighters' thermal protective clothing exhibit greater thickness compared to woven fabrics. Consequently, under specific thermal exposures, moistened or wetted nonwoven fabrics are observed to provide superior thermal protective performance compared to their woven counterparts [8].

3 Mechanism of Heat Transfer

To understand the thermal properties of the textile system, it is necessary to assess the effects of various heat transmission mechanisms that may be in play. The three methods are thermal conductivity, thermal convection, and radiant heat transfer.

3.1 Conduction

Fibers and air coexist within textile yarns and fabrics, resulting in materials that are neither homogeneous nor isotropic. Nevertheless, when aiming to measure and calculate the thermal conductivity of textiles through abstract and applied means, it is logical to treat a fabric as a uniform and consistent substance in heat transfer. Additionally, in typical clothing scenarios where the thickness of a fabric is considerably smaller than its width and length, it is feasible to simplify the thermal conduction as a linear problem. Given those norms, the dynamic heat transfer via the insulating material is delineated in the following expression [32]:

$$\frac{\partial T}{\partial t} = \frac{\lambda}{cp} . \frac{\partial 2T}{\partial x^2}$$

where T temperature (°K); t time (s); λ conductivity (W m^{-1} K^{-1}); *p* mass density (kg m^{-1}); c specific heat (W S kg^{-1} K^{-1}); and x direction of heat transfer.

3.2 Convection

Convection, a fundamental heat-transfer mechanism, entails the conveyance of energy through the movement of the heat-transfer medium—in this context, atmosphere. When chilly air flows past a heated body, it displaces the temperate air in proximity and replaces it with colder air. Notably, research indicates the absence of convection within clothing insulation, even at low densities. In FEM, convective heat transfer is established as a constraint. The thermal flux due to convection be stated as follows [33]:

$$q = h(T_r - T_x)$$

where q heat flux (W m^{-2}); h film coefficient (W m^{-2} K^{-1}); T_r surface temperature of the fabric (°K); and T_x; temperature of the ambient atmosphere (°K)

3.3 Radiation

The thermal radiation emitted from a clothed human body to the surroundings. In this case, the heat flux by radiation at the external surface of the textile assembly is regulated by the resulting mathematical expressions [34].

$$q = h\left(T_r^4 - T_x^4\right)$$
$$\lambda\frac{\partial y}{\partial x} = h(T_r - T_x) + \sigma\varepsilon\left(T_r^4 - T_x^4\right)$$

where q heat flux (W m^{-2}); σ Stefan–Boltzmann constant which is 5.6703×10^{-8} W m^{-2} K^{-4}; ε emissivity of the surface; h film coefficient (W m^{-2} K^{-1}); T_r surface temperature of the fabric (°K); and T_x temperature of the ambient atmosphere (°K); and λ conductivity (W m^{-1} K^{-1}).

4 Effect of Clothing on Thermal Comfort

Apparel plays a crucial role in preserving heat equilibrium by influencing both heat loss from the epidermis and moisture loss from the skin. Yet, there is no one-size-fits-all apparel system suitable for every occasion. A clothing system effective in one climate may not be suitable for another [35].

In cold climates, clothing and textiles require good thermal insulation properties. The thermal insulation rests on several features, including thickness and sum of layers, drape, fiber density, flexibility of layers, and the effectiveness of closings. Thermal insulation significance of clothing, once worn, is not solely determined by the insulation of specific attire but by the entire outfit. Air gaps among clothing layers can significantly contribute to the overall thermal insulation importance, assuming the gaps are not great enough to allow air mobility and subsequent heat loss through convection [36]. The snugness of fit of a dress plays a vital role in its insulation value, along with the material from which it is made. The fabric's resistance to the movement of heat is critically important for achieving thermal comfort. From Fig. 1 it can be seen that in summer, heat primarily transfers from the environment to the human body when outdoors. Conversely, in winter, heat is consistently released from the human body to the surroundings.

5 Factors Affecting Clothing Comfort

Thermal comfort in clothing primarily revolves around the sensations related to temperature and moisture. This aspect predominantly engages thermal receptors in the skin, intertwining with clothing's transfer properties like heat and moisture

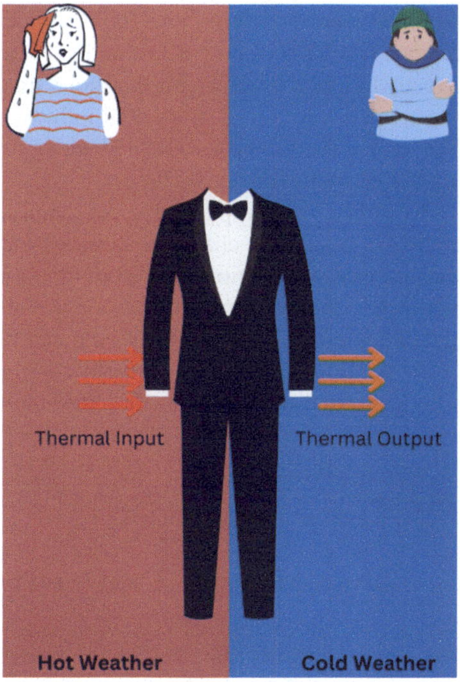

Fig. 1 Movement of heat during different weather conditions

transfer, as well as air permeability [37]. Clothing serves as a protective barrier against cold or heat, ensuring body thermal comfort across a wide spectrum of human activities. Tactile interactions, encompassing moisture and thermal dynamics amid apparel material and human skin, dictate an individual's comfort level in specific environmental conditions and activities. The fabric type, blend composition, tactile and thermal insulation conduct, and moisture management capabilities all contribute to overall comfort [38].

Kothari [39] states that diverse attributes of fibers, yarns, fabrics, and garments suggestively impact well-being and must be considered in creating suitable gear. Textile technologists play a crucial role in developing fabrics that prioritize comfort. The assortment of fiber composition, yarn selection, and techniques employed in fabric construction, along with the treatments applied at the end, all play a crucial role in determining physical comfort. These factors influence thermal retention or dissipation, breathability, resistance to water, buildup of static charge, protection against ultraviolet rays, and other aspects of comfort. The quality of fabrics for clothing relies heavily on aesthetic performance, comfort-related properties, and wear-related attributes.

Within the realm of properties that impact comfort, the characteristics of fibers, such as their type, fineness, cross-sectional shape, crimp, length, and surface attributes are of paramount significance [40]. Additionally, factors related to fabric structure, including yarn linear densities, sett, weave patterns, and crimp levels, play a crucial role in determining fabric properties such as coverage, thickness, bulk density, mechanical properties, and surface characteristics, all of which contribute directly to fabric comfort. Furthermore, treatments that modify fabric properties and appearance can exert a substantial influence on the overall performance of clothing textiles. Fabric properties, coupled with garment design and size, collectively impact several garment attributes [39].

To attain thermo-physiological comfort, garments must exhibit appropriate thermal conductivity characteristics and adequate permeability to moisture vapor, as well as ample ventilation. Fabric configurations can be engineered to elevate garment comfort, with a primary emphasis on thermal and mechanical attributes. It is widely agreed upon that the transfer of heat and moisture vapor within garments emerges as the paramount factor in garment comfort. Scholars such as [41] emphasize the body's thermal management to define the system in which comfort must be kept, while others like Hollies [42] stress the position of 'contact comfort' in addressing a clothing system.

6 Fibrous Materials Used for Thermal Protection

The various hazards that fire-fighters face while dealing with fire and thermal environments were discussed by Ali and Mohammed [43]. Fire-fighters often encounter fatal and nonfatal injuries, such as skin burns, strains, bruises, and fume or vapor inhalation [44]. It is assessed that a significant number of fire-fighters suffer injuries every year across different countries, with some even losing their lives in the line of duty.

Abbott [45] and Foster [46] reported that the thermal environmental conditions can be categorized as monotonous, hazardous, or emergency. Alternative thermal environments, like flashover conditions, can be extremely dangerous, with temperatures exceeding 600 °C and high heat flux range of 21–210 kW/m^2. Fire-fighters have only a few seconds to react in such situations [47, 48]. Routine and hazardous thermal environments involve lower levels of radiative heat exposure, with heat fluctuation ranging from 4 to 21 kW/m^2 [45]. Despite not causing immediate thermal degradation of protective materials, widely occurredburn accidents arise in these exposure circumstances [49].

Barker [47, 48] stressed that in order to reduce or avert lethal and non-lethal injuries, fire-fighters must don protective attire with flame-resistant and thermally insulating characteristics. The efficiency of this protective clothing rest on factors like fabric properties, air gap size, moisture content, and dress design. Proper fibrous materials play a vital role in improving the thermal protective act of the clothing against different heat exposures.

The research regarding performance specifications of thermal safeguarding attire provides an essential foundation for understanding the necessary features and characteristics that such clothing must possess to safeguard fire-fighters and workers exposed to thermal hazards. The key aspects that are often emphasized in this literature include flame retardance, thermal insulation, and protection against radiant heat, convective heat, and contact heat.

6.1 Thermal Protective Fibers

Fire-retardant fibers can be generally categorized into two categories: chemically modified flame-resistant fibers and inherently flame-resistant fibers.

1. Chemically Modified Flame-Resistant Fibers: These fibers, which include both natural and synthetic fibers, are treated with flame-retardant chemicals during a textile finishing process. The flame-retardant properties are achieved through the application of these chemicals, which act as a protective barrier against flames and reduce the material's flammability. Examples of chemically modified flame-resistant fibers include cotton treated with flame-retardant chemicals and flame-retardant polyester [50].
2. Inherently Flame-Resistant Fibers: Inherently flame-resistant fibers have a unique molecular structure that imparts high resistance to decomposition at elevated temperatures. These fibers do not require the application of additional flame-retardant chemicals because their inherent properties provide flame resistance. Examples of inherently flame-resistant fibers include meta-aramid (e.g., Nomex) and para-aramid (e.g., Kevlar). These fibers have high thermal transition temperatures, making them capable of withstanding intense heat without breaking down [51].

The burning behavior of fibers is influenced by various thermal properties, including softening temperature, melting temperature, and rate of heat release. The frequency of heat release is a critical parameter that determines the burning hazard, indicating how quickly the material will burn and the severity of burns it may cause. The transition temperatures and thermodynamic parameters of the commonly fibrous materials are compared to inherently flame-resistant fibers to chemically modified flame-resistant fibers, the former typically exhibit higher thermal transition temperatures. This is because the inherent molecular structure of these fibers is designed to withstand higher temperatures, providing enhanced flame resistance. LOI is an additional important parameter used to assess flammability. Inherently flame-resistant fibers often have greater LOI values, usually above 29%, indicating better flammability resistance.

Therefore, both types of fire-retardant fibers play a crucial role in developing protective clothing and equipment for fire-fighters and other individuals who work in high-risk environments where exposure to flames and heat is a concern.

Here are some examples of the fibers showing their thermal properties [52]:

1. Cotton fibers do not possess inherent flame resistance. They typically soften and begin to pyrolyze at around 350 °C and ignite at the same temperature.

Cotton has a relatively low Limiting Oxygen Index (LOI) of 18.4%, indicating its susceptibility to combustion. Chemically modified flame-resistant cotton fibers are treated with chemical finishes to enhance their flame resistance, with a heat release rate of 19 kg/J.

2. Viscose fibers soften at around 350 °C and ignite at approximately 420 °C. They have an LOI of 18.9% and typically undergo additive introduction or copolymeric modifications to improve flame resistance. This treatment results in a similar heat release rate of 19 kg/J.
3. Nylon fibers have a softening temperature of 50 °C and pyrolyze at 215–431 °C, with ignition occurring at approximately 450 °C. They possess an LOI of 20–21.5% and have a relatively high heat release rate of 39 kg/J.

7 Evaluation Method and Performance Standards

Based on laboratory simulation techniques, the protective capabilities of fabric or apparel can be assessed. In recent decades, numerous related assessment standards and procedures have been established. There are typically two processes involved in the assessment of fibrous materials for prospective usage in thermal protective apparel. The ability of the fibrous materials to withstand flame and maintain thermal stability is first investigated. To determine whether a material is suitable for thermal protection, a flammability test is conducted to assess the performance of the material as per the ASTM and ISO standards [75–85]. The capability of protective clothing to withstand heat transmission, including conduction, convection, and radiation, should be confirmed by evaluating its thermal protective performance.

The ability to sustain combustion and ease of ignition are two factors that determine flammability. The type of igniting source and the moment it comes into contact, the alignment of the fabric, and the ignition location (such as at the edge, face, top, or bottom of the fabric), the surrounding temperature and humidity levels, the speed of the air, and the structural characteristics of the fabric all have an impact on how fabrics burn [53]. Various standardized test methods have been created to evaluate flammability. The most popular evaluation techniques are the smoke concentration test, the vertical flammability test, the 45-degree flammability test, and the LOI.

Wearing thermally protected clothes can prevent heat from entering the body from the surrounding environment. The level of thermal protection that clothing offers has a direct impact on worker safety. Therefore, the choice of protective materials is greatly influenced by the evaluation of thermal protective performance. According to ASTM and ISO standards [75–85] the tabletop tests and full-scaled manikin tests are two examples of laboratory simulation techniques that can be used to assess the thermal protection qualities of fabric or clothing. A flame, thermal radiation, and heated surface are frequently used to imitate thermal environments for testing. When exposed to heat, the thermal sensor at the back of the specimen captures thermal histories that can be used to determine the rate at which heat is transferred through the fabric [54].

There has been a lot of research done in the area of thermal protective clothing to give higher thermal protection, but the development of some smart materials and new technology opens up new opportunities for thermal protective clothing enhancement. These materials are easily affected by environmental and social factors, changing their physical characteristics.

Additionally, intumescent treatments are a relatively new development in flame-retardant technology. When activated by too much heat or flame, the majority of flame retardants, which are typically thin and low-insulating coatings on lining fabrics, immediately generate an inert insulative char to protect the wearer [55]. To reduce inhalation of harmful fumes and pollutants, the flame retardant, the equipment used to apply it, and the product itself must be ecologically friendly. Furthermore, in addition to flash fire and high-intensity thermal radiation, fire-fighters are frequently exposed to thermal risks such as contact with hot objects, hot liquid splashes, and hot steam [56, 57]. The shift in protective highlight may be caused by the differentiation of heat sources. The differentiation of heat sources may cause the protective highlight to alter. Increasing the heat insulation of protective materials is one way to guard against hot objects, while reducing and preventing the ingress of hot liquid and steam is a crucial technique for preventing scalds or vapor burns. Consequently, the next research should create thermal protective attire that satisfies various functional needs.

8 Design of Safety Clothing for Thermal and Fire Resistance

Kothari [38, 44] stated that heat and moisture transport within clothing can manifest through various channels, either in a trans-planar direction, traversing the diverse clothing layers directly or laterally parallel to both the skin and fabric layers. It has long been recognized that a well-designed garment can enhance ventilation effects, facilitating the transmission of heat and moisture through openings, mostly over wearer-induced pumping effects. Nevertheless, in heat- and flame-resistant clothing, openings in cuffs are typically minimized to prevent the ingress of environmental heat [58]. While some ventilation effects may occur at the lower end of a jacket, particularly in two-piece protective clothing, their occurrence is likely limited in overalls. As a result, the primary mode of heat and moisture transfer in heat- and flame-resistant clothing is trans-planar. Consequently, the structure of various fabric layers plays a crucial role in evacuating excess heat and moisture (sweat). Ideally, body-produced sweat should evaporate near the skin, as evaporative cooling efficiency significantly diminishes when evaporation occurs in outer clothing layers. Liquid moisture can traverse between layers, and the layer worn above the underwear can significantly impact sweat distribution in the underlying layers. This liquid transport is primarily influenced by the fabrics' hydrophilicity [59]. A substantial amount of sweat accumulates within the clothing layers, emphasizing the need for well-coordinated layer adaptations to prevent steam formation. Protective clothing design is pivotal for heat and flame protection, as evidenced by efficient energy transfer reduction in simulated flash fire measurements with air gaps. Recent studies highlight higher heat transfer to limbs, likely due to smaller air gaps, emphasizing the importance of clothing fit for optimal protective performance [60]. Striking a balance, the garment should incorporate air gaps to enhance protection while avoiding excessive looseness to prevent snagging on obstacles.

9 Upcoming Developments

Without a doubt, the overwhelming majority of advancements in the realm of heat- and flame-resistant attire aim to mitigate the heat strain experienced by those who wear it. Three primary avenues of development are readily discernible:

- Enhancement
- Enhancing the capacity for heat (and moisture) retention, partly by employing materials with adaptable properties; and
- Advancement of alert mechanisms employing non-intrusive sensors integrated within the textiles.

9.1 Development of New Materials

In essence, air stands out as an excellent insulator, prompting numerous endeavors to enhance the air content within clothing layers [61]. Advancements in three-dimensional spacer fabrics have been notable in this regard. Innovations like 'super insulators,' such as aerogels, which boast lower thermal conductivity than air (typically below 0.020 W m^{-1} K^{-1}), have been applied in building insulation [62]. Despite numerous patents combining aerogels with textiles, practical implementations remain limited, likely due to the high cost or brittleness of these materials, complicating their incorporation into textile components. Nevertheless, advancements in processing technologies and the burgeoning market for such materials are likely to encourage the broader use of aerogels in textile applications. Materials with dynamic properties under heat exposure can contribute to reducing the wearer's heat stress [63]. These materials offer moderate thermal insulation under normal working conditions but can enhance insulation dramatically in emergency situations with extremely high temperatures. Horrock and Price [53], covered in their book, the flame-retardant materials, including intumescent materials that promote char formation when exposed to heat and flame. Additionally, structures like shape memory alloys and fibers undergo morphological changes when heated, enabling them to increase the thickness of a multilayer assembly when a critical temperature is reached when integrated into clothing [64].

9.2 Problems Related to Thermal Protective Clothing

The fundamental purpose of heat- and flame-resistant clothing is to shield the body from potentially dangerous heat exposure. Consequently, the thermal insulation is intentionally high, hindering the required metabolic heat dissipation. The interplay of heightened physical exertion, elevated clothing insulation, and rising ambient temperatures leads to a rise in the body's core temperature. Additionally, the clothing acts as an obstacle to the transport of water vapor, causing a portion of the body's generated sweat to accumulate within the fabric. This poses a distinct challenge for the clothing, as its thermal characteristics undergo alterations when wet, markedly affecting its capacity for heat and moisture transfer [65–68].

9.3 External Cooling Systems

The passive and active cooling garments have been proposed to enhance heat dissipation from the body and mitigate thermal stress [69]. Active cooling systems typically involve incorporating ventilators into clothing to enhance convective heat exchange or employing a system that circulates cooling liquids to conductively remove heat between the skin and the liquid. In liquid-cooled garment, the tubing system is sewn into the garment's inner surface that is in contact with the skin. The cooling liquid is provided by an external water circulation system [70]. The vest equipped with coolants has given beneficial effects on heart rate, skin temperature, sweat rate, and overall thermal strain. Additionally, research has explored the use of phase change materials in heat protective clothing to alleviate heat strain. Some studies have examined the impact of post-exertion cooling, with forearm immersion in cold water [71]. The flame retardancy of the system has to be kept in mind while implementing these cooling systems under heat- and flame-protective clothing. Any additional device increasing the overall equipment weight may potentially diminish the positive effects of external cooling [72].

9.4 Evaporation and Vapor Formation

The utmost effective way for the build to dissipate excess heat is through sweat evaporation. Water has an enthalpy of vaporization of approximately 2400 kJ kg^{-1} at 34 °C. Still, heat and flame protecting wear typically comprises multiple layers, limiting the passage of water vapor through the fabric [73]. This vapor flow depends on the partial water vapor pressure gradient amid the skin and the atmosphere. To achieve evaporative cooling, the vapor pressure in the environment would not surpass that of the skin. Yet, at high atmospheric temperatures, the fractional water vapor pressure may exceed that on the skin, reaching 5.4 kPa at 45% RH at 51 °C or

27% at 61 °C [67, 68]. Beyond these thresholds, the body's moisture cannot escape the clothing, conversely, water vapor may move from the environment to the body. Studies have demonstrated that the most effective way to prevent this exchange is through the usage of water vapor-tight clothing. Below definite heat exposure conditions, moisture can evaporate, forming exceedingly energetic vapor that may reach the body and cause steam burns. Research using radiography has explored the creation and transportation of steam in multilayer fabric mixtures, clearly illustrating the potential risk of steam burns [74].

References

1. Gilligan (2010) The prehistoric development of clothing: archaeological implications of aThermal model. J Archaeol Method Theory 17(1):15
2. Li Y (2001) The science of clothing comfort. Text Prog 31(2):1
3. Fourt L, Hollies NRS (1970) Clothing: comfort and function. Marcel Dekker, New York
4. Bajaj P (1992) Fire-retardant materials. Bull Mater Sci 15(1):67
5. Bajaj P, Sengupta AK (1992) Protective clothing. Text Prog 22(1):1
6. Kahn SA, Patel JH, Lentz CW, Bell DE (2012) Firefighter burn injuries: predictable pat-terns influenced by turnout gear. J Burn Care Res 33(1):152
7. Farnworth B (1983) Mechanisms of heat flow through clothing insulation. Text Res J 53(12):717
8. Matusiak M, Sikorski K (2011) Influence of the structure of woven fabrics on their thermal insulation properties. Fibres Text East Eur 19(5):46
9. Varshney RK, Kothari VK, Dhamija S (2011) A study on thermophysiological comfort properties of fabrics in relation to constituent fibre fineness and cross-sectional shapes. J Text Inst 101(6):495
10. Murakami H, Hegemier G, Maewal A (1978) A mixture theory for thermal diffusion in unidirectional composites with cylindrical fibers of arbitrary cross section. Int J Solidsstruct 14(9):723
11. Hearle JWS (2001) High performance fibres. Woodhead Publishing, Cambridge
12. Matsudaira M, Tan Y, Kondo Y (1993) The effect of fiber cross-sectional shape on fabricmechanical properties and handle. J Text Inst 84(3):376
13. Shekar RI, Kotresh TM, Subbulakshmi MS, Vijaylakshmi SN, Prasad SK (2009) Thermal resistance properties of paratrooper clothing. J Ind Text 39(2):123
14. Sun G, Yoo HS, Zhang XS, Pan N (2000) Radiant protective and transport properties of fabrics used by wildland firefighters. Text Res J 70(7):567
15. Vallabh R, Banks-Lee P, Mohammadi M (2008) Determination of radiative thermal conductivity in needle punched nonwovens. J Eng Fiber Fabr 3(4):46
16. Bozkurt E, Kaya E, Tanoglu M (2007) Mechanical and thermal behavior of non-crimp glassfiber reinforced layered clay/epoxy nanocomposites. Compos Sci Technol 67(15–16):3394
17. RweiS P, LinY T, SuY Y (2005) Study of self-crimp polyester fibers. Polymer Eng Sci 45(6):838
18. McGregor BA, Postle R (2008) Mechanical properties of cashmere single jersey knitted fabrics blended with high and low crimp superfine merino wool. Text Res J 78(5):399
19. Hearle JWS (1991) Understanding and control of fibre structure. J Appl Poly Sci 47:1
20. Bogaty H, Hollies NRS, Harris M (1957) Some thermal properties of fabrics part I: theeffect of fiber arrangement. Text Res J 27(6):445
21. Ghassemi MRM, Mojtahedi RSR (2011) Investigation on the physical and structural properties of melt-spun multifilament yarns, drawn yarns and textured yarns produced from blend of PP and oxidized PP. Fiber Poly 12:789
22. Hatch KL (1993) Textile science. West Publishing Company, New York

23. Rengasamy RS, Kawabata S (2002) Computation of thermal conductivity of fibre from ther-mal conductivity of twisted yarn. Ind J Fibre Text 27:342
24. Abdel-Rehim ZS, SaadM M, El-ShakankeryM HanafyI (2006) Textile fabrics as thermalinsulators. Autex Res J 6(3):148
25. Qashou HV, Tafreshi BP (2009) An investigation of the radiative heat transfer through nonwoven fibrous materials. J Eng Fiber Fabr 4(1):9
26. Gibson PW, Lee C, Ko F, Reneker D (2007) Application of nanotechnology to nonwoventhermal insulation. J Eng Fiber Fabr 2(2):32
27. Barker RL, Heniford RC (2011) Batting materials for use in firefighter turnout suit. J Eng Fibres Fabr 6(1):1
28. Mao N, Russel SJ (2007) The thermal insulation properties of spacer fabrics with a mechanically integrated wool fiber surface. Text Res J 77(12):914
29. Frydrych G, Dziworska JB (2002) Comparative analysis of the thermal insulation properties of fabrics made of natural and man-made cellulose fibres. Fibres Text East Eur 10(4):40
30. Fan JT, Luo Z, Li Y (2000) Heat and moisture transfer with sorption and condensation inporous clothing assemblies and numerical simulation. Int J Heat Mass Transf 43(16):2989
31. Celcar D, Meinander H, Gersak J (2008) A study of the influence of different clothing materials on heat and moisture transmission through clothing materials, evaluated using asweating cylinder. Int J Cloth Sci Tech 20(2):119
32. Yang SM, Tao WQ (1999) Heat transfer. Higher Education Press, Beijing
33. Incropera FP, DeWitt DP (2002) Fundamentals of heat transfer, 5th edn. Wiley, Somerset
34. Gagge AO, Buston AC, Bazett HC (1941) A practical system of units for the description of heat exchange of man with his environment. Science 94:428–430
35. Kandha Vadivu P (2014) Effect of clothing materials on thermoregulatory responses of the human body, textiles history, properties and performance and applications. Nova Science Publishers Inc, London
36. Psikuta A, Frackiewicz-Kaczmarek J, Frydrych I, Rossi R (2012) Quantitative evaluation of air gap thickness and contact area between body and garment. Text Res J 82(14):1405–1413
37. Jeffries R (2005) Functional aspects of high-performance clothing. Book of Abstracts, Fashion the future, British Textile Technology Group, Shirley Publication, pp 126–128
38. Kothari VK (2004) Fabric comfort. In: Proceedings of the seminar on comfort in textiles, IIT Delhi, New Delhi, India
39. Kothari VK (2006) Thermo-physiological comfort characteristics and blended yarn woven fabrics. Ind J Fib Text Res 31(1):179–186
40. Bhat P, Bhonde HU (2006) Comfortable clothing for defense personnel. Asian Text J 2006:73–77
41. Rees WH (1972) Materials and clothing in health and disease—the biophysics of clothing material. Lewis Pub, London
42. Hollies NRS, Goldman RF (1977) Clothing comfort: interaction of thermal, ventilation construction and assessment factors. Michigan, Ann Arbor Science publishers Inc. Ann Harbor, USA
43. Ali AH, Mohmmed R (2015) A review of the firefighting fabrics for flashover temperature. Int J Eng Sci Res Tech 4(3):247–257
44. Karter MJ, NFP Association (2009) Patterns of firefighter firegroundinjuries. National Fire Protection Association Quincy
45. Abbott NJ, Schulman S (1976) Protection from fire: nonflammable clothing—a review. Fire Technol 12(3):204–218
46. Foster J, Roberts G (1995) Measurements of the firefighting environment—summary report. Fire Eng J 55:30–34
47. Barker, R. (2005). A review of gaps and limitations in test methods for first responder protective clothing and equipment. Report Presented to National Personal Protection Technology Laboratory.
48. Barker RL (2005) A review of gaps and limitations in test methods for first responder protective clothing and equipment. National Institute for Occupational Safety and Health, pp 1–98

49. Guowen S, Paskaluk S, Sati R (2010) Thermal protective performance of protective clothing used for low radiant heat protection. Text Res J 81(3):311–323
50. Jackson RH (1978) PBI fiber and fabric—properties and performance. Text Res J 48(6):314–319
51. Wang CC, Chen CC (2005) Some physical properties of various amine-pretreated Nomex Aramid yarns. J Appl Polym Sci 96(1):70–76
52. Liu X-Y, Tao Y-K, Wang X-L, Yu W-D (2009) The thermal property of high-performance fiber. J Xi'an Polytech Univ 23(2):267–271
53. Horrocks AR, Price D (2001) Fire retardant materials. Woodhead Publishing, London
54. Lawson LK, Crown EM, Ackerman MY, Dale JD (2004) Moisture effects in heat transfer through clothing systems for wildland firefighters. Int J Occup Saf Ergon 10(3):227–238
55. Shalev RLB (1983) Analysis of heat transfer characteristics of fabrics in an open flame exposure. Text Res J 53(8):475
56. Lu Y, Song G, Zeng H, Zhang L (2013) Characterizing factors affecting the hot liquid penetration performance of fabrics for protective clothing. Text Res J 84(2):174–186
57. Su Y, Li J (2016) Development of a test device to characterize thermal protective performance of fabrics against hot steam and thermal radiation. Meas Sci Technol 27(12):1–9
58. Havenith G, Bröde P, Den Hartog E, Kuklane K, Holmer I, Rossi RM, Richards M, Farnworth B, Wang X (2013) Evaporative cooling: effective latent heat of evaporation in relation to evaporation distance from the skin. J Appl Physiol 14:778–785
59. Adler MM, Walsh WK (1984) Mechanisms of transient moisture transport between fabrics. Text Res J 54:334–343
60. Keiser C, Becker C, Rossi RM (2008) Moisture transport and absorption in multilayer protective clothing fabrics. Text Res J 78:604–613
61. Nocker W, Seibert J (2005) Firefighter garments with non-textile insulation. In: Yutaka T, Tadakatsu O (eds) Elsevier ergonomics book series, vol 3, pp 379–382
62. Cheuvront SN, Goodman DA, Kenefick RW, Montain SJ, Sawka MN (2008) Impact of a protective vest and spacer garment on exercise-heat strain. Eur J Appl Physiol 102:577–583
63. Sankaran V, Younes A, Engler T, Cherif C (2012) A novel processing solution for the production of spatial three-dimensional stitch-bonded fabrics. Text Res J 82:1531–1544
64. Koebel M, Rigacci A, Achard P (2012) Aerogel-based thermal superinsulation: an overview. J Sol Gel Sci Technol 63:315–339
65. Bröde P, Havenith G, Wang X, Candas V, den Hartog EA, Griefahn B, Holmér I, Kuklane K, Meinander H, Nocker W, Richards M (2008) Non-evaporative effects of a wet mid layer on heat transfer through protective clothing. Eur J Appl Physiol 104:341–349
66. Lotens WA, Vandelinde FJG, Havenith G (1995) Effects of condensation in clothing on heat-transfer. Ergonomics 38:1114–1131
67. Rossi R, Indelicato E, Bolli W (2004) Hot steam transfer through heat protective clothing layers. Int J Occupat Saf Ergonom 10(3):239–245
68. Rossi RM, Gross R, May H (2004b) Water vapor transfer and condensation effects in multilayer textile combinations. Text Res J 74:1–6
69. Chinevere TD, Cadarette BS, Goodman DA, Ely BR, Cheuvront SN, Sawka MN (2008) Efficiency of body ventilation system for reducing strain in warm and hot climates. Eur J Appl Physiol 103:307–314
70. Xu XJ, Gonzalez J (2011) Determination of the cooling capacity for body ventilation system. Eur J Appl Physiol 111:3155–3160
71. Kim JH, Coca A, Williams WJ, Roberge RJ (2011) Subjective perceptions and ergonomics evaluation of a liquid cooled garment worn under protective ensemble during an intermittent treadmill exercise. Ergonomics 54:626–635
72. Smolander J, Kuklane K, Gavhed D, Nilsson H, Holmér I (2004) Effectiveness of a light-weight ice-vest for body cooling while wearing fire fighter's protective clothing in the heat. Int J Occupat Saf Ergon 10:111–117
73. Desruelle AV, Schmid B (2004) The steam laboratory of the institute de medicine naval du service De Sante Des Armies: a set of tools in the service of the French navy. Eur J Appl Physiol 92:630–635

74. Sati R, Crown EM, Ackerman M, Gonzalez J, Dale D (2008) Protection from steam at high pressures: development of a test device and protocol. Int J Occupat Saf Ergon 14:29–41
75. ASTM F1060-08 A (2008) Standard test method for thermal protective performance of materials for protective clothing for hot surface contact. ASTM International, West Conshohocken, PA
76. ASTM F1930-15 A (2015) Standard test method for evaluation of flame-resistant clothing for protection against fire simulations using an instrumented manikin. ASTM International, West Conshohocken, PA
77. ASTM F1939-15 A (2015) Standard test method for radiant heat resistance of flame-resistant clothing materials with continuous heating. ASTM International, West Conshohocken, PA
78. ASTM F2700-13 A (2013) Standard test method for unsteady-state heat transfer evaluation of flame-resistant materials for clothing with continuous heating. ASTM International, West Conshohocken, PA
79. ASTM F2702-15 A (2013) Standard test method for radiant heat performance of flame-resistant clothing materials with burn injury prediction. ASTM International, West Conshohocken, PA
80. ASTM F2703-13 A (2013) Standard test method for unsteady-state heat transfer evaluation of flame-resistant materials for clothing with burn injury prediction. ASTM International, West Conshohocken, PA
81. ASTM F2731-11 A (2011) Standard test method for measuring the transmitted and stored energy of firefighter protective clothing systems. ASTM International, West Conshohocken, PA
82. ISO/TC 94 ISO 6942 (2002) I. Protective clothing-protection against heat and fire method of test: evaluation of materials and material assemblies when exposed to a source of radiant heat. Geneva, Switzerland: ISO, 2004
83. ISO/TC 94 ISO 9151 (1995) I. Protective clothing against heat and flame determination of heat transmission on exposure to flame. Geneva, Switzerland: ISO, 2004
84. ISO/TC 94 ISO/DIS 13506 (2004) I. Protective clothing against heat and flame test method for complete garments prediction of burn injury using an instrumented manikin. Geneva, Switzerland: ISO
85. ISO/TC 94/SC13 ISO 17492 (2003) I. clothing for protection against heat and flame determination of heat transmission on exposure to both flame and radiant heat. Geneva, Switzerland: ISO, 2004

E-Textiles and Wearables for Healthcare

Rashmi Thakur and Md. Vaseem Chavhan

1 Introduction

As the global population steadily increases, there is a noticeable demographic shift characterised by greater life expectancy and a growing ageing population. Simultaneously, expectations for improved healthcare are on the rise, creating mounting pressure on medical care systems. To address these challenges, researchers are actively exploring the potential of smart textile systems. These innovative systems seamlessly integrate advanced functionalities into various medical textiles, which include surgical implants, dressings, garments, and beddings, to name a few. The overarching aim is to establish a seamless relationship between textiles and technology within the medical environment. These smart textile systems play a versatile role, from facilitating therapeutic care to aiding in diagnostics and monitoring vital signs. The smart textile systems offer adaptability to diverse locations, enabling remote monitoring of patients through wireless communication technologies. This capability holds significant advantages, particularly in streamlining and freeing up hospital resources. By allowing for the earlier transition of patients to their home environment, healthcare facilities can operate more efficiently. Throughout this transition, patients remain under careful monitoring through telecommunications systems, empowering medical professionals to collect and interpret data remotely [1].

Smart Textiles majorly encompass shape memory polymer, phase change materials and conductive. This chapter aims at summarising the existing healthcare solutions aided specifically through e-textiles and wearables, made thereof. Application of e-textiles and Wearables spans a large sector as already briefed in prior chapters including Military, Sports, Entertainment, Healthcare, Wellness, and Leisure. It is

R. Thakur (✉)
Department of Fashion Technology, National Institute of Fashion Technology, Mumbai, India
e-mail: rashmi.thakur@nift.ac.in

Md. V. Chavhan
Department of Knitwear Design, National Institute of Fashion Technology, Hyderabad, India

S. Sundaramoorthy et al. (eds.), *Textile Materials for Good Health and Wellbeing*, SDGs and Textiles, https://doi.org/10.1007/978-981-97-6496-9_17

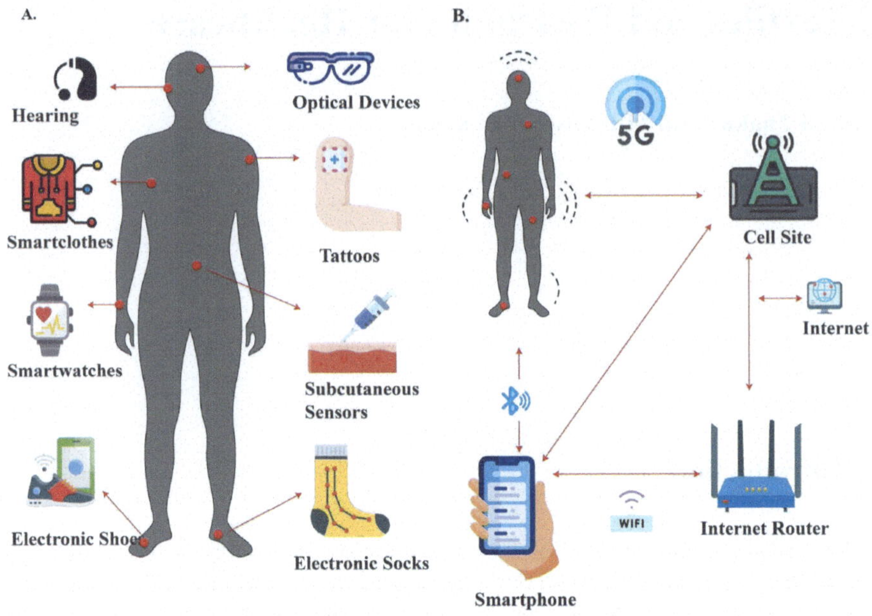

Fig. 1 Application areas of e-textiles for healthcare

important to disseminate the information to the readers in a form which makes them connect to the topic as a consumer as well. Thus, the chapter elaborates different forms of e-textiles and wearables based on various healthcare applications (Fig. 1). Electronic textiles can incorporate materials like metallics, optical fibers, and conductive polymers for electrical conductivity, sensing, and data transmission. Organic polymers offer elastic, resilient, lightweight, mechanically flexible, easy-to-process, and inexpensive alternatives.

2 Monitoring Breathing and Sweating Activity in Human

Monitoring the breathing and body perspiration-related vitals is one of the preliminary checks for fitness among humans whether for sports activities or for daily well-being. Being one of the primary diagnostic steps, it has wider scope of application and therefore the system is desired to be in wearable form. E-textiles have been found to perform well in monitoring the thoraco-abdominal pattern and the variation in it. They can monitor parameters like breathing frequency, breath by breath respiratory period, and volume variations of the whole chest wall with a good agreement to the existing clinical solution [2].

Continuous and real-time study of temperature and sweating profile can also be utilised for diagnosing symptom-based ailments associated with these vitals, like hyperthermia, infection, inflammation, etc. This calls for development of a next to skin solution, provided by e-textiles. Jiang et al. [3] reports fabrication of textile-based Near Field Communication antenna by using silver coated conductive threads on a cotton fabric. The antenna is enabled by an NFC sensor transponder, having an inbuilt temperature sensor, which is connected to an external humidity sensor. A Dickson charge pump and a voltage regulator was used for developing a voltage boosting rectifier which is connected between the NFC antenna and the sweat (humidity) sensor for providing power supply to the sensor wirelessly. The maximum reach for reading sensor data is 6 cm. The measurement accuracy reported for this system is $\pm$ 0.14 °C and $\pm$ 0.2% for temperature measurement and sweat measurement, respectively. However, the challenge lies in maintaining consistency in measurement while the body location where the flexible sensors are mounted, bends or stretches. Therefore, continuous attempts are made among researchers to reduce the circuit size and select a body location with minimal movement activity for placing the system [3].

Another study reports utilisation of tribo-electric all-textile sensor array named as "TATSA", for monitoring arterial pulse and respiratory signals respectively. Such a sensor along with being comfortable offers high pressure sensitivity, quick response time, and machine washability of up to 40 washes. This array of sensors is prepared by knitting a full cardigan stitch fabric using conductive and nylon yarns. This knit structure can then be integrated into clothing at suitable locations. Cardigan knit stitch structure was selected among the rest as the contact between the conductive yarn and nylon yarn is in surface form rather than a point contact form thus enabling better tribo-electric effect owing to larger contact area. The conductive yarn had stainless steel in the core wrapped with Terylene fibre on the sheath. TATSA was proposed further for development of a monitoring system to assess cardio-vascular ailment and sleep apnoea syndrome [4].

Wang et al. [5] also report exploration of knitted sensors integrated into a seamless garment for monitoring breathing conditions. The prototype developed and studied had 60 such flexible sensors integrated into it. The fabric was prepared on a circular knitting machine where each loop was made up of two yarns. One was the silver plaited nylon filament yarn and other was nylon/spandex core-spun yarn. The conductive yarn constituted the conductive element of the sensor, responsible for monitoring the health condition. The conductive yarn offered a low resistance of 0.516 Ω/mm and acted as a piezo-resistive sensor for measuring breathing amplitude and frequency obtained from the variation in sensor resistance. Through this the system was capable of classifying breathing patterns as slow breathing, normal breathing, fast breathing, and shallow breathing [5].

Another research studies the feasibility of using cover stitched textile-based stretch sensors on an existing garment (shirt in this case) for monitoring the breathing pattern. A 23 cm long sensor strip developed was cover stitch at chest level, on a polyester/ spandex, 84/16 blended, knitted T-shirt, insulated by stitch-less bonded 3 M film. Snap-style connectors were used to connect with the conductors present at both ends of the sensor. The change in resistance of the stretch sensor was measured through analog circuit and analyzed to study the breathing pattern. The results taken from 3 subjects performing eight different activities were compared with Respiratory Inductive Plethysmography (RIP)-based breathing system, which were found to be similar with minor error percentage [6].

Zhao et al. [7] came up with an interesting proof of concept, wherein biosensors were developed to detect sweat. The textile-based biosensors were prepared using conducting yarn which were integrated with zinc-oxide nanowires (ZnO NWs). These were utilised for detecting sodium and lactate in human perspiration during physical exercise. The system was used in the form of a wearable headband. It can monitor human sweat wirelessly. The test conducted on a healthy volunteer both in static and dynamic state was found to be in good agreement with the standard test method, thus claimed to be robust in signal capture and transmission [7].

Lugoda et al. [8] reported the usage of temperature sensing yarn for creating temperature sensing garments such as armbands, gloves, and socks. It was established through the series of experiments conducted on developed socks that the sensing technique is dependent on the pressure applied on the temperature sensing yarn owing to its compressible and flexible nature. With the application of pressure, yarn geometry changes thus leading to change in the scale of temperature sensing. This highlights the issues existing with e-textile-enabled wearables, thus motivating to address the same [8].

3 Human Body Motion Monitoring

Ailments leading to or caused by poor body posture further leads to mobility restrictions in humans. Diagnosis of most of such ailments is majorly dependent on symptom analysis such as gait and motor skill study. Patients suffering from Motor Skill Disorder, Developmental Coordination Disorder, and other neurological disorders leading to Parkinsonism, Epilepsy, etc., require their body motion activities to be tracked both for improvement of motor skills and for gait correction. In order to make the data capturing continuous and yet less intervening, during daily activity, smart textiles could be the most sought-after solution. Also, smart textiles have found their relevance for sports training or posture correction at workplace for the personnel to address the increasing occurrence of musculoskeletal disorder. Various research in this domain or related fields have been reported by scientists across the world.

Musculoskeletal disorders (MSD) developed due to long hours of working in poor posture, such as bent, turned, or twisted trunk, lead to moderate to severe low back pain and has become the most common MSD encountered by the majority. Though the awareness regarding the importance of maintaining correct posture has increased across the workforce, maintaining the same during work hours is difficult and lacking. Therefore, it is important to develop a feedback mechanism for individuals to send posture correction alarms during their daily activities at work or wherever required. Such attempts have been made by the researchers through the posture monitoring smart wearable vest. One such textile-based wearable to track the movement of the trunk when the wearer bends forward or laterally or turns has been developed using inductive sensors. The sensor is formed by sewing the copper wire onto an elastic fabric. After multiple iterations of varied sewing patterns, the zig zag pattern was found to provide the most optimised result as the pattern enabled increase in inductance value thus leading to better resolution. The wearable was tested on the subjects and was found to be able to distinguish the three movements of the trunk that are forward bend, lateral bend, and twist. Owing to the working principle of inductive sensor, magnetic interference of signals was anticipated to occur when in close vicinity of mobile phones. The same was evaluated and it was found not to interfere with the forward bend signal while the prototype was worn by the wearer but affected in unworn condition [9].

Smart wearable in the form factor of Socks has been further utilised by Amitrano et al. [10] for remote monitoring of biomedical signals. Sensoria socks is one such product existing commercially as well. The system consists of the sensor, data transmission unit, advanced signal processing desktop, and an Android-based application for real-time signal visualisation. The wearable is in the form of a Socks made up of 03 pressure sensors using e-textile which is placed at 03 different points of the foot arch. As per the human anatomy the force transmitted to the foot in standing positions is passed on from the lower leg onto the foot through the ankle by distributing them into three forces in a tripod style. Accordingly, the three sensors are placed one under the heel, second one onto the 1st metatarsal, and third one under the 5th metatarsal as located in Fig. 2.

These textile-based sensors are prepared out of conductive nonwoven fabric of size 2 cm × 4 cm by utilising its piezo-resistivity functionality, where electrical resistance is proportionately reduced with the application of force. These textile sensors are secured against abrasion and other wear by covering them with non-conductive fabrics. These three sensors are connected to the output unit with the help of a conductive ribbon of 5 mm thickness. The conductive path length is minimised by studying the shortest path to reduce noise and interference during signal capture and transmission. In this manner the sock acquires the plantar pressure signals along with angular velocities of the lower limbs which on post processing provides overview of patient's gait and postural assessment. Such wearable provides an alternate more comfortable

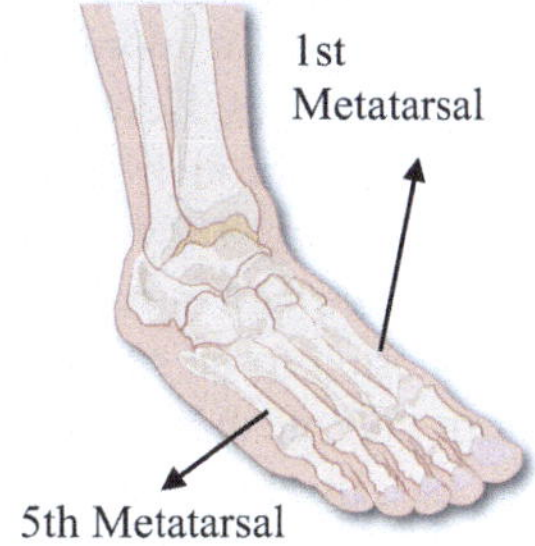

Fig. 2 Metatarsal locations in foot

and less intrusive solution in a domestic environment for continuous monitoring of patients. This can also be used for ergonomic assessment of personnel at the workplace [10]. E-textile enabled socks have been developed with sensory capability to capture plantar pressure and acceleration signals derived from the lower limbs motion. This attempt is to replace the clinical instrument otherwise used in hospital set-up. These signals captured through the socks were compared with the readings obtained from clinical instruments and were found to be similar, thus concluded to be reliable [11].

Another research reports usage of fabrics embedded with textile sensors to enable classification of human body movements. The product reported is in the form of a muscle activity pants through which initial trials were conducted for recognition of four different human motions namely sprinting, running, jumping and walking. Herein textile-based stretch sensors embedded in fabric collects reading from stretch points which act as data for the classification model. The model is based on an Adaptive Neuro Fuzzy Inference System (ANFIS) which integrates the features of Fuzzy Inference System (FIS) and Artificial Neural Network Model (ANN) and was claimed to be better in accuracy than both. The digitisation of the motion data from the signal captured by the stretch sensors, embedded in fabrics, is carried out by the processing unit, which is then fed to the microcontroller for extracting relevant data points to classify the motions [12].

Li et al. [13] have attempted to utilise an accelerometer for capturing the walking and running activity along with step counting. The study reports to have used the smallest commercially available accelerometer sensor of size 2 mm $\times$ 2 mm $\times$ 0.95 mm. This has been integrated within the woven structure thus making it hidden within the textile substrate. The form factor of this wearable is in a sleeve form, which can be worn in the arm and leg to measure arm and knee joint bending angle. The measurement of elbow joint angles ranging from 0° to 180° by this developed wearable was compared with a custom built goniometer and a commercially available bending sensor. The results were in good agreement with an error of 1% or lesser. Further, knee joint angles were measured on 03 subjects with 03 replications each for walking, running, step counting (stair climbing). The angle measurement data captured was then utilised for identification of activity type and for step counting [13]. With a different approach [14] have carried out a study to provide textile wearable solutions for monitoring skin strains along with human motions by utilising nano-composites. Herein graphene-based nanosheet films are applied onto the kinesiology tape through spray painting to develop a self-adhering strain sensor with high sensitivity. They were found to be capable of capturing the muscle engagement during functional movements, through subject testing. Further, spatial strain sensing was reported to be successful in two different configurations, i.e. both single strip pattern and mesh pattern. This research concludes with the possibility of utilising these tape meshes to measure patterns of engagement of different muscle groups for different physical activities [14].

Further, these textile-based wearable sensors have found their applications in monitoring progress of motor rehabilitation and sport training along with everyday activities monitoring. One such attempt has been reported by the researchers wherein warp-knit-based wearable sensor has been developed to function like a goniometer. A commercially existing conductive fabric named Electrolycra was selected based on the elasticity and conductivity offered by it. The fabric is prepared by warp knitting process using yarns made up of fibers which are turned conductive by silver particle coating. The process relies on the piezo-resistive property of the conductive fabric for it to be utilised as strain sensor and placed at elbow joint, in this case, to measure motion in two degrees of freedom-flexion–extension and prono-supination. This solution is reported to be suitable for situations where measurement accuracy could be compromised over the possibility of non-intrusive way of tracking signals continuously even through joint movements (flexion/ extension) [15].

Li et al. [16] have reported a dyeing method to fabricate a textile-based sensor. This study utilises interlock knitted polyester fabrics, which was plasma treated before functionalization and thereafter dip-coated using reactive silver ink. Thus coated (dyed) fabric was then connected to external copper wire on its two ends with the help of silver paste, for it to be utilised as a silver-based textile strain sensor. Thus, developed sensors were attached onto a stretchable adhesive tape and fixed to the corresponding positions for signal capture with attention paid to avoid friction and perspiration. The sensor was developed to monitor and analyse the kineses of the human body. The developed sensor with silver coating (dyeing) was reported to perform better over other conductive solutions like carbon nanotube and graphene. It was found to have excellent durability for strain-electric response. The captured data could be classified through Principal Component Analysis for identification of basic kineses such as squat, jumps, relax, walk, jog, and extend [16].

An attempt was made by researchers to develop textile-based sensors for detecting the activities such as breathing, walking, and other activities by capturing the minor changes in the resistance caused due to different movements of the muscles in humans. The study involved development and evaluation of sensors with different combinations of stitching pattern and stitch length with conductive sewing threads in terms of their sensitivity, response time, and working range. The sensor consisted of a multi-layer structure, made up of fabric, flexible conductive sheet, and a semi-rigid material. The developed textile-based sensors worked on the principle of piezo resistivity capable of monitoring step counting through hamstring muscle and breathing rate through ribcage. These were also evaluated post 10 cycles of washing and were found to be stable in performance post washing thus making them fit for various wearable applications [17].

Zhang et al. [18] report development of a self-powered activity monitoring textile-based wearable. The sensor is prepared out of silver yarn and polytetrafluoroethylene yarn to form a fibre shaped coaxial tribo-sensor. This ensured ease of integration in the clothing with desired flexibility, satisfactory breathability, and good washability. The sensor was found to offer outstanding finger-touching sensitivity, application for which could be as wireless controllers, active sensors, and for human–machine interactions such as to control home appliances wirelessly. Further calorie consumption and gait recognition can also be realised through this self-powered tribo-sensor [18].

Vu and Kim [19] have attempted to devise a textile-based sensor using 76/24 polyester/spandex woven fabric. The fabric was treated with ink having 1% by weight of a single wall carbon nanotube (SWCNT) through pad and dry process. The treated fabric was later printed with silver paste (Ag) through screen printing and developed into strain sensors of varied shapes through laser cutting. These sensors were integrated within a glove for monitoring the angle of finger motions. It was also attempted to classify the activity of speaking, coughing, and swallowing by placement of this strain sensor onto the skin of the neck and track the pharynx motion. The conversion of fabrics to conductive form by using SWCNT/Ag combination was reported to be better at performance than only SWCNT treatment [19].

While the section reports numerous examples existing as a solution for activity monitoring, very less attention has been paid to the fact that all of them require garments to be tight-fitting in to obtain sufficient signal-to-noise ratio (SNR). This makes them uncomfortable limiting the usage mainly for the applications such as active sports performance monitoring. One of the studies has attempted to solve this problem by adopting the principle of tribo-electric charge measurement induced by folding and compression of textiles. Successful capture of signals with high SNR through this principle would enable development of natural fit clothing with signal monitoring capability. However, attaining high SNR is a challenging task here, owing to the fact that this principle requires a large sensing surface, adding the body-coupled noise which is required to be suppressed in order to detect the weak tribo-electric signal generated. This research thus aimed at developing the textile-based sensor ensemble in a manner to enable maximum SNR. The developed fabric-based tribo-electric joint sensing system was named as Tribexor which could be integrated with natural fitting clothing and was capable of sensing sweat level, joint velocity, and joint flexion–extension. It consists of a textile patch which is optimised to collect electrical charge generated by the textile and a multi-stage amplifier circuit for voltage amplification of the tribo-signal and noise rejection. The layered ensemble of the sensor is provided in the schematic diagram given below for better understanding. A single layer of plain-woven fabric with cotton lycra blend, which functions as a dielectric, is used to cover the electrodes and the functionalized fabric. The additional layer of conductive silver coated nylon fabric serves as a Faraday cage, to protect the tribo-electric device from the impact of electric fields. The blue layer, composed of a two-ply gauze fabric, helps in the separation of tribo-electric layers. The whole ensemble is designed in a manner to be used as a stand-alone product in the form of a sleeve or alternatively to be integrated into a shirt. The accuracy achieved for detecting flexion and extension for elbow and knee joints was 95.1% followed by 88% accuracy for elbow angular velocity estimation and 83% accuracy for knee angular velocity estimation. The textile-based sensor thus developed was proven to be capable of activity classification, perspiration measurement, and comfort level detection for HVAC systems [20].

4 Electrocardiogram (ECG) Monitoring

The use of wearable systems for monitoring vital parameters has gained wide popularity in several medical fields. The electrocardiogram (ECG) is a crucial parameter for monitoring an individual's physiological state. Existing ECG monitoring systems, whether stationary or wearable, often fall short in providing satisfactory comfort for the wearer, as they are not seamlessly integrated into standard clothing. The attempts made to make ECG monitoring mobile, portable, yet comfortable are expected to be majorly used for the case of patients having chronic health conditions requiring continuous monitoring of ECG, blood pressure, and heart rate variability. Such a situation is conventionally addressed through halter-type systems which interferes with daily activity of Unlike halter-type systems, the textile-based sensors and the monitoring system designed thereof are to accommodate the daily activities of patients. Majority of e-textile enabled ECG monitoring systems focus on the development of wearable textile electrodes designed for an ECG measurement that prioritises user comfort. The electrode material comprises electrically conductive textiles, resulting in a textile composite that ensures both optimal user comfort and high-quality ECG measurements.

Vojtech [21] developed a textile electrode made of electrically conductive yarn made up of silver nano-coated polyester and cotton fibre blend. Thus, making the yarn not only is conductive but also anti-allergic and antibacterial owing to the presence of silver nanoparticles. These sensors are embedded on a textile substrate to be used in the wearable form as a T-shirt. This alternative utilises two lead ECG signals principle for monitoring the same. The system is resistant to washing and ironing and offers required comfort with sufficient quality of ECG measurements through captured signals [21].

Ambulatory ECG monitoring can be achieved using them in wearable forms with reduced number of surface leads. This is where smart garments have a role to play through utilisation of conductive textile electrodes to monitor ECG continuously and comfortably. A study took up T-shirt as the carrier using 13 textile electrodes for an ECG measurement system. The textile electrodes were prepared out of silver yarns and hydrogel pads with an ability to release water vapour to moisturise the skin for better signal capturing. It is important to highlight here that the smart T-shirt is CE-marked medical device with a capability to monitor ECG on 15 leads. To verify the working of the system readings were collected on subjects for four modes of activity three in rest positions, i.e. supine, seated, standing, and fourth as walking. The readings were comparable to that of conventional halter readings [22].

Mestrovic et al. [23] developed textile electrodes of area 20 mm × 20 mm which were integrated through knitting of conductive yarns. For this commercially available conductive yarn, of varied type, i.e. silver coated nylon yarn, stainless steel yarn, and silver coated copper filaments were used. These electrodes were strategically placed at required locations in a manner which ensured optimum skin pressure. The silver-nylon yarn was a multifilament twofold yarn where the core is nylon 6,6 and the sheath is made of silver coating. The stainless steel yarn was a multifilament yarn as well having 550 monofilaments in it. For silver coated copper yarns two variants were used in the form of continuous single filament and continuous multifilament. However, the silver coated copper yarn was found to be prone to breakage during knitting and was therefore twisted with wool and polyester yarn for mono and multifilament yarn respectively. The aim of the study was to study the feasibility of providing dry electrodes for ECG capture. Validation for the same was carried out by using the electrodes on the subject without preparing the skin with gel or any other moisture laden agent and they were found to be suitable for the same in dry form Mestrovic et al. [23].

A life shirt was developed to have inbuilt alarm-based system for detecting arrhythmia. This was attained by attaching and ECG sensing patch using textile-based conductive fibre. The selection of conductive textile was based on its knittability and ability to capture measurable signals upon connection with sensors. The sensor synchs with mobile devices enabling the user to monitor their state at any point of time and at any location. To provide uninterrupted power supply to the ECG sensor, the T-shirt is also developed with an option of wireless charging. The data captured is stored for later reference enabling remote diagnostic consultation [24].

Continuous efforts are being made by researchers to develop graphene-based conductive textiles to provide economical choice, as compared to commercially existing silver/silver chloride electrodes. In one of the attempts the synthesised Graphene oxide prepared through Hummer's method was applied on nylon fabric through dip coating, followed by reduction reaction on drying of the coating. Nylon fabric selected here was based on its minimum surface roughness among the available options. The resistance of thus developed conductive textile was reported to be in the range of 3 kΩ to 5 kΩ. The developed textile electrode was used in the form of ECG sensors by using snap on connectors to attach these textile leads onto the chest band fabric. The ECG waveforms from this textile-based system were found to correlate with waveforms gathered from conventional gel-based silver/silver chloride electrode [25].

Few researchers have worked on development of graphene-based conductive textile electrodes prepared using pad-dry-cure method. The developed resistive type sensors prepared out of reduced graphene oxide (rGO) and Poly(3,4-ethylenedioxythiophene) Polystyrene sulphonate (PEDOT: PSS) were found to be suitable for measuring ECG, pulse rate, and pressure response. The performance of these electrodes was found to be exceptional when compared with metal-based ECG electrodes along with the added advantage of flexibility and stretchability [26].

Another research in similar direction reports the development of graphene-based textile electrodes for capturing ECG by integrating the same in the clothing. Herein, the exhaust dyeing methods has been used instead of pad-dry-cure method. A one step process was developed to dye and reduce the graphene oxide on a textile substrate at 90-degree centigrade. Later the dyed and reduced knitted fabric was coated with conductive polymer PEDOT: PSS through dip coating. Thus, developed textile electrode was found to be stable till 20–30 washes and 120–130 flexing cycles. The ECG signals captured were found to be of high quality for heart rate in both dry and wet conditions [27].

Hafid et al. [28] have carried a feasibility study on integrating conductive textile in an intarsia knit structure. The developed electrodes were positioned strategically in a sensorized T-shirt. The measurement of ECG and heart rate through this developed system was compared with that of commercially existing chest strap. These measurements were carried out on the subject, while they carried out varied activities in static and dynamic conditions. The results were found to be comparable with chest strap both in dynamic and static condition [28].

Maglietta Interattiva Computerizzata (MagIC) is a system consisting of an ensemble inclusive of vest, textile-based sensors for ECG and respiratory activity monitoring, electronic board, signal pre-processing unit, and wireless data transmission unit. The vest is made up of cotton and spandex on which are mounted the two woven electrodes made up of conductive fibers for capturing ECG. Vest also consists of a textile-based transducer for measuring respiratory frequency which is related to change in volume of the thorax. The subjects were tested in moving and static conditions while at home, at office, driving, cycling and also while in microgravity condition during a parabolic flight. The device was able to give consistent results in both rest and moving condition of the subject. It also utilised an accelerometer for motion detection along with ECG monitoring. The data is remotely collected through the electronic board and data transmission unit which is connected through conductive yarns only [29].

A study reported development of textile-based resistive pressure sensors to detect physiological signal such as acoustic vibrations and hand motions with a response time of as short as 20 ms. This was prepared by drop coating a three-dimensional spacer knit textile with carbon nanotubes. The spacer textile consisted of two layers which were interconnected with a monofilament yarn acting as a spacer thus providing compression resistance and high air permeability, thus making it capable of being used as a pressure sensing device. It was found to be capable of measuring the change in pressure over a range of 200 Pa to 50 kPa. The carbon nanotube inks sensors could also detect various vibrations when attached directly to the human skin, thus creating real-time pressure spatial distribution of pressure mapping as well. The challenge lying with such a development is the limitation of scaling up the solution [30].

One of the research projects attempted to utilise e-textile-based wearable device for carrying out lower extremity Venous Occlusion Plethysmography (VOP). This is a diagnostic method to assess and then detect abnormal blood circulation in lower limbs and enable diagnosis of any related medical conditions. The study explored five different types of conductive textiles for proving their scope to be used as electrodes as an alternate to conventional set-up used for LEVOP. These were namely Argenmesh, Ripstop Silver, Stainless-Steel Mesh, Silver Mesh, and Stretch. Barring Argenmesh and Rip Stop Silver all other fabrics were developed through knitting. It was also observed that yarn which were all 100% metallic (Argenmesh and Stainless-Steel Mesh) faced challenges during knitting in flexing, handling, and breakage. The other yarn variants, i.e. silver mesh, ripstop, and stretch, designed especially for e-textile exploration offer better capability towards washability, flexing, twisting, etc. The readings from proposed LEVOP textile electrodes were found to be in good agreement with the standard Ag/AGCl electrodes. These samples used for testing were all made from nylon and silver fibre with blend % varying for each type of fabric [31].

Exploring the conductive materials in the form of coating, dyeing, etc., for developing e-textile-based wearables have been majorly attempted as a single stage approach. However, [32] report an exploration of the fabric structure during knitting to develop a textile-based wearable. Commercially available conductive yarn was used to prepare the electrode portion of the wearable by knitting the electrode areas with this yarn. This region was knitted in a manner that the region is voluminous in nature to have protruding area to enable better contact of electrodes with the skin and provide a stabilised position within the clothing. This was attained in a jacquard knitting machine. The yarn Elitex is a nylon yarn with silver coating in a textured multifilament form. The electrical connections were also integrated directly during the knitting process in the direction of wales. Regions required to be insulated were done through silicone coating. While the ensemble is stated to be in the form of a shirt, it may be noted that this is in the form of a body-hugging vest. This is a common practice among the researchers to quote the wearable product form to be called a shirt, for knitted vests which ensures required fit for electrodes to be placed for signal capturing, they all are mostly in the form of vests. The clothing here was made up of nylon/ spandex blend. This ensemble is one of the examples of complete integration of e-textiles into textile to provide an unobtrusive diagnostic method. However, such structures too have issues related to comfort to a certain extent and its possibility to be used daily could also be deliberated upon [32].

5 Electromyography Monitoring

Electromyography (EMG) is one of the useful diagnostic tools for monitoring muscle activity. However, the usage of the same is difficult for untrained users and the clinical solutions existing are invasive and expensive in nature. Researchers in the field of smart textiles have proposed an alternate solution by developing textile-based surface EMG (sEMG) which can be integrated into clothing. However, using textiles for such signal capturing has certain challenges due to variation in electrical properties of conductive textiles used as electrodes, lack of precise fabrication technologies while dealing with textiles such as embroidery, sewing, printing, etc., and lack of uniformity in design selection. Therefore, there has been continuous research in this area to improve the design of the textile sensors for addressing these issues stated and come up with a textile-based sEMG.

Researchers have attempted to study the effect of design variables on the quality of signal captured by textile-based sMEG, by optimising the set of design variables to trade-off between high manufacturing consistency and low electrode impedance. These textile-based electrodes were found to provide similar sensitivity to variation in force when compared with gel-based electrodes. The developed electrodes were taken for subject testing to detect muscle fatigue while running on different surfaces [33].

Graphene coated fabrics were prepared to be used as electrodes for surface electromyography (sEMG). The signal-to-noise ratio (SNR), skin–electrode impedance (SEI), and cross correlation of these electrodes were compared to that of commercially available wet silver/silver chloride (Ag/AgCl) electrodes. Skin–electrode impedance values of graphene-based textile electrodes were found to be within the range of commercial electrodes. Further The performance of textile-based sEMG in static settings was found to be at minimum 94% of SNR obtained from clinical setting. The cross-correlation readings from both also proved similarity of signal morphologies. The textile-based sEMG sensors were used as a textile elastic band along with a mobile acquisition circuit to validate its usage as an alternative to gelled Ag/AgCl electrodes and possibility of being used as a smart garment application [34].

Similar study was reported wherein a polyester fabric coated with edge-oxidised graphene oxide was used as a sensor to develop an EMG compression sleeve. The measurements carried out using the same were found to be better than that of commercial Ag/AgCl wet electrodes with respect to signal-to-noise ratio (SNR). The testing was carried out on eight test subjects. The existing commercial solution has a shortcoming of being non-reusable when compared with the graphene-based textile system of measurement [35]. Graphene coated fabrics have been explored in other research as well, for muscle activity monitoring and sEMG [36].

Muscle fatigue monitoring, through varied methods, is paid utmost attention in sports, for injury prevention. While attempts are being made to replace existing commercial solutions, which majorly consist of gel-based wet electrodes, with textile-based surface EMG, one of the studies has proposed usage of e-textile headband for indirect monitoring of muscle fatigue. This is to address the two major shortcomings existing with the e-textile sEMG solutions. Firstly, the motion artefacts created by the wearer (especially in the case of sports) interfere with sensor data. Secondly, the sensors mounted on the limbs (both lower and upper) disturb the limb motion, thus affecting performance involving physical movement of limbs. The e-textile-based headband estimates biceps fatigue. The e-textile headband consists of 5 sets of PEDOT-PSS (poly(3,4-ethylenedioxythiophene) polystyrene sulfonate) textile electrodes. While the centre electrode acts as a reference electrode one pair of electrodes each is located on both left and right temporal. These electrodes are connected to a Wi-Fi transmitter for data acquisition and transmission. A strong correlation was reported to exist between frowning and jaw clenching muscle activity with the physical efforts made when exercising. This method utilised both EMG signals and motion artefacts for studying muscle fatigue. While the EMG signals of the temporal muscles were used as an indicator for jaw clenching, motion artefacts were used as an indicator for frowning. With the increased fatigue in biceps the root mean square values calculated through the reading collected by the headband sensor was found to increase [37].

sEMG signals obtained from the remnant muscles above the amputation are one of the most convenient control signals for active prostheses. These require systems which can offer easy application for daily placement/repositioning of the electrodes by the user. The same is feasible to be offered by textile-based sEMG electrodes for recording high-density EMG signals to enable myo-electric control. One of the studies names it as Smart Fabric and Interactive Textile (SFIT) systems. The SFIT system is integrated in the sleeve form and connected through stainless steel yarns. The sleeve is a multi-layer structure; with electrodes being placed in the internal layer while external layer ensures right pressure on the electrodes by acting as support. SFIT system uses textile electrodes which enables usage of large numbers of electrodes in grid format thus helping in high-density EMG signal capture. Further the usage of SFIT does not require expertise or any clinical condition and was reported to be user friendly during application [38].

6 Electro-Oculography Monitoring

Electro-oculography (EOG) is a unique approach to analyse eye movement features which can then be applied in the area of medical diagnosis, wearable medical sensors, neurosciences, interfaces for human–computer interaction, etc. EOG records the bio potential signals induced due to eyeball movement. Conventionally existing EOG solutions, like EMG conventional electrodes, consist of wet gel-based electrodes. Wet electrodes reduce the portability and hinder the development of wearable EOG solutions for easy eye movement tracking. Thus, making it difficult to design and develop long-term EOG monitoring devices and applications. To address these challenges, smart textile researchers have proposed solutions by replacing the gel-based electrodes with e-textile electrodes. For this purpose, graphene-coated textile electrodes have been designed and developed to provide a dry monitoring solution for EOG. The performance of the developed electrodes was compared with commercially available Ag/AgCl wet electrodes and strong correlation of up to 87% was reported between the readings of the two. Further, comparison of both conventional and e-textile electrodes was carried out for a use case of blink detection and was found to perform at par with conventional measurement system [39–41].

6.1 Other Applications

Though the healthcare sector is dominated by e-textile-based wearables for majority of healthcare monitoring, smart textiles based on shape memory polymers (SMP) have found their applications in pain management through compression-based products (stockings or sleeves). However, they still have their limitations which appear over prolonged usage of the SMP-based textile solutions. Therefore, e-textiles have provided alternate and comfortable solutions by replacing the rigid components of a transcutaneous nerve stimulation (TENS) system with textile-based electrodes. But, currently the full integration of e-textiles in these solutions is being explored [42].

One of the studies reports development of a textile-based glove sensor which could be successfully applied to a 3D printed prosthetic hand, which exhibited stable responses to various types of grasps. The successful functioning of this sensor further provides options to explore its application for integrated prosthetic hand and in medical robotic hands. The textile sensor consisted of silver nanowire (AgNW) deposited over the polyester spandex fabric substrate. Herein polydimethylsiloxane (PDMS) films acted as interface for both planarization and passivation layers. The glove was mounted with five strain sensor units, one each of each finger to track their motion [43].

In the realm of therapeutic interventions, e-textiles provide innovative solutions for targeted drug delivery, wound management, and rehabilitation. By incorporating drug-eluting fibres or smart textiles with therapeutic functionalities, these textiles can deliver medications precisely to affected areas, promote wound healing, and assist in physical rehabilitation exercises.

Smart textiles have also been able to provide a novel solution for monitoring the status of incontinence. The ensemble is in the form of an incontinence pants prepared by advanced seamless knitting technology by the use of conductive yarns. The conductive yarn acts as electrodes which are connected through wires to an alarm system. The multi-level engineering (fiber, yarn, fabric, and clothing) possibility of textiles further enables achieving desired level of flexibility and tactile comfort. This smart incontinence pants is capable of real-time monitoring and sensing of incontinence status of the wearer and alerting the caregiver when required, thus, reducing the workload of patient care takers. Successful utilisation of this solution in its form is going to provide further potential applications in bio-functional healthcare products and solutions [44].

7 Conclusions

Exploration of e-textiles within the realm of healthcare unveils a multitude of promising avenues for improving patient well-being and transforming traditional medical practices. Seamless integration of advanced textile engineering with healthcare technologies is going to revolutionise patient monitoring, diagnostics, and therapeutic interventions. Through the development of smart fabrics embedded with sensors and microelectronics, e-textiles offer unprecedented capabilities for real-time health monitoring. These textiles have the potential to continuously assess vital signs, detect anomalies, and provide timely alerts, thereby facilitating early intervention and reducing the risk of adverse health events. Moreover, the fusion of e-textiles with telemedicine platforms enables remote patient monitoring. Furthermore, e-textiles hold promise in the realm of diagnostics, offering non-invasive methods for gathering physiological data with minimal discomfort to patients. However, as we continue to advance in the field of e-textiles for healthcare, it is imperative to address challenges related to durability, washability, and user acceptance. Collaborative efforts between textile engineers, healthcare professionals, and regulatory bodies will be crucial in overcoming these hurdles and realising the full potential of e-textiles in revolutionising healthcare delivery.

Acknowledgements Authors would like to acknowledge the contribution of Ms. Komal Sanghvi, Graduate in Fashion Design from Institute of Design and Media (ITM IDM), Mumbai, for figures drawn by her.

References

1. Black S (2007) Book Chapter "Trends in smart medical textiles." In: Van Langenhove L (ed) Smart textiles for medicine and healthcare: materials, systems and applications. Woodhead Publishing Limited, London
2. Massaroni C, Venanzi C, Silvatti AP, Presti DL, Saccomandi P, Formica D, Giurazza F, Caponero MA, Schena E (2018) Smart textile for respiratory monitoring and thoraco-abdominal motion pattern evaluation. J Biophotonics. https://doi.org/10.1002/jbio.201700263
3. Jiang Y, Pan K, Leng T, Hu Z (2020) Smart textile integrated wireless powered near field communication body temperature and sweat sensing system. IEEE J Electromagn RF Microw Med Biol 4(3):164–170. https://doi.org/10.1109/JERM.2019.2929676
4. Fan W, He Q, Meng K, Tan X, Zhou Z, Zhang G, Yang J, Wang ZL (2020) Machine-knitted washable sensor array textile for precise epidermal physiological signal monitoring. Sci Adv 6(11):eaay2840. https://doi.org/10.1126/sciadv.aay2840
5. Wang J, Soltanian S, Servati P, Ko F, Weng M (2020) A knitted wearable flexible sensor for monitoring breathing condition. J Eng Fibers Fabr. https://doi.org/10.1177/1558925020930354
6. Ramos-Garcia RI, Silva FD, Yashvanth K, Sazonov E, Dunne LE (2016) Analysis of a cover-stitched stretch sensor for monitoring of breathing. In: 10th international conference on sensing technology (ICST), Nanjing, China, pp 1–6. https://doi.org/10.1109/ICSensT.2016.7796258
7. Zhao C, Li X, Wu O, Liu X (2021) A thread-based wearable sweat nanobiosensor. Biosens Bioelectron. https://doi.org/10.1016/j.bios.2021.113270
8. Lugoda P, Riley TH, Oliveira C, Morris R, Dias T (2018) Developing novel temperature sensing garments for health monitoring applications. Fibers. https://doi.org/10.3390/fib6030046
9. Patiño AG, Khoshnam M, Menon C (2020) Wearable device to monitor back movements using an inductive textile sensor. Sensors 20(905):1–17. https://doi.org/10.3390/s20030905
10. Amitrano F, Coccia A, Ricciardi C, Donisi L, Cesarelli G, Capodaglio EM, D'Addio G (2020) Sensors, design and validation of an E-textile-based wearable sock for remote gait and postural. Assessment 20(6691):1–20. https://doi.org/10.3390/s20226691
11. Amitrano F, Donisi L, Coccia A, Biancardi A, Pagano G, D'Addio G (2020) Experimental development and validation of an E-textile sock prototype. In: IEEE international symposium on medical measurements and applications, Bari, Italy, pp 1–5. https://doi.org/10.1109/MeMeA49120.2020.9137302
12. Vu C, Kim J (2018) Human motion recognition using e-textile sensor and adaptive neuro-fuzzy inference system. Fibers Polym 19(12):2657–2666. https://doi.org/10.1007/s12221-018-8019-0
13. Li M, Torah R, Nunes-Matos H, Wei Y, Beeby S, Tudor J, Yang K (2020) Integration and testing of a three-axis accelerometer in a woven e-textile sleeve for wearable movement monitoring. Sensors 20(5033):1–15. https://doi.org/10.3390/s20185033
14. Lin Y-A, Zhao Y, Wang L, Park Y, Yeh Y-J, Chiang W-H, Loh KJ (2020) Graphene K-tape meshes for densely distributed human motion monitoring. Adv Mater Technol 2000861:1–9. https://doi.org/10.1002/admt.202000861
15. Grassi A, Cecchi F, Maselli M, Röling M, Laschi C, Cianchetti M (2017) Warp-knitted textile as a strain sensor: characterization procedure and application in a comfortable wearable goniometer. IEEE Sens J 17(18):5927–5936. https://doi.org/10.1109/JSEN.2017.2736944

16. Li Y, Li Y, Su M, Li W, Li Y, Li Y, Li H, Qian X, Zhang X, Li F, Song Y (2017) Electronic textile by dyeing method for multiresolution physical kineses monitoring. Adv Electron Mater. https://doi.org/10.1002/aelm.201700253
17. Choudhry NA, Rasheed A, Ahmad S, Arnold L, Wang L (2020) Design, development and characterization of textile stitch-based piezoresistive sensors for wearable monitoring. IEEE Sens J 20(18):10485–10494. https://doi.org/10.1109/JSEN.2020.2994264
18. Zhang X, Wang J, Xing Y, Li C (2019) Woven wearable electronic textiles as self-powered intelligent tribo-sensors for activity monitoring. Global Chall 1900070:1–8. https://doi.org/10.1002/gch2.201900070
19. Vu C, Kim J (2020) Highly sensitive e-textile strain sensors enhanced by geometrical treatment for human monitoring. Sensors 20–2383:1–12. https://doi.org/10.3390/s20082383
20. Kiaghadi A, Baima M, Gummeson J, Andrew T, Ganesan D (2018) Fabric as a sensor: towards unobtrusive sensing of human behavior with triboelectric textiles. In: SenSys'18: proceedings of the 16th ACM conference on embedded networked sensor systems, November 2018, pp 199–210. https://doi.org/10.1145/3274783.3274845
21. Vojtech L, Bortel R, Neruda M, Kozak M (2013) Wearable textile electrodes for ECG measurement. Biomed Eng 11(Special Issue 5):410–414
22. Fouassier D, Roy X, Blanchard A, Hulot J-S (2019) Assessment of signal quality measured with a smart 12-lead ECG acquisition T-shirt. Ann Noninvas Electrocardiol 00:e12682. https://doi.org/10.1111/anec.12682
23. Mestrovic MA, Helmer RJN, Kyratzis L, Kumar D (2007) Preliminary study of dry knitted fabric electrodes for physiological monitoring. In: 3rd international conference on intelligent sensors, sensor networks and information, Melbourne, VIC, Australia, pp 601–606. https://doi.org/10.1109/ISSNIP.2007.4496911
24. Park JA, Han HJ, Heo JC, Lee JH (2017) Computer aided diagnosis sensor integrated outdoor shirts for real time heart disease monitoring. Comput Assist Surg (Abingdon) 22(sup1):176–185. https://doi.org/10.1080/24699322.2017.1389396
25. Acar G, Ozturk O, Yapici MK (2018) Wearable graphene nanotextile embedded smart armband for cardiac monitoring. IEEE Sens. https://doi.org/10.1109/ICSENS.2018.8589800
26. Shathi MA, Chen M, Khoso NA, Rahman MT, Bhattacharjee B (2020) Graphene coated textile based highly flexible and washable sports bra for human health monitoring. In: Materials and design, 193, August 2020, p 108792
27. Shathi MA, Minzhi C, Khoso NA, Deb H, Ahmed A, Sai WS (2020) All organic graphene oxide and poly(3,4-ethylene dioxythiophene)-poly(styrene sulfonate) coated knitted textile fabrics for wearable electrocardiography (ECG) monitoring. Synth Met. https://doi.org/10.1016/j.synthmet.2020.116329
28. Hafid A, Gunnarsson E, Ramos A, Rödby K, Abtahi F, Bamidis PD, Billis A, Papachristou P, Seoane F (2023) Sensorized T-shirt with intarsia-knitted conductive textile integrated interconnections: performance assessment of cardiac measurements during daily living activities. Sensors (Basel). https://doi.org/10.3390/s23229208
29. Rienzo MD, Rizzo F, Parati G, Brambilla G, Ferratini M, Castiglioni P (2005) MagIC system: a new textile-based wearable device for biological signal monitoring. Applicability in daily life and clinical setting. In: IEEE engineering in medicine and biology 27th annual conference, Shanghai, China, pp 167–7169. https://doi.org/10.1109/IEMBS.2005.1616161
30. Kim K, Jung M, Jeon S, Bae J (2019) Robust and scalable three-dimensional spacer textile pressure sensor for human motion detection. Smart Mater Struct. https://doi.org/10.1088/1361-665X/ab1adf
31. Goy CB, Dominguez JM, Lopez MAG, Madrid RE, Herrera MC (2013) Electrical characterization of conductive textile materials and its evaluation as electrodes for venous occlusion plethysmography. J Med Eng Technol 37(6):359–367. https://doi.org/10.3109/03091902.2013.812689
32. Catarino A, Carvalho H, Dias MJ, Pereira T, Postolache O, Girão PS (2012) Continuous health monitoring using E-textile integrated biosensors. In: International conference and exposition on electrical and power engineering, Iasi, Romania, pp 605–609. https://doi.org/10.1109/ICEPE.2012.6463867

33. Shafti A, Manero RBR, Borg AM, Althoefer K, Howard MJ (2017) Embroidered electromyography: a systematic design guide. IEEE Trans Neural Syst Rehabil Eng 25(9):1472–1480. https://doi.org/10.1109/TNSRE.2016.2633506
34. Ozturk O, Yapici MK (2021) Surface electromyography with wearable graphene textiles. IEEE Sens J 21(13):14397–14406. https://doi.org/10.1109/JSEN.2020.3048924
35. Awan F, He Y, Le L, Tran L-L, Han H-D, Nguyen LP (2019) ElectroMyography acquisition system using graphene-based e-textiles. In: International symposium on electrical and electronics engineering, Ho Chi Minh City, Vietnam, pp 59–62. https://doi.org/10.1109/ISEE2.2019.8920937
36. Ozturk O, Yapici MK (2019) Muscular activity monitoring and surface electromyography (sEMG) with graphene textiles. IEEE Sens. https://doi.org/10.1109/SENSORS43011.2019.8956801
37. Nijima A, Isezaki T, Aoki R, Watanabe T, Yamada T (2018) Biceps fatigue estimation with an e-textile headband. In: ISWC'18: proceedings of the 2018 ACM international symposium on wearable computers, Singapore, pp 222–223. https://doi.org/10.1145/3267242.3267281
38. Farina D, Lorrain T, Negro F, Jiang N (2010) High-density EMG e-textile systems for the control of active prostheses. In: Annual international conference of the IEEE engineering in medicine and biology, Buenos Aires, Argentina, pp 3591–3593. https://doi.org/10.1109/IEMBS.2010.5627455
39. Golparvar AJ, Yapici MK (2017) Wearable graphene textile-enabled EOG sensing. IEEE Sens J. https://doi.org/10.1109/ICSENS.2017.8234242
40. Golparvar AJ, Yapici MK (2018a) Electrooculography by wearable graphene textiles. IEEE Sens J 18(21):8971–8978. https://doi.org/10.1109/JSEN.2018.2868879
41. Golparvar AJ, Yapici MK (2018) Graphene-coated wearable textiles for EOG-based human–computer interaction. In: IEEE 15th international conference on wearable and implantable body sensor networks (BSN), Las Vegas, NV, USA, pp 189–192. https://doi.org/10.1109/BSN.2018.8329690
42. Papaiordanidou M, Takamatsu S, Mazinani SR, Lonjaret T, Martin A, Ismailova E (2016) Cutaneous recording and stimulation of muscles using organic electronic textiles. Adv Healthc Mater. https://doi.org/10.1002/adhm.201600299
43. Heo JS, Shishavan HH, Soleymanpour R, Kim J, Kim I (2020) Textile-based stretchable and flexible glove sensor for monitoring upper extremity prosthesis functions. IEEE Sens J 20(4):1754–1760. https://doi.org/10.1109/JSEN.2019.2949177
44. Liu R, Wang S, Lao TT (2012) A novel solution of monitoring incontinence status by conductive yarn and advanced seamless knitting techniques. J Eng Fibers Fabr 7(4):50–56

Herbal Infusions: Weaving Nature's Pharmacy into Textiles

Hariharasudan Natarajan and Periyasamy Siddhan

Abstract This chapter focuses on using different plant-based extracts and infusing them with textile substrate which in turn can be used for various medical applications besides antimicrobial effect especially in assisting wound healing either directly or indirectly, leading to a sustainable future. The chapter deliberates on methods of extracting medicinal compounds from plants that can assist the healing process either directly or indirectly, methods and techniques used for integrating the extracts into the textile material for durability and efficacy. Further on the section has a detailed discussion on aromatic textiles using plan-based scents and its used and application in holistic healing. A new frontier in medical textiles—adaptable textile, stimuli-responsive textiles and controlled drug release textile material and incorporating herbal extracts with this material are conversed. Added to these, natural dyes and the method for dyeing them create a sturdy effect of coating for various medical applications including colour therapy that can assist ailments in eco-friendly healing which will be the new era of healing.

Keywords Adaptable textiles · Aromatic textiles · Antimicrobial textiles · Herbal wound dressings · Natural dyes plant extracts · Textile infusion

1 Introduction

Plants have been used for centuries to treat diverse ailments and they are still widely used today [1]. Plant extracts are made by soaking the herbs in an appropriate solvent and the solvent extracts contain the active ingredients from the herbs that are further filtered and concentrated. Herb extracts are often more potent than the herbs themselves and they can be easier to contain and store. They are also more consistent in terms of their active ingredient content. Herb extracts can be used in a variety of ways, including internal/oral consumption, external application and aromatic infusion.

H. Natarajan (✉) · P. Siddhan
Central Silk Technological Research Institute, Central Silk Board, Bangalore, India
e-mail: hari.csb@nic.in

S. Sundaramoorthy et al. (eds.), *Textile Materials for Good Health and Wellbeing*, SDGs and Textiles, https://doi.org/10.1007/978-981-97-6496-9_18

The practice of extracting herbal compounds from plants for incorporation into medical textiles dates back to the beginning of human civilization. Throughout history, various cultures have utilized the healing powers of nature's resources. This ancient practice has been modernized and incorporated into medical textiles, providing a harmonious combination of traditional wisdom and modern technology.

By combining the wisdom of medical compounds from plants, traditional herbal practices with the accuracy of modern extraction techniques, medical textiles not only provide protection and healing, but also bridge the gap between traditional and modern textile and manufacturing techniques. In a future, where nature's pharmacy is met with modern fabric engineering, medical textiles offer the potential for holistic healing and improved well-being. When extracted, these valuable compounds are incorporated with a fabric, creating a medical textile, providing its wearer with its therapeutic benefits. Antimicrobial herbal extracts are used in wound dressings to protect against infections and hasten healing. Compounds for aromatherapy derived from plants such as Lavender or Chamomile can be encapsulated in textiles, providing long-term relaxation and stress-relieving benefits. Textiles with natural dyes that contain medicinal compounds offer a comprehensive healing approach. For example, natural dyes such as turmeric and indigo can have an anti-inflammatory effect; improve skin health and relieving irritations [2]. However, as with any innovative endeavour, there are certain challenges that must be addressed when integrating natural compounds into textile materials, which include the stability of the bioactive molecules in herbal compounds, the release kinetics of these compounds and the potential effect on fabric durability during manufacturing process.

1.1 Extracting Medicinal Compounds from Plants

On careful separation of the bioactive molecules, the medicinal compounds can be extracted from plants, which can include essential oils, polyphenols, flavonoids and other compounds with therapeutic properties. There are a variety of extraction methods available to entrap these biological and healing substances, such as steam distillation, solvent extraction, cold pressing, etc. Steam distillation involves the use of pressurized steam to vapourize essential oils, which are then condensed into a liquid. This method is known for its ability to preserve the aromatic compounds present in plants. Solvent extraction, on the other hand, involves the use of solvents such as ethanol, methanol or water to dissolve the desired compounds and extract them from the plant material. The solvent is then evaporated, resulting in a concentrated extract that does not require high temperatures. Cold pressing is a popular extraction method for citrus fruits, where the plant/biological material is pressed to yield juices/oils that maintain their natural fragrances and therapeutic properties.

Further, novel methods like supercritical fluid extraction typically involve the use of supercritical CO_2 as a solvent. This method provides a gentle and effective way to extract compounds without the need for high temperatures, thereby preserving the fragile nature of the biological molecules. Traditional methods of infusion and decoction involve the steeping of plant material into water or other fluids to produce herbal tea or solution extracts. These methods are less concentrated than other methods; however, they provide a direct approach for extracting and transferring medicinal properties to textile materials.

1.2 Integrating Herbal Extracts into Textile Fibres

The combination of natural healing resources and cutting-edge textile engineering has resulted in novel concepts where the incorporation of herbal extracts in textile fibres is made possible. This innovative approach not only alters the characteristics of fabrics, but also marks the beginning of a new age of functional textile textiles that provide a variety of health and well-being benefits. By incorporating the healing compounds of herbal extracts into textiles, we are embarking on a journey in which nature and technology work together to improve human health and well-being [3].

1.2.1 Methods and Techniques of Integration

Incorporating herbal extracts into textiles is one of its kind and relatively complex process that necessitates careful consideration of how to balance the effectiveness of the extracts with the longevity of the fabric. Various techniques and approaches have been developed to accomplish this goal.

Solution dyeing method involves the addition of herbal extracts in a dye bath, along with appropriate solution like water, ethanol, methanol, etc. [4]. As the dye is applied to the fabric, both its colour and its medicinal properties are absorbed into the fibres, resulting in a fabric that bears both the visual and medicinal properties of the herbal extracts, that enables the therapeutic potential of the extracts within the textile structure.

Pad-dry-cure method, on the other hand, involves saturating a fabric with a solution of herbal extracts, after which the fabric is padded to remove excess solution further to which the material is dried to remove any excess moisture. This is followed by a curing process at appropriate temperature to enable the extracts to be fixed onto the fibres, thus prolonging the duration of the beneficial effects of the herbal extracts.

Microencapsulation is a sophisticated process that involves the encapsulation of herbal extracts in microscopic particles or capsules, which are then adhered to the textile surface. This enables the controlled and permanent release of compounds over time, thus conserving the textile properties while simultaneously providing therapeutic benefits.

Nanocoating is another form of nanotechnology that enables the formation of very thin protective layers on the surface of textiles [2]. These coatings contain herbal extracts and act as reservoirs for therapeutic compounds. Nano coating allows for the gradual and appropriate release of extracts when the fabric is exposed to the skin. Herbal extracts can be incorporated into textile fibres to provide a range of functional benefits, from anti-inflammatory properties to aromatherapy benefits.

2 Applications on the Medical Infused Textiles

2.1 Antimicrobial Properties

Herbal extracts, which possess natural antimicrobial qualities, can be beneficial in medical application, as they can act as a protective barrier against harmful micro-organisms. This makes them an ideal choice for medical clothing/apparel and wound dressings, as well as for use in healthcare applications.

2.2 Herbal-Based Aromatherapy

Herbal extracts can be used for providing a calming and therapeutic aroma, which can be infused into textiles to create a harmonious release of calming scents. This can help to reduce stress and promote relaxation.

2.3 Skin Health

Textiles that are integrated with herbal extracts with skin-soothing properties, can help to improve skin health by reducing inflammation and providing comfort. This is especially beneficial for sports apparel and medical textiles.

The integration of herbal extracts in textile fibres is expected to continue to revolutionize functional textiles with advances in technology and scientific knowledge. Future developments in sustainable extraction methods, optimization of encapsulation and controlled release mechanisms will further refine the integration process, paving the way for a new co-relation between nature and fabric engineering. This innovative combination has the potential to transform the way textiles are viewed, from passive materials to dynamic agents for wellness, comfort and healing.

3 Aromatic Textiles: Scented Therapy Through Fabric

Aromatherapy is a centuries-old form of therapeutic art and practice that utilizes aromatic herbal extracts to influence and promote psychological, physical and emotional well-being. Now, with the advent of modern medical textiles, aromatherapy has the potential to transform health care sector, providing patients along with the medical professionals with the sensory drive to improved physical and emotional well-being.

The connection between aromatherapy and the brain is deeply rooted. While inhaling, the aromatic compounds in essential oils activate the olfactory nerves, which are connected directly to the brain's limbic system, which is responsible for emotion, memory and physiological responses. For example, the scent of lavender is known to induce relaxation through the influence of neurotransmitters such as serotonin and gamma-amino-abundant acid (GABA) which are responsible for regulating mood and stress.

3.1 Role of Aromatherapy in Medical Textiles: Scented Healing for Enhanced Well-Being

Aromatherapy is based on the volatile compounds found in essential oils derived from different varieties and parts of aromatic plants, which have distinctive scents and therapeutic effects. When inhaled, these scents trigger reactions in the brain, which can have an impact on emotion, mood and physiology. This is supposed to be due to the link between the body's olfactory and limbic systems, which are regulating emotions, memory and stress responses. This shift in healthcare practices is marked by the inclusion of aromatherapy textiles in medical field, which can provide a calming atmosphere and reduce stress and anxiety.

Microencapsulated fabrics, when infused with essential oils, can emit therapeutic scents over a long period of time, creating a harmonious and reassuring atmosphere that can reduce stress, anxiety, pain, and mood and promote sleep. Lavender, for example, can be incorporated into textiles, such as hospital gowns and bedding, to create a tranquil atmosphere amid medical environments. The therapeutic aroma creates a serene ambiance, helping patients and healthcare providers alike manage stress.

Some essential oils have analgesic or muscle-relieving properties. For example, textiles infused with eucalyptus, chamomile and peppermint have analgesic and muscular-relieving properties and can be used on bedding, in patient clothing and in recovery rooms to reduce discomfort and improve overall wellness. When textiles are infused with essential oils, patients not only do experience physical comfort, but they also experience psychological relief. For example, recovery rooms and patient clothing can become sensory experiences that contribute to pain relief.

Aromatherapy's effect on sleep quality has been well-documented. Textile infused with essential oils that induce sleep, such as Chamomile or Lavender, can create sleep-inducing environments in healthcare facilities. Not only does this improve sleep quality for patients and residents, but it also aids in the healing process, making it a natural approach to improve sleep quality meanwhile textile-infused essential oils have a positive effect directly on mood and behaviour of humans.

For example, citrus scents, like lemon or orange are known to be energizing and uplifting and can be incorporated into textiles. These scents can be used for a mental health clinic, relaxation area and space where emotional support is predominant. Spaces designed for relaxation or mental health support can benefit from the positive effects of these scents.

The use of textiles infused with extracts for aromatherapy can have a significant impact on the emotional well-being of paediatric patients. Children often find medical facility to be intimidating and the use of gentle and calming scents can help to create a more comfortable environment in which they can feel more at ease and supported throughout their treatment at the facility. This can lead to a significant transformation in children's wards.

3.2 Holistic Healing and Future Perspectives

Medical textiles infused with aromatherapy provide a comprehensive approach to healthcare, recognizing the strong correlation between sensory experiences and overall health. As healthcare facilities become more patient-focused, textiles infused for aromatherapy provide an invaluable tool for improving the healing environment. Future developments may include the inclusion of customizable and on demand scent profiles, allowing for the sensory experience to be tailored to the individual's preferences. Ultimately, these textiles bridge the gap between clinical effectiveness and compassionate care, creating an environment where healing embraces the body, mind and senses.

Lavender essential oil is a well-known and popular choice for aromatherapy that focuses on relaxation. Its gentle floral aroma has been found to be particularly effective in reducing stress, anxiety and insomnia. Peppermint essential oil, also referred to as *Mentha Piperita*, is characterized by its refreshing and invigorating scent and is frequently used to boost energy levels, improve concentration and reduce headaches, nausea and other symptoms. Eucalyptus essential oil has a refreshing and medicinal aroma. Its medicinal properties for respiratory wellness are well-known and it is often used for congestion and improved breathability. Chamomilla essential oil, which is derived from the plant *Matricaria chamomilla*, has been found to have calming properties and is used for relaxation, stress reduction and sleep aid. It is suitable for both adults and children.

Essential oils with bright, uplifting scents, such as Lemon, Citrus, Rosemary, Geranium, Frankincense and Ylang Ylang, are known to improve mood, energy and cleanliness behaviour. Geranium and frankincense are particularly well-known for their grounding and calming properties, while Ylang Ylang and geranium are known for their floral and slightly sweet scents. Ylang Ylang and geranium essential oils are well-known for their mood balancing properties and are used to reduce stress, improve concentration and improve mental clarity. Sandalwood (*Santalum album*) essential oils are also popular for their calming effects and are often used in meditation, sleep and relaxation.

Essential oils such as Bergamot, Clary sage and Tea Tree are known to have a range of effects on mood, anxiety and emotional well-being. Bergamot has a citrusy, uplifting aroma, while Patchouli is an earthy, musky scent with a calming effect. Clary sage has a herbal, slightly sweet scent, while Tea Tree has a fresh, medicinal aroma with powerful antimicrobial, antiseptic and immune support properties. Jasmine is a deeply floral, romantic fragrance that can be used to enhance mood, confidence, sensuality and relaxation.

Essential oils for calming the mind, promoting relaxation, emotional balance, reducing anxiety, can be found in a variety of sources, including Lemongrass, Neroli, Cedarwood and Basil. Lemongrass is a bright, citrusy scent that invigorates and energizes the body, while Neroli is derived from citrusy orange blossoms. Cedarwood is a warm, woody aroma with grounding and calming effect, while Ginger is a spicy, warming fragrance that invigorates the mind, promotes digestion and provides comfort during times of mental and physical discomfort. Lastly, Marjoram is a nature-derived essential oil that is known to promote emotional release, improve focus and ease muscular discomfort.

Aromatherapy involves the use of a wide variety of herbs and herbal extracts, each of which has its own distinct set of therapeutic benefits and properties. As a result, aromatherapy has a high potential to be adapted for meeting the specific requirements and preferences of any person.

3.3 Infusing Textiles with Aromatic Healing: The Scented Fabric Experience

Aromatherapy integration into medical textiles gives advantage of the long-lasting release of essential oil aromas. Microencapsulation is a technique as described previously that involves the encapsulation of essential oil droplets into microscopic capsules. As the fabric is worn or is used, the capsules gradually diffuse the contented aroma, creating a healing atmosphere.

Application of textiles for aromatherapy offers great potential, however, there are some challenges that must be addressed. Stability during textile integration, long-term scent release and any potential allergies to certain scents must be considered. Additionally, the patient's and healthcare professionals' preferences and sensitivities should be taken into account when selecting scents.

Encapsulation is a more advanced technique that prolongs the aromatic effects in a variety of applications, such as personal care products, textiles, air fresheners and more. This method involves the entrapping of essential oil molecules into microscopic capsules, allowing for a more controlled and consistent release of therapeutic scents over a longer period of time. This approach increases the efficacy of aromatherapy.

3.3.1 Microencapsulation Process

The process of microencapsulation involves the encapsulation of small droplets of essential oils within protective shells composed of materials such as polymers and lipids. The capsules serve as reservoirs for the essential oil, gradually releasing its aroma as it interacts with the environment.

In order to ensure the stability of essential oils to be encapsulated, a number of steps must be planed and taken care of. The first step is to select an appropriate encapsulation shell material, such as a natural or synthetic polymer or lipid-based material. The essential oil is then mixed with the material to form an emulsion, which ultimately breaks down into small droplets. This emulsion is then added to an appropriate solution that enables microscopic capsulation. The second step is to harden and solidify the capsules, either by cooling or by chemically cross-linking them, to prevent leakage. Finally, the capsules are washed and dried to remove any excess moisture, further increasing their stability during storage and application.

This method of encapsulation can be used to incorporate the essential oil into a variety of products, including different/various textile, personal care and air freshener applications. As the product is utilized, the capsules gradually release their contents, providing a long-term aromatherapy benefit.

3.3.2 Advantages of Encapsulation for Aromatherapy

Encapsulation for aromatherapy offers a wide range of applications and advantages, including increased longevity, controlled release and enhanced aromatic experience. Traditional diffusing methods can quickly evaporate essential oils, resulting in uneven scent distribution, while microencapsulation allows for a consistent, controlled release. Additionally, encapsulated essential oils are able to be incorporated into a variety of products, such as personal care items and textiles, allowing a smoother integration into daily life. Additionally, microencapsulation reduces the intensity of some essential oils, while still maintaining their therapeutic effects.

3.4 Challenges and Considerations

Integration of aroma with material to create medical textiles is a unique combination of traditional healing methods and modern textile engineering that provides a comprehensive approach beyond physical healing. Medical facilities are increasingly recognizing the significance of emotional well-being in the healing process and textiles infused with aromatherapy are a shining example of innovation. However, there are some important considerations that must be taken into account when integrating extracts and textiles for aromatherapy. The integration of essential oils requires careful investigation and development to ensure their stability and longevity of scent on processing. Additionally, certain scents may cause allergic reactions, so it is important to carefully select and test the essential oils for suitability. Lastly, microencapsulation necessitates specialized equipment, knowledge and consideration of factors including the selection of encapsulating material, compatibility of essential oils, capsule stability and the durability of the capsules.

Aroma-infused fabrics for the future have opened up new possibilities and have the capability to transform healthcare sector into environments of both clinical effectiveness and sensory comfort, providing a holistic healing experience for patients, care-takers and healthcare professionals. As technology progresses, researchers are continuing to investigate encapsulation methods and materials that provide enhanced scent release, longevity and personalization. Eco-labelled aromatherapy has a high potential to significantly improve well-being and transform everyday routines into spaces of sensory solace and therapeutic support.

4 Antimicrobial Weaves: Nature's Defence Mechanisms in Textile Form

4.1 Unveiling Nature's Healing Arsenal for Infection Control

As the fight against infectious diseases intensifies and antibiotic resistance becomes a major concern, the search for natural antimicrobial substances has gained significant traction. Drug-resistant pathogens are on the rise and pharmaceutical industries are looking for more sustainable, safe ways to control infections. Nature has a rich repertoire of compounds with powerful antimicrobial properties. These natural antimicrobial substances offer a holistic, sustainable approach to infection prevention. The variety of compounds in nature's healing arsenal, derived from plant, mineral and other sources, provide a unique way to fight infections while reducing the risk of developing resistance. From aromatic plant essential oils to bioactive molecules from herbal sources, there is a wealth of therapeutic potential in the natural antimicrobial world.

4.2 Power of Nature's Pharmacy: Natural Antimicrobial Agents

Antimicrobial agents discussed in the chapter are natural compounds derived from a variety of sources that contain a range of bioactive compounds. These compounds are designed to protect against pathogens and while they are effective against microbes, they have a gentler effect on human cells than synthetic antibiotics, thus making them ideal for infection control [5]. A few of the main categories of natural antimicrobials are listed.

4.2.1 Essential Oils

Essential oils are highly concentrated extracts derived from plants, which capture their aromatic and therapeutic properties. These essential oils are composed of a wide range of volatile compounds, including terpenes, phenols and other compounds, which are known to possess potent antimicrobial properties. A prime example of this is Tea Tree oil, which contains terpenes such as thyme and cinnamon is renowned for its powerful antibacterial and antifungal properties. Additionally, oregano oil, which contains carvacrol, as well as thymol, has broad spectrum antimicrobial activity against a variety of pathogens.

4.2.2 Plant Extracts

Extracts from a variety of plants contain bioactive molecules with intrinsic antibacterial and antimicrobial properties. Garlic, for example, is a well-known allicin-rich plant that has been studied and shown to possess antibacterial, antiviral and antifungal properties. Additionally, other plants such as Echinacea and Neem, as well as grapefruit seed, have been studied and demonstrated to be effective in the treatment of infections.

4.2.3 Antimicrobial Peptides

An Antimicrobial Peptide (AMP) is a short chain of amino acids that plays a critical role in innate immunity against a variety of target pathogens. AMPs attack pathogens through a variety of pathways, making them difficult for microbes to become resistant to. AMPs are derived from a variety of organisms, such as humans, insects and amphibians and offer a novel approach to infection control.

4.2.4 Silver Nano Particles

Silver has a long history of antimicrobial use in natural and alternative medicine where it exhibits enhanced antimicrobial properties due to its nano size and larger surface area. It is employed in a variety of applications, such as medical dressings and coatings for medical textiles [6]. Nano-sized silver particles can be integrated with medical textiles, devices and wound dressings to reduce microbial growth in textile materials.

4.2.5 Honey

Honey, a highly esteemed natural sweetener, has outstanding antimicrobial properties. The antimicrobial properties of honey are largely attributed to its low water content and acidity, as well as the release of the bioactive compounds like hydrogen peroxide. These factors contribute to its efficacy in wound healing and the control of infection. Manuka honey is particularly renowned for its antibacterial properties and its wound healing capabilities.

4.2.6 Propolis

Propolis is a natural resin produced by bees through the collection of plant resins. It is used to seal the hives made by the bees. Propolis is composed of a complex combination of compounds with antibacterial, antiviral, anti-inflammatory and antifungal properties. Its therapeutic potential is widely used in natural medicinal products and topical applications, ranging from oral health to skin care.

4.3 Benefits and Applications

The application of natural antimicrobial agents in infection control offers several advantages. Natural antimicrobial agents offer a potential solution to the increasing concern of antibiotic resistance, as they possess complex mechanisms that make it difficult for pathogens to become resistant. Furthermore, they have a wide range of effects beyond antimicrobial properties, as they can be effective against a variety of pathogens. Furthermore, these natural antimicrobial agents are biodegradable, with a lower environmental impact than any synthetic chemicals, which is in line with sustainable practices. This holistic approach to healing, which takes into account not only the physical aspects of infections, but also the emotional and mental health aspects, that is beneficial for overall health and well-being. The following are some of the natural antimicrobial agents that offer a variety of therapeutic possibilities for infection control, demonstrating the potential of natural antimicrobial pharmacies.

As previously discussed, Tea Tree oil is a well-known source of antibacterial, antifungal and antiviral properties. Oregano oil is also known for its antimicrobial properties due to the presence of thymol. Garlic is another source of antimicrobial properties, as it contains allicin. Echinacea contains immune-boosting compounds that help to combat infection. Grapefruit seed extract is also known to be antimicrobial, as it is able to fight off bacteria, fungi and viruses. Manuka Honey is also well-known to have antimicrobial properties due to its low water content and acidity. Propolis, which is derived from bees, has antiviral, antibacterial and antifoliate properties. Silver nanoparticles are also known for their increased antimicrobial activity when used in medical textiles and dressings [7]. Cinnamaldehyde-rich cinnamon essential oil is known to have antimicrobial effects, while Lavender essential oil is also known to possess antiviral and antibacterial properties. Lemon is another source of anti-inflammatory properties due to its citric acidity and hydrogen peroxide content. Manjistha (*Rubia cordifolia*) is used in Ayurvedic medicine for its antibacterial and anti-inflammatory effects.

Ginger, Turmeric, Chitosan and Neem leaf extract are all antimicrobial and antiviral substances that are known to have anti-inflammatory and antifungal properties. Goldenseal contains berberine, which has anti-inflammatory and immune-boosting properties, while Clove essential oil contains eugenol. Neem leaf extract has powerful antiviral, antibacterial and antifoliative properties, while Lemongrass is an essential oil with a refreshing citrus aroma. Additionally, Myrrh essential oil has antibacterial and anti-inflammatory properties, as well as antiseptic and respiratory health properties and is traditionally used for wound healing. Bamboo fibres contain a natural antimicrobial substance called bamboo kun, which can resist microbial growth when applied to textiles.

4.4 Enhancing Textile Performance with Natural Antimicrobials

In a world where global health challenges, hygiene and environmental concerns which are paramount, the demand for antimicrobial textiles has surged as a game changer in health care sector and even in day-to-day clothing. These textiles are designed to inhibit the growth and proliferation of micro-organisms, deter the proliferation of pathogens on fabric surfaces [8], thwarting the spread of infections and preventing the emergence of odours. Commercially used synthetic antimicrobial agents have been instrumental and widely used in achieving the objectives, but the appeal of natural alternatives lies in their potential to address these challenges and compatibility with sustainability practices and consumer preferences for eco-conscious choices.

4.5 Benefits and Applications of Antimicrobial Textiles

Antimicrobials are a fundamental component of healthcare and play an essential role in the healthcare sector in terms of infection control and prevention of healthcare-related infections. Additionally, antimicrobials can help to reduce odours caused by micro-organisms in textiles, making them an ideal choice for sportswear, activewear and underwear. Furthermore, antimicrobial fabrics are eco-friendly, as they are compatible with sustainable practices and can be washed multiple times to maintain their antimicrobial properties. Finally, antimicrobial fabrics can be used in everyday applications, such as beds, towels and daily clothing, to promote hygiene in routine activities.

4.6 Challenges and Considerations

Medical textiles based on natural antibiotics offer a range of advantages with potential to revolutionize the use of natural antimicrobial substances. However, challenges include maintaining uniformity, achieving standardization, establishing optimal concentrations of extract, controlled antimicrobial release, maintaining textile properties after treatment and managing potential allergic and toxic reactions. Current researches are focused on optimizing integration methods and prolonging the effects of antimicrobial substances. Further investigations are required to gain a comprehensive understanding of their mechanism of action, synergies and potential constraints.

4.7 Future prospects of Nature's Defence Mechanisms in Textile Form

The use of textiles is no longer limited to aesthetics and comfort, but is now embracing a comprehensive approach to health and prevention of infection. In the exponentially evolving world of textile development, the combination of natural antimicrobial agents with textiles has initiated a transformation that goes beyond simply improving functionality; it is a step towards a more health-conscious future, a step towards a healthier and more sustainable way of living. Utilizing the natural antimicrobial qualities of natural compounds to create versatile textiles that not only enhance user experience, but also promote hygiene, safer environment, the combination of textiles and natural antimicrobial compounds transcends the concept of materials; it elevates them to an integral part of hygiene, health and safety. As technology progresses and scientific knowledge advances, the potential for the use of antimicrobial fabrics to transform our surroundings and promote wellness is becoming increasingly attractive and promising.

This novel collaboration between nature and technology is a major development in fabric cum medical engineering. The potential of natural antimicrobial compounds has been identified in a variety of areas, ranging from medical to household use as stated previously. Further investigation of their combination therapies, prevention approaches and preservatory applications could lead to a paradigm shift in infection control. As science progresses, this harmonious partnership between man and nature's pharmacy could open the door to a more secure, sustainable and efficient approach to infection control and wellness. Research is continuing to investigate optimal integration techniques and the durability of antimicrobial effects.

5 Herbal Dressings: Wound Healing Enhanced by Medicinal Textiles

5.1 Herbal Ingredients for Enhanced Wound Care: Nature's Remedies for Healing

The emergence of herbal-based solutions to wound care is revolutionizing the approach to promote effective healing and reduce complications. The medicinal properties of herbs have long been appreciated for their medicinal properties and their revival in wound care reflects the wide range of bioactive compounds involved in healing. This resurgence is not simply a trend, but rather a paradigm shift in which nature's gifts are embraced in a scientific context. Drawing on the wealth of natural remedies, herbal ingredients are becoming increasingly effective in wound management due to the combination of traditional wisdom and modern scientific knowledge. By embracing the therapeutic properties of herbs, wound care is undergoing a holistic development that accelerates healing, reduces inflammation and promotes tissue regeneration with overall well-being.

Aloe-vera is a symbol of natural healing, with its polysaccharide and enzyme-rich gel providing a combination of anti-inflammatory, cell-stimulating and collagen-promoting properties. Its moisture-retaining properties help to prevent wound dehydration and scabbing, thus aiding in wound healing. Additionally, the fragrant petals of Calendula contain flavonoids and triterpenoid-carotenoids, which combine to produce remarkable antiseptic and antimicrobial applications. Comfrey, on the other hand, is known for its allantoin-rich compounds, which promote wound healing by stimulating cell growth and collagen synthesis, as well as stimulating skin-soothing, reducing pain and restoring damaged skin. Lavender is also a natural remedy for wound healing.

Tea Tree oil contains a wide range of antimicrobial compounds, which are effective in fighting bacterial, fungal and viral infections, thus providing protection against wound care related issues. Echinacea, on the other hand, is a polysaccharide and alkamide-based compound that helps to regulate the immune system and promote tissue repair and inflammation. Turmeric, the compound of a golden hue, is known for its antioxidant and anti-inflammatory properties, as well as its ability to reduce oxidative stress, minimize inflammation and actively promote tissue regeneration. Additionally, Gotu Kola, a compound found in plantain leaves, helps to promote wound healing while also reducing inflammation and infection. All of these compounds work together to improve immune and accelerate wound healing.

5.2 *Benefits and Mechanisms of Herbal Wound Care*

Herbal ingredients possess a range of anti-inflammatory properties, creating a supportive environment for wound healing, reducing pain and swelling. Additionally, many herbal compounds possess inherent antimicrobial properties, which can reduce the risk of infection and its associated complications. Furthermore, certain herbs stimulate the production of collagen, which is essential for wound closure and the formation of healthy tissue. Parallelly, hydration and moisture balance are key components of wound healing, as herbal ingredients prevent excessive wound dehydration and promote the optimal moist environment for optimal healing. Finally, certain herbs can help to reduce scar formation by aiding the development of healthy tissue.

5.2.1 Applications and Considerations

The primary method of use for herbal ingredients in wound care is topical application, which typically takes the form of a cream, ointment, gel, powder or poultice. Nevertheless, it is important to take into account individual sensitivities and allergies, as well as potential interactions with traditional wound care treatments.

5.3 *Creating Medicinal Dressings with Herbal Textiles*

A novel combination of herbal knowledge and textile innovation has resulted in the development of medicinal dressings made from herbal textiles in the field of wound care. This convergence combines the natural healing powers of herbal ingredients with the functional abilities of textiles to offer improved wound care solutions [9]. Immersing nature's remedies in the fabric that supports wound healing provides a comprehensive approach for healing, soothing, protecting and accelerating healing process.

5.4 *Genesis of Medicinal Dressings: Integrating Nature and Technology*

The integration of traditional knowledge of herbs with modern textile engineering has resulted in the development of nature based medicinal dressings. These dressings are composed of textiles that have been infused carefully with chosen herbal ingredients, providing a dynamic synergy that addresses both physical wound care and physiological needs. Textiles that are infused with herbal extracts or essential oils or plant compounds are used to create a range of benefits, including accelerated healing, minimization of scarring and protective benefits [10].

Textile-infused dressings deliver a calming and moisturizing effect of aloe extracts directly onto the wound site, providing an environment that is conducive to healing and reducing discomfort. Aloe has anti-inflammatory properties and is known to be effective in wound healing. When combined with the essential oil of lavender, the dressing is further enhanced by its antimicrobial properties, analgesic properties and antiseptic properties. Turmeric contains curcumin which when incorporated into dressings is highly potent in anti-inflammatory properties, antioxidant properties and tissue regenerative properties. Tea Tree oil is used in wound dressings to provide natural protection against infections due to its broad spectrum of antimicrobial properties.

Medicinal dressings provide an alternative route for herbal compounds to reach the wound site, optimizing their effectiveness while reducing systemic exposure. These dressings contained with herbal ingredients help to promote a supportive healing environment by reducing inflammation, avoiding infection and stimulating tissue regeneration. As a result of their prolonged antimicrobial activity and their support for wound healing, these dressings find potential of being used more frequently. Furthermore, medicinal dressings take into account not only physical healing, but also the emotional and psychological well-being of the patient, resulting in a holistic healing experience. Despite their potential, the challenges associated with the development of medicinal dressings include consistency in distribution of herbal ingredients within textile surface, optimizing release rates and managing potential allergic reactions.

5.5 Shaping the Future of Wound Care: A Promising Horizon

Medical dressings offer the potential to transform the way one can care for the wounds. Textiles are no longer confined to their traditional roles, but instead become active partners in the health, hygiene and recovery. As scientific knowledge advances, herbal textiles and modern wound care technologies are being combined to create a revolutionary approach. With each woven thread, the healing bounty of nature is woven into the material of the dressings, merging the essence of traditional knowledge with the science of modern medicine and each dressing serves as a proof of nature's ability to heal. Further research is essential to understand the mechanisms and clinical effectiveness of herbal components in wound care. The combination of traditional knowledge with modern medical practices has the potential to revolutionize this field.

The incorporation of herbal components into wound care is a traditional practice that combines traditional knowledge with modern science. Utilizing the benefits of natural herbal remedies to enhance wound care goes beyond traditions and treatments. Herbal ingredients possess antiseptic, antimicrobial and regenerative capabilities, providing a natural pathway for faster healing, fewer complications and improved overall health. As we further explore the intricate web of nature's pharmacology, wound care has the potential to embrace the harmonious combination of traditional and contemporary practices.

6 Adaptable Textiles: Responsive Medicine-Infused Fabrics

6.1 Stimuli-Responsive Textiles for Targeted Drug Delivery

Stimuli responsive textiles, commonly referred to as "smart textiles," are the materials that are able to adapt and modify their characteristics in response to either external or internal stimuli or conditions. Stimuli responsive textiles are capable of sensing one or few changes in the material or around it and adapting accordingly, creating a dynamic interface between the textured material and the wearer. The combination of various advanced textile and pharmaceutical technologies has led to the development of this new field, stimuli-responsive textiles for directed drug delivery. By combining the functional capabilities of textile materials with the precision required for drug administration, this innovative approach to therapeutic intervention offers the potential to optimize drug release, minimize adverse reactions and improve treatment efficacy. By enabling the self-regulation of drug delivery, stimuli-responsive textiles have the potential to revolutionize the complete medical landscape.

At the intersection between textiles and herbal remedies, a revolutionary breakthrough has been made. This breakthrough has led to the development of stimuli-responsive, herbal-based textiles for drug delivery. By combining the healing powers of nature with fabric that responds to specific cues in the body, textiles offer a novel approach to therapeutic intervention. These textiles promise to optimize drug release, reduce adverse reactions and improve treatment effectiveness. Not only does this innovation have the potential to revolutionize drug delivery, but it is also likely to contribute to the development of personalized medicine.

A few groundbreaking discoveries have been made at the interface between textiles and herbal medicinal products. This discovery has resulted in the creation of stimuli-responsive herbal textiles for the purpose of drug delivery [11]. Combining the healing capabilities of nature with a fabric that reacts to specific signals in the human body, textiles provide a novel therapeutic intervention. These textiles are expected to optimize drug release, minimize adverse reactions and enhance treatment efficacy. This innovation not only has the potential to transform drug delivery, but also has the potential to contribute to the advancement of personalized medicine.

6.2 *Advantages of Stimuli-Responsive Textiles in Targeted Drug Delivery*

Stimulus-responsive textiles facilitate the precise delivery of drugs to specific areas of the body, thereby reducing systemic exposure and reducing adverse reactions. These textiles may be designed to release drugs at predetermined rates and intervals, thus enhancing treatment effectiveness and patient comfort. Additionally, stimuli-responsive textiles enable treatments to be tailored to specific areas, such as wound areas or affected tissues, thus maximizing therapeutic impact. Furthermore, these textiles may reduce frequency of drug administration by responding to physiological changes, thereby increasing patient compliance to treatment plans. Comparison of stimuli-resistant textiles may provide real-time feedback regarding drug release, allowing for timely adjustments and customized treatment optimization.

6.3 *Incorporating Herbal Agents for Controlled Therapeutic Release*

The art of controlled therapeutic release is based on the artful combination of textiles and herbal agents, which incorporates the essence of natural healing properties within the fibres in the fabric. This integration is accomplished through a combination of advanced techniques, enabling the controlled, sequential and purposeful liberation of herbal compounds.

Therapeutic fabrics are revolutionizing the field of medicine, as the healing process of nature is incorporated into textiles to facilitate controlled therapeutic release. This novel approach is a perfect combination of ancient herbal wisdom and advanced material science, providing a platform where textile material acts as sophisticated carriers of therapeutic compounds. Through complex engineering, these fabrics react to physiological cues, facilitating the controlled and targeted liberation of herbal compounds in a way that optimizes treatment results, minimizes adverse reactions and expands the boundaries of personalized medicine and therapeutic interventions.

Herbal agents are incorporated into textiles through a variety of processes, such as encapsulation, coating or impregnation. All of these processes are designed to facilitate controlled release of herbal agents. This process involves the gradual release of herbal compounds from textile material over a predetermined period of time at a consistent rate. This approach involves textiles that are designed to react to external stimuli, such as temperature fluctuations or exposure to chemicals, in order to release herbal agents when activated. These textiles respond to the body's physiological signals and release herbal agents when desired conditions are met, resulting in a highly personalized and targeted therapeutic effect.

Therapeutic textiles utilize a controlled release mechanism to deliver herbal agents precisely to specified areas, thereby reducing systemic exposure and improving treatment accuracy. Maintaining a consistent, controlled release profile optimizes the delivery of herbal compounds, thereby maximizing their therapeutic efficacy. The controlled release of herbal agents reduces sudden peaks in concentration of herbal compounds, which can result in adverse effects, thereby enhancing the safety and comfort of the treatment experience. By delivering herbal agents in a controlled and steady manner, patients receive a consistent therapeutic effect, resulting in improved overall comfort and well-being. Patients can benefit from a customizable and adaptive therapeutic approach that is tailored to their individual health conditions.

Despite the immense potential of these textiles, challenges remain in terms of consistent and predictable herb compound release, durability of herb coatings and biocompatibility issues. Advancements in nanotechnologies, material sciences, herbal medicine and biomedical engineering are expected to address these issues. Overcoming the challenges of engineering accurate release profiles, maintaining stability of herb coatings or encapsulations and assuring biocompatibility is essential for the widespread use of controlled therapeutic release fabrics.

With each fibre and herb infusion, the present gets one step closer to a future where textiles are actively contributing to our health and well-being and tailored treatments are created to meet our individual needs. These re-imagined textiles combine the best of nature, engineering and medicine, transforming the way we experience healing and heralding a major step forward in the development of therapeutic fabrics.

The integration of herbal agents into controlled therapeutic release is a momentous event in which science and nature come together to create a story of healing. Textiles that carry herbal wisdom become active agents for health restoration and this innovation opens the door to reimagining medical treatments, enabling a harmonious combination of tradition and innovation [12]. With each fibre, controlled therapeutic release creates a web of possibilities, in which precision, efficacy and patient-oriented care come together to define the future of medical interventions.

7 Natural Dye-Infused Textiles: Colourful Medicine for Ailments

7.1 Colouring Fabric with Nature's Healing Palette: Utilizing Natural Dyes with Medicinal Properties

Since its inception, the art of textile dyeing has undergone a dramatic and drastic transformation as the use of natural dyes containing medicinal properties has become the primary focus of many innovations. This remarkable combination of colour and healing has led to the incorporation of herbal-based natural pigments into textiles, providing not only a vibrant range of hues, but also the therapeutic qualities of nature [13]. By utilizing the various benefits of plants, textiles have achieved a harmonious combination of aesthetic appeal and health-promoting benefits, creating a comprehensive approach to well-being through fashion [14].

Incorporating medicinal properties into textiles through natural dyes is a combination of art form, science and engineering. These pigments are sourced from plants, each with its own unique therapeutic properties, are fermented, boiled, etc., to extract the variety of colours and medicinal properties found in nature [15].

Dye-infused textiles offer a unique aesthetic appeal that is apprehended by eco-minded consumers, combining a classic charm with contemporary sensibilities. By wearing textiles dyed with natural colours which also inhibit medicinal values, the skin can interact with the therapeutic compounds, potentially providing therapeutic benefits and promoting overall well-being [16]. Additionally, the colours and scents of such textiles can stimulate the senses, resulting in an immersive and healing experience. These materials are often sustainably sourced and have a minimal environmental impact, in line with eco-friendly practices, as natural dyes are more environmentally friendly than synthetic dyes [17].

Turmeric, which is renowned for its bright golden yellow hue, contains curcumin which is discussed to have powerful anti-inflammatory, antioxidant and healing properties. Indigo, which is used to dye textiles, is also known for its anti-inflammatory and soothing properties. Madder root, which is found in *Madder tinctorum*, produces red and pink shades and contains compounds that are known to be antibacterial and antiseptic, making it suitable for fabrics that come into contact with skin. *Lawsonia inermis*, also known as henna, contains the compound lawsone, which is known to be soothing and pleasant, can be used to promote comfort, skin and hair care with textiles. Safron is not only used for culinary purposes, but it also contains antioxidants and is examined to offer a range of health benefits besides its bright colour.

The Chamomile, Hibiscus, Lavender and Walnut species offer a variety of colour extracts which are also consumed as herbal tea, each with its own unique characteristics. The Chamomile flower has a light yellow hue and is rich in compounds such as chamazulene, which is known to have antiseptic and calming properties. The Hibiscus flower has a vibrant red hue and is packed with antioxidants, making it a potential health benefit. The Walnut hull is brown in colour and contains the flavonoid juglone. The Annatto seeds, Onion skins and Lavender are all sources of antioxidants and flavonoids, with Lavandula being particularly well-known for their antioxidant properties.

Rose, Beet-root, Nettle and Lemon peels offer a range of health benefits when ingested. Rose's petals are pink and red, while beetroot's leaves are brown and contain beta-vulgaris, which are known to have anti-inflammatory and antioxidant properties. Nettle's leaves are green in hue and contain nutrients such as vitamins and minerals, while lemon peels are yellow in hue and contain citric bio-flavonoids, which are believed to support the immune system. Lemon peels are also known for their digestive properties and soothing properties.

Black Walnut, also known as *Juglans nigra*, is a species of hardwood tree whose hulls are typically dark brown or black in colour. It is traditionally used for its antiparasitic properties and its peel can produce shades of yellow. Additionally, pomegranate, also known as *Punica granatum*, has compounds such as ellagic acid which is known to have antioxidant properties that are often associated with health benefits.

The natural dyes continue to add to the beauty of textiles, both in terms of their aesthetic colours and the therapeutic properties of the plant compounds. Incorporating these dyes into fabrics can create a harmonious blend of aesthetics and well-being. The use of natural colours with medicinal properties transcends fashion to create a lifestyle that values both aesthetics and well-being [18]. Textiles become more than just garments; they become pitchers for self-care and a way to connect with the natural world's healing rhythms. By imbuing fabrics with the therapeutic properties of plants, they are able to transfer healing energy.

7.2 Colour Therapy and Healing Benefits of Herbal-Dyed Textiles

Herbal-dyed textiles offer a unique combination of aesthetic appeal and health benefits. Not only do they aesthetically enhance the body, but also engage the senses and may promote physical and emotional well-being. Herbal-dyed fabrics, which are dyed with extracts from natural sources, contain a wide range of healing properties as briefed that can be transmitted through the skin. These compounds can be used to soothe the skin, reduce inflammation and provide antimicrobial and antioxidant benefits. For example, compounds such as curcumin from turmeric, chamomile's chamazulene and lavender's linalool can be absorbed by the skin and lead to improved health and well-being.

Colour therapy (also referred to as chromotherapy) reveals a captivating world where colour's vibrational frequencies interact with human anatomy and physiology to facilitate holistic healing. Drawing on ancient wisdom and backed by modern scientific knowledge, this practice utilizes the profound influence of colours on human health. Warm colours such as red and orange evoke warmth and dynamic energy. These colours stimulate the adrenaline glands, resulting in increased energy and vitality. These bright colours are commonly used to combat exhaustion and boost motivation. Blue and green colours evoke tranquillity and calmness. They activate the peripheral nervous system, resulting in reduced stress and anxiety levels. These colours are used to promote inner peace and relaxation.

Colours have the capacity to evoke feelings. Bright colours, such as yellow, are associated with optimism, which can improve mood and reduce feelings of depression or sadness. Yellow is commonly used to evoke feelings of joy and emotional contentment. Colour therapy is based on the idea that each colour corresponds to an individual chakra, which is the energy centre of the body as said in the literatures on traditional medicine (Ayurveda) and spiritual science. Applying the appropriate colours for the correct chakras can help to restore balance and promote harmony. Some colours have analgesic properties, such as blue and violet, which are linked to pain relief and can be used to reduce physical discomfort. When exposed to these colours, it is witnessed that the body will release natural painkillers known as endorphins.

Colours are associated with cellular regeneration and healing, particularly green and turquoise. These colours are believed to activate the heart chakra and promote physical rejuvenation and the body's immune system. Additionally, colours influence cognitive function. For example, indigo and purple hues stimulate the mind, improving creativity and concentration. These colours are often used during activities that require creative thinking and increased mental clarity. Furthermore, colours carry emotional associations and colour therapy provides a safe and controlled path for any person to express and release their own emotions. Working with colours that resonate with specific emotions can help individuals to process and heal their emotions. Certain colours also empower the wearer. For example, gold and yellow, which are often associated with the solar charka and penta-chakra respectively, can help to boost self-confidence and increase self-esteem.

Colours have long been associated with spiritual symbolism, with violet and white being symbols of spirituality and higher awareness. These colours are thought to be beneficial for spiritual development, intuition and connecting with the divine. Additionally, colours can influence sleep patterns, with soft and calming colours such as lavender and pastel being known to promote improved sleep quality. Calming colours such as green and blue have been found to have a calming effect, reducing stress levels and fostering a sense of serenity. Colour therapy utilizes specific colours to target specific areas of the body, allowing healing energy to be directed towards physical restoration. Colour therapy serves as a mirror to our inner self, allowing one to become more aware of our own psychological and emotional states. Colour therapy strengthens the bond between mind, body and spirit, promoting holistic well-being.

Aromatic textiles, such as lavender and mint, are dyed with herbs that impart their fragrances to herbal dyeing textiles promote sustainability through the use of natural dyes, which are eco-friendly, biodegradable and environmentally friendly. In addition to their environmental benefits, herbal dye textiles offer additional value through potential healing properties. Wearing herbal dye textiles is more than just wearing garments; it is an act of conscious acceptance of nature's bounty. These textiles promote holistic well-being and encourage wearers to become more connected to the Earth's natural resources and to appreciate its nurturing properties. Traditional herbal dyeing techniques have a deep cultural and historical significance in various cultures [19]. By wearing herbal dye textiles, individuals are able to embrace these traditions and feel a tangible connection to the past. By choosing herbal dye textiles individuals are able to make conscious decisions about what they wear on their bodies, empowering them to live a mindful life.

Herbal-dye textiles combine colour, fragrance and healing properties in a harmonious way. These textiles symbolize a deep bond between people and nature, immerse themselves in a harmony of wellness, a harmony that is in harmony with the natural rhythms and the intent to live a balanced life.

Colour therapy is a form of vibration therapy that encompasses the entire spectrum of human experience, from physiological reactions to emotional shifts. Colour therapy is a keystone to a holistic approach to healing, encompassing the full range of human well-being. It is a reminder that healing transcends the physical realm, encompassing the emotional, physiological and spiritual dimensions of our lives.

Within each hue, there is a message that adds to the harmony of wellness. Colour therapy invites us to discover the transformative potential of colour and to embrace a life of healing harmony.

The potential of natural dyes and their medicinal properties is immense; however, challenges remain in terms of consistent colour and scalability, as well as the standardization of the inclusion of natural compounds in textiles.

8 Eco-Friendly Healing: Sustainable Approaches to Medicinal Textiles

In the current climate of environmental awareness and a growing need for a holistic approach to well-being, the integration of sustainable practices with medicinal textiles is a noble concept. As we work to bring healing into harmony with our planet's health, the notion of environmentally friendly healing through sustainable methods of using medicinal textiles is gaining traction. This new paradigm shift re-imagines the healing process as a multi-faceted endeavour that encompasses the health of the entire planet, not just the individual.

The concept of eco-friendly healing is rooted in an ethos of interdependence, which recognizes the delicate equilibrium between humans and nature. It promotes a mindful approach to well-being, which not only addresses the underlying causes of ailments, but also reduces the environmental impact of the treatment process. The eco-friendly approach to healing begins with the cultivation and harvest of medicinal plants in a sustainable manner. Through organic farming practices, the use of permaculture methods and the cultivation of agro-forestries, biodiversity is conserved and soil degradation is prevented, thus providing a continuous supply of healing herbs.

Eco-friendly healing is an ethos of interdependence that recognizes the delicate equilibrium between human well-being and the natural world. It focuses on a mindful approach to healing, which not only addresses the underlying causes of ailments, but also reduces the environmental impact of healing. Eco-friendly healing begins with sustainable cultivation and extraction of medicinal herbs. Organic farming, permacultures and agro pro-forestry promote biodiversity and prevent soil degradation, thus providing a sustainable source of healing herbs for the long term.

Recycling of textiles that are eco-friendly are made from materials that can be biodegraded and renewable, including Cotton, Bamboo, Tencel and more. These materials break down naturally, reducing the environmental impact of synthetic fabrics. The use of herbal dyes and environmentally friendly printing techniques on these materials eliminates the need for chemical dyes, as well as water-intensive manufacturing processes. This approach not only ensures that textiles have healing properties, but also meet eco-friendly standards. Furthermore, the adoption of zero or minimum waste production processes reduces the impact of textile production on the generation of waste and the landfill contribution, thus adhering to sustainable production principles.

Manufacturing medicinal textiles domestically reduces the carbon footprint of production, processing and transportation. Promoting local artisans and manufacturers also contributes to the resilience of communities and economic sustainability. Recycling or upcycling of discarded textiles reduces textile waste and conservation of resources. Textile design for durability ensures that products can be used for a long period of time, resulting in fewer replacements and reduced resource consumption. Efficient healing through sustainable solutions to medicinal textiles has a positive impact on individual well-being as well as global ecological health. When we embrace practices that are in line with sustainability of the Earth's resources and ecosystems, not only do we experience healing within ourselves, but we also contribute to the health of the planet.

8.1 A New Era of Healing

The emergence of eco-friendly healing in medicinal textiles marks a new era of consciousness and healing. It stresses that the decisions we make in our quest for wellness are inextricably linked to the decisions we make for the well-being of our planet. When we align our healing journey with environmental practices, we are not only nourishing our own vitality, but we are also actively contributing to the renewal of a harmonious, interconnected world—a world where healing is part of the intricate web of nature and humanity.

In the quest for holistic health, the incorporation of natural medicinal products into textile applications has become increasingly popular as a means of improving therapeutic experiences. This combination of nature's healing potential and cutting-edge textile technology offers great potential, however, it is important to examine both the effectiveness and safety of these applications. Natural medicine research examines the complex web of plant-based compounds, fabric engineering techniques and their implications for human health [20]. By unravelling the complexities of efficacy and safety in textiles, researchers navigate the intricate web of nature's potential.

Several researches have been carried out worldwide to identify and quantify the bioactive compounds present in medicinal plants. Establishing the phyto-chemicals that are responsible for therapeutic effects is the basis of efficacy evaluation. By examining the bioavailability of the medicinal compounds when integrated into textiles, it is possible to determine how well these compounds are absorbed by the body and interact with it. Textile engineering methods facilitate the controlled release of the herbal compounds. The objectives of these researches are to optimize the mechanisms to guarantee a consistent and lasting therapeutic effect. By optimizing the properties, it is made possible to direct the medicinal compounds to specific parts of the body, improving the targeted therapeutic effects. Studies conducted in-vitro and in-vivo environment to assess the physiological responses elicited by herbal and textile interactions. The outcome of the studies reveal the potential health benefits associated with the wearing of medicinal textiles and human trials evaluate their efficacy in real-world situations, providing valuable information on user experiences, dosing and overall effectiveness.

Herbal textiles are subjected to research and analysis to determine how they interact with various skin types, as well as to identify any potential irritation, allergy, or sensitivity. Studies are conducted to assess the potential toxicity associated with herbal compounds when exposed to the skin for extended periods of time, in order to ensure the safe use of textiles in everyday life. Furthermore, research is conducted to identify the potential allergic reactions associated with herbal textiles, in order to guarantee that therapeutic benefits are not jeopardized by undesirable side effects. Meanwhile, long-term studies conducted to determine the safety of long-term use of medicinal textiles ensure that long-term use does not result in unexpected complications. Research conducted to assess the environmental impact of herbal textiles, with a focus on biodegradation and ecological friendliness have proven results, in order to contribute to overall well-being.

Research in the field of natural medicine combines with textile applications works between science, technology and holistic wellness, embodying a rigorous exploration of natural resources and a dedication to responsibly utilizing them. By combining efficacy with safety, a network of innovation that improves human health, enhances therapeutic outcomes and is in line with environmentally conscious practices is witnessed. The path to investigate the role of natural medicine in textiles is a complex one, one that will ultimately enrich our knowledge of the healing relationship between nature and human creativity.

8.2 Future Horizons: Innovations in Natural Medicine Integration in Textiles

The future of integration of natural medicine/herbal extracts into textiles is set to be a thrilling journey at the intersection of ancient healing knowledge and modern technology. As we currently stand at the precipice of transformative opportunities, the integration of natural medicine with textiles that herald innovations that will revolutionize the realm of wellness and therapeutic experience. Here, we look forward to the captivating future where fabrics act as conduits for holistic health and healing.

Nanotechnology allows medicinal compounds to be encapsulated at the nanoscale enabling drug delivery to specific areas of the body to optimize therapeutic effects and minimize systemic exposure which can be enhanced with sensor-embedded textiles with micro-processors to detect physiological alterations in the user. Textiles that emit herbal compounds when exposed to stress or discomfort can provide tailored healing responses. By combining 3D printing with herbal compounds, a complex and personalized medicinal textile could be created. Textiles could be designed to adapt to the user's needs and provide personalized therapeutic benefits. Textiles could respond differently to each user based on genetic predisposition. Wearable bioreactor textiles could contribute to the health of oneself and microbial balance as wearable health technology becomes more integrated.

The application of quantum physics to textiles along with natural healing compounds could lead to the harnessing and transmission of delicate energies for healing purposes. These fabrics could interact with the natural energy fields of the body, thus promoting holistic well-being. Researchers may develop cutting-edge textiles with intrinsic bioactive properties, which could be naturally infused with compounds that provide ongoing therapeutic advantages without additional treatments. Advancements in genetic analysis and data collection could open the door to textiles specifically tailored to an individual's health requirements, utilizing the potency of particular medicinal compounds. Additionally, textiles could be integrated with virtual reality environments, creating multi-sensory healing spaces that integrate visual, auditory and tactile sensations for enhanced therapeutic benefits. Sustainability is likely to be a key factor in the future integration of natural medicine into textiles, with textiles being engineered to be more environmentally friendly through the use of eco-friendly dyes and circular fashion practices.

Artificial Intelligence driven analysis of wearable health device data could provide guidance on the optimal herbal compounds to be included in textiles, as well as guidance on the best time and place to wear certain textiles for optimal benefit [21]. Furthermore, research into traditional medicinal practices across the globe could uncover new herbal resources to be incorporated into textiles, thus broadening the range of healing compounds available. Innovation is boundless and collaborations between science and herbals from various backgrounds, as well as designers and engineers from various disciplines, could lead to breakthroughs that bring together different perspectives. Integrating natural medicine in textile products for a broader wellness eco-system could create a smooth integration of clothing, wearable technology, nutrition and mindfulness practices [22].

The end result of the integration of natural medicine in textiles is a world where people are able to take control of their health and well-being through the textiles they wear. Textiles become more than just garments; they become active participants in a personal journey towards vitality and equilibrium. In this innovative tapestry, natural medicine integration within textiles transforms from an idea to a dynamic reality, improving well-being, allowing for individualized healing journeys and reimagining the relationship between the fabrics that touch us. As we enter this new era, we bring together the threads of history and innovation, crafting a fabric that aligns with the healing rhythms of nature and human creativity.

References

1. Rockefeller JD (2015) An introduction to Ayurveda and its benefits. J.D. Rockefeller Book Club, Scotts Valley, California, USA

2. Natarajan H, Alagundi PM (2023) Metal-containing coatings on textiles—concepts and properties. Int J Adv Res Sci Eng Technol 10(10):21156–21163
3. Gupta C, Prakash D, Gupta S (2013) Functional clothing: a novel concept for disease therapeutics. Afr J Basic Appl Sci 5(4):160–166. https://doi.org/10.5829/idosi.ajbas.2013.5.4.73225
4. Mukherjee S, Kanakarajan S (2017) Extraction, optimisation and dyeing of silk yarn using natural dye from Cosmos sp. Int J Develop Res 7(7):13865–13871
5. El-Shafei A, Shaarawy S, Motawe FH, Refaei R (2018) Herbal extract as an ecofriendly antimicrobial finishing of cotton fabric. Egypt J Chem 61(2):317–327. https://doi.org/10.21608/ejchem.2018.2621.1209
6. Parani M, Lokhande G, Singh A, Gaharwar AK (2016) Engineered nanomaterials for infection control and healing acute and chronic wounds. ACS Appl Mater Interfaces 8(16):10049–10069. https://doi.org/10.1021/acsami.6b00291
7. Machado TB, Pinto AV (2003) Antimicrobial activity of selected plants for medical textiles. Int J Antimicrob Agents 21(3):279–284. https://doi.org/10.1016/S0924-8579(02)00349-7
8. Elayarajah B, Rajendran R, Balakumar C, Venkatrajah B, Sudhakar A, Janiga PK (2011) Antibacterial synergistic activity of ofloxacin and ornidazole treated biomedical fabrics against nosocomial pathogens. Asian Text J 1(2):87–97. https://doi.org/10.3923/ajt.2011.87.97
9. Deepak K, Abhishek S, Raghav V, Dharmendra G, Abhishek K (2011) Herbal textiles: green business, green earth. Colourage 58(4):54–60
10. Venkatesh CS (1976) Our tree neighbours. National Council of Educational Research and Training, New Delhi
11. Kolte PP, Shivankar VS, Ramachandran M (2015) Herbal clothing—an Ayurveda doctor. REST J Emerg Trends Model Manuf 11:26–31
12. Parvin F, Islam S, Urmy Z, Ahmed S (2020) A study on the textile materials applied in human medical treatment. Eur J Physiother Rehab Stud 1(1):56–79. https://doi.org/10.5281/zenodo.3779236
13. Bechtold T, Mussak R, Mahmud-Ali A, Ganglberger E, Geissler S (2006) Extraction of natural dyes for textile dyeing from coloured plant wastes released from the food and beverage industry. J Sci Food Agric 86:233–242. https://doi.org/10.1002/jsfa.2360
14. Mahanta D, Tiwari SC (2005) Natural dye yielding plants and indigenous knowledge on dye preparation in Arunachal Pradesh. North East India Curr Sci 88:1474–1480
15. Hasan KMF, Deb H, Rahman MDM, Morshed MN, Sai A, Genyang C (2016) A review on antibacterial colouration agent's activity, implementation and efficiency to ensure the ecofriendly and green textiles. Am J Polym Sci Eng 4(1):39–59
16. Gulrajani GML, Jain AM (1922) Application of natural dyes on bleached coir yarn. Indian Text J 102(78):1
17. Hana K (2015) Natural dyes: Their past, present, future and sustainability. In: Křemenáková D, Militký J, Mishra R (eds) Recent developments in fibrous material science. Kanina-O.P.S, London, pp 59–71
18. Rangari NT, Kalyankar TM, Mahajan AA, Lendhe PR, Puranik PK (2012) Ayurvastra: herbal couture technology in textile. Int J Res Ayurveda and Pharm 3(5):733–736
19. Mader LS (2011) Ayurvastra: dyeing fabric with medicinal ayurvedic plants. Am Bot Council 92:16–18
20. Jain M (2010) Ayurvedic textiles: a wonderful approach to handle health disorders. Colourage 57(10):45–52
21. Loret SJB, Sherimon PC (2022) Healthcare system using deep learning. In: Jain V, Chatterjee JM, Kumar P, Kose U (eds) Healthcare monitoring and data analysis using IoT: technologies and applications, Chap 13. IET, pp 237–255. https://doi.org/10.1049/PBHE038E_ch13
22. Jyothirmai S, Panda S (2016) Ayurvastra—herbal clothing (A new technology to heal naturally). IJARIIE 2(4(3044)):1166–1171

Scope of Sustainable Materials in Medical Textiles

V. R. Giridev and S. Sudha

1 Introduction

Medical textiles is one of the first to be recognised as a field of technical textiles and growing field of technical textiles sector that includes personal protective equipment, prosthetics and bio-scaffolds used for tissue engineering. It serves its application throughout the body that provides solution for various complications in operative procedures, orthopaedic and cardiovascular surgeries, tissue engineering and procedures to manage obesity, dentistry and veterinary requirements [1]. The Textile Institute defined "Medical textile as a textile structure designed for usage in a various medical application" [2, 3]. In the field of medicine, safety of the patient is of prime importance and moreover, the adequate personal protection to safeguard the medical practitioners is required in order to protect them from infections. It is one of the sectors of higher demand that is driven by technology, demand and sustainability. Sustainability is one of the key factors of the current scenario. This is because the consumption of synthetic polymers increases rapidly in medical textiles. It is mainly due to its light weight, and resistance towards chemicals and environmental conditions [4]. But the major problem with respect to the usage of synthetic materials is the waste management for manufacturing wastes and post consumer wastes. These wastes are disposed by burning which leads to the release of dangerous gases into the atmosphere or by burying the waste in the soil which leads to slow degradation. Hence alternative strategies of the waste are to be developed. With this background, the various sustainable strategies towards the waste management of medical textiles are discussed in this chapter.

V. R. Giridev (✉) · S. Sudha
Department of Textile Technology, Anna University, Chennai, India
e-mail: vrgiridev@gmail.com

S. Sundaramoorthy et al. (eds.), *Textile Materials for Good Health and Wellbeing*, SDGs and Textiles, https://doi.org/10.1007/978-981-97-6496-9_19

2 Medical Textiles

2.1 Categories

Based on the application of textiles in medical field, it is categorised as,

Implantable materials—It is a kind of medical device used to replace the damaged biological structure in a body to support or enhance the structure [5, 6].

- Sutures—natural and synthetic materials, absorbable and non-absorbable
- Vascular grafts—woven, warp knitted, weft knitted, membrane, polyurethane polymer
- Re-absorbable polymer

Non-implantable materials: The type of textile material that comes in contact with a wound but not absorbed into the body, e.g. wound dressing, pressure garments, healthcare/hygiene textiles

- Absorbent: hydrogel, alginate, collagen, odour absorbent
- Semi-permeable: film, hydrocolloid, polyurethane foam
- Secondary dressing: non-adherent, low-adherent, adherent
- Vapour permeable
- Pressure garments: elastic and non-elastic, compression bandage

Extracorporeal devices (Sastri [7]): The textile material that is used as a substitute for the essential organs that are not functioning properly. It is applied for various purposes of filtration, purification and blood circulation

- Artificial and bio-artificial kidney, liver and pancreas

Healthcare/hygiene textiles: The textile materials applied for the hygienic applications in order to avoid infection and spread of diseases thereby improves the care in the hospital and operation theatre. E.g. wipes, masks, surgical gowns

- Disposables: wipes, panty shields
- Clothing: mask, uniform
- Infection control
- Antimicrobial/antistatic
- Dental products

Intelligent medical and healthcare textiles—The textile material that can sense and react to the external conditions of mechanical, thermal, magnetic, chemical or electrical stimuli.

- Chromic materials—thermo-chromistic, photo-chromistic, halo-chromistic, electro-chromistic
- Phase changing materials—shape memory polymers, thermo-regulating
- Intelligent textile systems—passive smart, active smart (or) adaptive function

Textiles in Healthcare Environment

- Furnishing fabrics
- Textiles used in fixtures and fittings

Environmental Hygiene Control Component

- Air filters

2.2 Need for Sustainability in Medical Textiles

The advancements in data mining, artificial intelligence and robotics create a range of new opportunities in healthcare applications. In addition, the expansion of emerging economies creates a demand for the better healthcare among the raising population. The growing demand in healthcare applications leads to the increased use of single-use textile-based materials that in turn cause significant environmental harm. Medical textile products are made out of woven, knitted or nonwoven structure based on the application using natural and synthetic fibres. Among the fibres used in the medical textile products, synthetic fibres play a major role. Medical textile products consist of many chemical substances to provide the required and desired properties. These chemicals create significant risks throughout the product's lifecycle with respect to manufacture, utilisation and disposal. During manufacturing, the use of certain hazardous chemicals creates a threat to the health of the workers exposed to it. In turn, the remnant residues in the finished medical textile products pose a potential risk to the users of the products. It is also observed that the disposal of the single-use medical textile products creates a significant impact on the air in the environment through incineration, e.g. increase in carbon emissions.

The fibres of polyamide, polyester, polypropylene polytetrafluoroethylene, polyethylene, ceramic fibres, polysulphone, glass fibres, carbon fibres, etc., made in different forms of woven, nonwoven or braided structure are used in medical textile applications [8–10]. These structures can be used as implantable materials, non implantable materials, extracorporeal devices and healthcare and hygiene textile products in the field of medical textiles.

The use of synthetic fibres creates a variety of health and environmental issues throughout the lifecycle like pollution, greenhouse gas emissions, present persistently in marine and terrestrial habitats [11]. Hence, it is high time to find the right alternative for synthetic fibres to have reduced health and environmental issues such as greenhouse gas emissions and adverse change in land use. It also reduces the exposure of hazardous chemicals to the workers in the manufacturing unit.

3 Sustainability in Implantable Materials

During surgery, sutures used for the tissue ligation should have good mechanical stability, tissue compatibility, thermal stability and tensile strength [12]. The knot security, tensile strength and tissue reactivity are major drawbacks with the commercially available suture materials that lead to the development of novel suture materials. Natural-based materials are materials that have good biocompatibility of lower cost and are eco-friendly in nature. Banana fibre is a light weight fibre comprising of cellulose, hemi-cellulose, lignin and pectin [13]. Himadri Kalita [14] developed banana fibre into a biomaterial used for the prevention of infection in post operative wound. In this research, pseudo stem-based fibre of banana was treated with clotrimazole, chloramphenicol and growth factors using hydrogel system. This surface modified suture shows excellent tensile strength and physico-chemical properties and exhibits sustained release of drugs for upto 144 h. This suture shows decrease in infection and inflammatory signs at the wound site. It also exhibits significant antimicrobial activity against *E. coli, S. aureus* and *C. Albicans* in the in vitro and in vivo conditions.

Eri silk waste-based biocompatible suture material was developed in order to avoid infection at the surgical site [15]. In general, many biomaterials were developed from *B. mori, A. assama and P. ricini* and among these *P. ricini* is the cheapest to rear without causing any hazards to the environment [16, 17]. More tonnes of fibrous waste materials are generated from the reeling of the silk from the cocoon every year [18]. This fibrous waste is utilised in this research for the development of cost effective suture with biocompatibility and higher tensile strength. The higher tensile strength of the suture is mainly due to the presence of higher percentage of C and H in the silk waste obtained from *P. ricini* [19]. The suture from *P. ricini* was obtained by reeling through a five-loop technique that is embedded with antimicrobial agents in a hydrogel base consisting of aloevera and gum acacia. This suture shows sustained release of drugs and biocompatibility towards mammalian cells. The presence of galactan, glucomannan and polysaccharide present in aloevera presents anti-inflammatory and antibacterial properties to the sutures [20]. This polymeric matrix helps in drug release through leaching process that leads to the diffusion into the wound area followed by biphasic release mechanism. Gum Arabic acts as a binding agent that exhibits antimicrobial activity and has wide application in therapeutic purposes [21]. This makes the *P. ricini*-based suture help in the wound healing process faster by inhibiting microbial colonisation on the surface of the suture. In addition aloevera hydrogel helps in the synthesis of collagen in the sites of wound, which plays a major role in the tissue regeneration in the healing area by the synthesis of connective tissue [22]. It is also found that the brady kinase enzyme present in aloevera helps in reducing the inflammation on skin application that in turn stimulate the immune defence system present in the site of wound that fasten the healing process [23].

According to the research of Raghuram Kandimalla et al. [24] ramie fibre is one of the right alternatives for the development of surgical sutures. In general, ramie is used for the manufacturing of sewing threads, packing materials, fishing nets and filter cloths. Ramie has tensile strength eight times greater than cotton and seven times of silk. It is found that ramie has silky lustre and is resistant to mildews, rotting, alkalis, light and inherent characteristics of strong bacterio-static efficiency [25]. The five-loop braiding technique is used to reel the degummed fine fiber filaments in order to develop the suture material. The developed sutures show exceptional tensile strength and minimal tissue reactivity. This type of suture shows noticeable antibacterial activity against different bacterial stains and increased efficiency in wound healing with no scar on the area of wound. From this research it is found that ramie fibre shows comparable biocompatibility, tensile properties and wound closure efficiency as commercial sutures.

An attempt was made to develop the artificial human skin with silk-based biomaterials engineered using non-weaving or electrospinning technologies that support human keratinocytes and fibroblasts [26]. Silk monofilament, multifilament fibres of knitted, woven, braided or combined structures to form wire-rope scaffolds having micro-porous structure was used as nerve guides. These structures have biocompatibility and required mechanical performance. Silk-based biomaterials were used for the vascular tissue regeneration to divert the flow and as stents. Similarly, nonwoven silk fibre mesh scaffolds made from silkworm cocoons exhibit anti-thrombic surface having increased resistance to blood flow and high shear stress. It also shows good biocompatibility and supports intercellular contact with endothelial cells [1].

4 Sustainability in Extracorporeal Implants

The artificial organs applied for purifying blood are called extracorporeal implants that benefit from using fibrous materials. The artificial lung consisting of micro-porous membrane having high gas permeability and low liquid permeability as natural lungs allow oxygen through patient's blood [27]. The artificial liver made of hollow fibres or membranes is used similar to artificial kidney. The use of silk-based hollow fibres and hollow fibres made of regenerated cellulose composite fibres were used as artificial organs. It is also observed that silk is considered to be a suitable substrate for the liver cell adhesion and considered to be an alternative to collagen [28].

5 Sustainability in Non-implantable Materials

The skin acts as an effective barrier to pathogens and also plays a major role in immunisation, protection and regulation of body temperature and sensory functions [29]. As the skin acts as an interface between the internal organs of the body and the environment, there is a possibility for it to get injured. This skin injury creates an issue of bacterial infection that complicates the process of healing [30]. Wound dressing requires high antibacterial activity which is essential for the wound healing process. These wound dressings may be in various forms as electrospun fibres, biofilms, porous sponges and hydrogels [31–35]. Among these, hydrogels maintain wound healing properties which reduces the temperature of the wound surface and promotes cell proliferation [36–38]. This can be achieved by using many natural-based materials as hydrogels that need more chemical or biological reactions.

Xiang [39] developed an innovative hydrogel textile of fish de-cellularised extracellular matrix using bio printing approach having the specialised features for wound healing. This was found to be having good biocompatibility, tissue cell induction function and mechanical properties. The collagen present in the fish skin has the special characteristics of biological activity due to its low thermal shrinkage, low thermal denaturation temperature and low intermolecular cross-linking. The pepsin was used to extract the fish skin, then acylated with methacrylic anhydride. It is further mixed with Alg in order to achieve the mixture having printable properties. These bio-printed textiles show better performance with respect to cell adherence and proliferation. It also has good air permeability due to its porous structure. It also has high specific surface area that makes it possible for loading active molecules that improve wound healing process by inhibiting bacterial growth, promoting cell proliferation and accelerating wound healing process.

In another research, extrusion pretreatment and hot water extraction were used to extract gelatin from processing waste. This type of extracted gelatine shows higher cell adhesion, cell growth and wound healing property [40]. The wound surface is protected by placing the physical barrier on the wounds. This wound dressing should also help in retaining moisture, absorbing secretions and detaching it comfortably with the help of hydrogels. These hydrogels are generally obtained by cross-linking of soluble or hydrophilic polymers through chemical or physical means. These fish gelatine gels were used because of their biodegradability, biocompatibility and less antigenicity. The combination of fish gelatine with konjac shows promising results in overcoming the drawbacks and reinforcing gel properties [41]. This gelatin of fish scale extracted from tilapia fish scales covering the bioresorbable phosphate-based glass fibre shows potential application in wound healing [42, 43].

Nonwoven fabrics are generally porous structures that are formed by arranging the fibres of differing diameters and length randomly that are used in the various fields of medicines as wound dressings [44]. Jolana Kubíckova et al. [45] developed a nonwoven textile from hyaluronan that can be used for the wound healing. An anionic, non-sulphated linear polysaccharide, hyaluronan is present naturally in human structure and is biocompatible, biodegradable and non-toxic in nature. It takes part in the wound healing process, tissue regeneration, hydration and lubrication of the skin [46, 47]. This hyaluronan is applied as a coating on the cotton fabrics that can be used as wound dressing and on carbon nonwoven fabric for the osteochondral repair. The hyaluronan coated nonwoven textiles seem to be soft, flexible and paper like in appearance. This nonwoven textile forms a protective gel layer on the wound by hydrating the textile gradually on placement over the wound. The properties of non-cytotoxicity and non-pyrogenicity of this structure make it suitable for the wound healing properties of different types of wounds.

Biomedical textiles and fibre-based implants are generally used for various external applications of wound dressings, orthopaedic bandages, pressure garments and prosthetic socks. Bombyx mori fibroin fibres on blending with silver nanoparticles can be used as an antibacterial wound dressing [48]. It is also found that Bombyx mori fibroin films in combination with titanium dioxide nanoparticles are used to inhibit bacteria [49]. The use of bioactive layer made of wax coated silk fibroin woven fabric-based non-adhesive layer along with a sponge of sericin and glutaraldeyde cross linked with silk fibroin/gelatin shows more reduction in wound size, epithelialisation and collagen formation [50–52].

6 Sustainability in Intelligent Medical and Healthcare Textiles

The sustainable alternative for invasive and non-invasive health monitoring that degrade naturally and dissolve completely in their physiology is by using biodegradable sensors and electronic devices [53, 54]. The health monitoring system is used to monitor various physiological aspects of strain, pressure, temperature, pH, oxygen, etc., to assess the information on tissue healing, post surgical infection detection and personalised treatments. The monitoring devices can be non-invasive or invasive based on the requirement and complications. Hence, it is highly required to have the resorbable devices as implants for diagnostic and healing applications to have improved patient's comfort, less comfort and reduced risk of removal surgery. It also reduces the reactions with implants as a foreign body due to the characteristics of biocompatibility, non-toxicity and biodegradability. The biomedical applications of silk and regenerated silk fibroin have increased due to its mechanical robustness, bioactivity, biocompatibility and immune-compatibility nature [55]. It is observed that silk has been obtained as the new generation of mechanically flexible and degradable implantable electronics that provides appropriate support [56]. The proteolytic enzymes of chymotrypsin, actinase and carboxylase break down the silk structure through adsorption at surface binding areas followed by ester bond hydrolysis [57]. Further the structure is degraded into non-inflammatory amino acids absorbed in vivo and used in cell metabolic activities. The ultrathin bio-interfaced electronic systems provide intimate contact and reduce invasiveness in integration with profiles of the human tissues [58]. Silk is also used in bacterial detection by supporting the bio-transfer of modified peptide, and the transfer of printed graphene nanosensors onto the biomaterials with tooth enamel. The dissolution of water soluble silk substrate produces the bio-transfer of nanosensors made of graphene [59].

Another approach to fabricating biodegradable, flexible and optically transparent, micro-patterned bioelectronics devices made of organic material by aqueous photolithographic processing of conducting polymers on silk substrate was carried out [60]. It is also observed that work was carried out in developing stretchable and skin comfortable electrodes using plasticisation of silk and thin film metallisation for wearable and implantable applications. This was obtained to have excellent performance in recording on-skin electrophysiological signals in comparison with that of the gel electrodes that are commercially available [61]. An attempt was also made to combine silk with grapheme to develop a substrate with electrically conductive path in recognising the self healing, and sensing the variation in strain, temperature and humidity in the form of applying as skin mountable electronic tattoos [62]. The plant-based fibres of cotton, jute, bamboo and banana shows increased k values due to the presence of huge amount of hydroxyl groups that impart polarity [63–65].

7 Sustainability in Healthcare/Hygiene Textiles

The textiles used for the surgical applications have shown unprecedented demand in the recent years because of the COVID-19 pandemic. This creates an increased demand for the medical textile goods such as gowns, gloves and masks for the protection of healthcare workers to protect and break the spread of the more infectious coronavirus. These products are generally plastic based and on usage by the public are not disposed to the environment properly. This adds mass to the plastic pollution that is non biodegradable and retains in the environment for several years and creates a threat to the land and marine life [66]. The incineration of the PPE in turn increases the carbon emission that is considered to be biohazard. Thus, the use of synthetic material in this environment creates a challenged environment. It is high time to move towards the reusable product than the disposable product with new technologies of sterilising and separating the infectious waste from the general waste to reduce the environmental footprints [67].

It is found that reusable gowns and wipes derive environmental benefits in comparison with disposable products. The use of washed, reusable napkins significantly reduces the impact than the single, heavier disposables. The disposables create impacts to change in climatic condition, smog, depletion of fossil fuel, acid rain, algae bloom, etc. The advancement of laundering technologies with respect to machineries and efficiency, the reuse of the medical textiles is an efficient method and in turn the impact on the environment will also be reduced. In order to wash a pound of healthcare textiles three gallons of water was needed in the earlier days, wherein, with the advanced technologies, only three quarters of gallons of water. It is also observed that reuse and recycling of wastewater is also made possible with the advanced technologies. With all these advancements, the laundering of healthcare textile products was made environmental friendly and cheaper means of usage.

It is also found that for the manufacturing of protective clothing biodegradable polymers such as polylactic acid [68] and recycled polymers can be used. The protective clothing based on recycling of the disposable plastic helps in reducing the environmental load and making medical textiles shift towards circular economy and sustainable protective clothing. The natural-based biopolymers extracted from animal or plant sources such as chitosan, sericin, cyclodextrin and alginate were considered to be the sustainable bioactive textiles [69]. Green-based technology of solvent free dry plasma was used for antimicrobial finishing on textiles that do not produce contaminated water or hazards to the fabric [70]. It is also observed that the use of UV radiation, enzymes and ultrasound energy was obtained to be a more environmental-friendly process that helps in adhesion of natural dye along with the antimicrobial effects. In addition the use of new fibre or fabric production techniques can be used for the development of protective clothing with increased comfort and protection from pathogens [71].

8 Challenges in Sustainability Concepts of Medical Textiles

Some of the challenges with respect to recycling of medical textiles are Sourav Kumar Das [10],

- Appropriate selection of fabrics based on the requirements as large amount of body fluids are involved in surgeries and of high risk.
- Proper follow-up of specification on manufacturing including visual and hygienic cleanliness, decontamination, disinfection and sterilisation would have to be developed, adapted and tested for the product.
- Nonwoven fabrics may generate lint as airborne particles create complications during surgeries and post operative problems

On considering all the challenges, the potential solution to be developed in order to reuse and recycle the medical textile products is by adapting appropriate eco-friendly technology in a cost effective manner.

References

1. Li G, Li Y, Chen G, He J, Han Y, Wang X, Kaplan DL. Silk-based biomaterials in biomedical textiles and fiberbased implants. Adv Healthc Mater. 2015 Jun 3;4(8):1134–51. https://doi.org/10.1002/adhm.201500002. Epub 2015 Mar 13. PMID: 25772248; PMCID: PMC4456268.
2. Grace Annapoorani S (2013) Recent developments in medical textiles implantable devices—an overview. Global Res Anal 2:12
3. Morris H, Murray R (2020) Medical textiles. Text Prog 52(1–2):1–127
4. Singsatit P, Pimpan V (2009) Recycling of medical gown nonwoven fabric manufacturing waste as a filler for high density polyethylene. J Met Mater Min 19:81–88
5. Azam Ali M, Shavandi A (2016) Medical textiles testing and quality assurance, performance testing of textiles, methods, technology and applications. Woodhead Publishing Series in Textiles, pp 129–153
6. Gad K (2014) Medical textiles. Encyclopedia of toxicology, 3rd edn, pp 182–190
7. Sastri VR (2022) Three-dimensional printing, wearables, medical textiles, adhesives, and coatings. In: Plastics in medical devices (third edition), properties, requirements, and applications, plastics design library, pp 381–421
8. Qin Y (2016) An overview of medical textile products. Med Text Mater 2016:13–22

9. Rajendran S, Anand SC, Rigby AJ (2016) Textiles for healthcare and medical applications. In: Handbook of technical textiles (second edition) volume 2: technical textile applications, pp 135–168
10. Das SK, Chinnappan A, Jayathilaka WADM, Gosh R, Baskar C, Ramakrishna S (2021) Challenges and potential solutions for 100% recycling of medical textiles. Mater Circular Econ 3:13
11. Ragitha VM, Edison LK (2023) Safety issues, environmental impacts, and health effects of biopolymers. In: Thomas SARA, Jose Chirayil C, Thomas B (eds) Handbook of biopolymers. Springer, Singapore
12. Pedro J, Serrano C, García-Fernández L, Barbeck M, Ghanaati S, Unger R, Kirkpatrick J, Arzt E, Funk L, Tur P (2015) Nanostructured medical sutures with antibacterial properties. Biomaterials 52:291–300
13. Ganan P, Cruz J, Garbizu S, Arbelaiz A, Mondragon I (2004) Stem and bunch banana fibers from cultivation wastes: effect of treatments on physico-chemical behaviour. J Appl Polym Sci 94:1489–1495
14. Kalita H, Hazarika A, Kandimalla R, Kalita S, Devi R (2018) Development of banana (*Musa balbisiana*) pseudo stem fiber as a surgical bio-tool to avert postoperative wound infections. RSC Adv 8:36791
15. Kalita H, Hazarika A, Kalita S, Kandimalla R, Devi R (2017) Antimicrobials tethering on suture surface through a hydrogel: a novel strategy to combat postoperative wound infections. RSC Adv 2017(7):32637
16. Dutta S, Talukdar B, Bharali R, Rajkhowa R, Devi D (2013) Fabrication and characterization of biomaterial film from gland silk of muga and eri silkworms. Biopolymers 99:326–333
17. Talukdar B, Saikia M, Handique J, Devi D (2011) Effect of organic solvents on tensile strength of muga silk produced by Antherae aassamensis. Int J Pure Appl Sci Technol 7:81–86
18. Javali UC, Padaki NV, Das B, Malali KB (2015) Developments in the use of silk by-product and silk waste. Advances in silk science and technology. Woodhead Publishing Series in Textiles, pp 261–270
19. Mai-ngam K, Boonkitpattarakul K, Jaipaew J, Mai-ngam B (2011) Evaluation of the properties of silk fibroin films from the non-mulberry silkworm *Samia cynthiaricini* for biomaterial design. J Biomater Sci Polym Ed 22:2001–2022
20. Silva SS, Oliveira MB, Mano JF, Reis RL (2014) Bio-inspired *Aloe vera* sponges for biomedical applications. Carbohydr Polym 112:264–270
21. Aderibigbe BA, Varaprasad K, Sadiku ER, Ray SS, Mbianda XY, Fotsing MC, Owonubi SJ, Agwuncha SC (2015) Kinetic release studies of nitrogen-containing bisphosphonate from gum acacia crosslinked hydrogels. Int J Biol Macromol 73:115–123
22. Grande D, Lee T, Ph D, Limpisvasti O (2010) The effect of recombinant human platelet-derived growth factor BB—coated sutures on rotator cuff healing in a sheep model arthrosc. J Arthrosc Relat Surg 26(11):1456–1462
23. Choi S, Chung MH (2003) A review on the relationship between aloe vera components and their biologic effects. Semin. Integr. Med. 1:53–62
24. Kandimalla R, Kalita S, Choudhury B, Devi D, Kalita D, Kalita K, Dash S, Kotoky J (2016) Fiber from ramie plant (*Boehmeria nivea*): a novel suture biomaterial. Mater Sci Eng C 62:816–822
25. Xi L, Qin D, An X, Wang G (2013) Resistance of natural bamboo fiber to microorganism and factor that may affect such resistance. BioResources 8:6501–6509
26. Min BM, Lee G, Kim SH, Nam YS, Lee TS, Park WH (2004) Electrospinning of silk fibroin nanofibers and its effect on the adhesion and spreading of normal human keratinocytes and fibroblasts in vitro. Biomaterials 25:1289
27. Horrocks AR, Anand SC (2000) Handbook of technical textiles. Elsevier, London
28. Cirillo B, Morra M, Catapano G (2004) Adhesion and function of rat liver cells adherent to silk fibroin/collagen blend films. Int J Artif Organ 27:60
29. Shi R, Geng H, Gong M, Ye J, Wu C, Hu X, Zhang L (2018) Long-acting and broadspectrum antimicrobial electrospun poly(ε-caprolactone)/gelatin micro/nanofibers for wound dressing. J Colloid Interface 509:275–284

30. Yu K, Lu F, Li Q, Chen H, Lu B, Liu J, Lan G (2017) In situ assembly of Ag nanoparticles (AgNPs) on porous silkworm cocoon-based wound film: enhanced antimicrobial and wound healing activity. Sci Rep 7(1):2107
31. Cheng H, Shi Z, Yue K, Huang X, Xu Y, Gao C (2021) Sprayable hydrogel dressing accelerates wound healing with combined reactive oxygen species-scavenging and antibacterial abilities. Acta Biomater 124:219–232
32. Qu J, Zhao X, Liang Y, Xu Y, Ma PX, Guo B (2019) Degradable conductive injectable hydrogels as novel antibacterial, anti-oxidant wound dressings for wound healing. Chem Eng J 362:548–560
33. Yang J, Wang K, Yu DG, Yang Y, Bligh SWA, Williams GR (2020) Electrospun Janus nanofibers loaded with a drug and inorganic nanoparticles as an effective antibacterial wound dressing. Mater Sci Eng C 111:110805
34. Yao X, Zhu G, Zhu P, Ma J, Chen W, Liu Z (2020) Omniphobic ZIF-8@hydrogel membrane by microfluidic-emulsion-templating method for wound healing. Adv Funct Mater 30(13):1909389
35. Zhang XX, Chen GP, Yu YR, Sun LY, Zhao YJ (2020) Bioinspired adhesive and antibacterial microneedles for versatile transdermal drug delivery. Research 8:3672120
36. Chaudhuri O, Cooper-White J, Janmey PA, Mooney DJ, Shenoy VB (2020) Effects of extracellular matrix viscoelasticity on cellular behaviour. Nature 584(7822):535–546
37. Qiao Z, Lian M, Han Y, Sun B, Zhang X, Jiang W (2021) Bioinspired stratified electrowritten fiber-reinforced hydrogel constructs with layer-specific induction capacity for functional osteochondral regeneration. Biomaterials 266:120385
38. Wang C, Lai J, Li K, Zhu S, Lu B, Liu J (2021) Cryogenic 3D printing of dual delivery scaffolds for improved bone regeneration with enhanced vascularization. Bioact Mater 6(1):137–145
39. Lin X, Zhang H, Zhang H, Zhang Z, Chen G, Zhao Y (2023) Bio-printed hydrogel textiles based on fish skin decellularized extracellular matrix for wound healing. Engineering 25:120–127
40. Huang C-Y, Wu TC, Hong YH, Hsieh SL, Guo HR, Huang RH (2018) Enhancement of cell adhesion, cell growth, wound healing, and oxidative protection by gelatins extracted from extrusion-pretreated tilapia (*Oreochromis* sp.) Fish Scale. Molecules 23:2406
41. Zhou L, Xu T, Yan J, Li X, Xie Y, Chen H (2020) Fabrication and characterization of matrine-loaded konjac glucomannan/fish gelatin composite hydrogel as antimicrobial wound dressing. Food Hydrocoll 104:105702–105735
42. Al-Nimry S, Abu Dayah A, Hasan I, Daghmash R (2021) Cosmetic, biomedical and pharmaceutical applications of fish gelatin/hydrolysates. Mar Drugs 19:145
43. Sghayyar HN, Lim SS, Ahmed I, Lai JY, Cheong XY, Chong ZW, Lim AFX, Loh HS (2020) Fish biowaste gelatin coated phosphate-glass fibres for wound-healing application. Eur Polym J 122:109386–109396
44. Ajmeri JR, Ajmeri CJ (2011) Nonwoven materials and technologies for medical applications. In: Bartels VT (ed) Handbook of medical textiles. Woodhead Publishing, Sawston, pp 106–131
45. Kubíčková J, Medek T, Husby J, Matonohová J, Vágnerová H, Marholdová L, Velebný V, Chmela J (2022) Nonwoven textiles from Hyaluronan for wound healing applications. Biomolecules 12:16
46. Litwiniuk M, Krejner A, Grzela T (2016) Hyaluronic acid in inflammation and tissue regeneration. Wounds 28:78–88
47. Voigt J, Driver VR (2012) Hyaluronic acid and wound healing. Wound Repair Regen 20:317–331
48. Gil ES, Panilaitis B, Bellas E, Kaplan DL (2013) Functionalized silk biomaterials for wound healing. Adv Healthc Mater 2:206
49. Xia Y, Gao G, Li Y (2009) Preparation and properties of nanometer titanium dioxide/silk fibroin blend membrane. J Biomed Mater Res B Appl Biomater 90:653
50. Kanokpanont S, Damrongsakkul S, Ratanavaraporn J, Aramwit P (2012) An innovative bi-layered wound dressing made of silk and gelatin for accelerated wound healing. Int J Pharm 436:141

51. Kanokpanont S, Damrongsakkul S, Ratanavaraporn J, Aramwit P (2013) Physico-chemical properties and efficacy of silk fibroin fabric coated with different waxes as wound dressing. Int J Biol Macromol 55:88
52. Wang R, Zhu J, Jiang G, Sun Y, Ruan L, Li P, Cui H (2020) Forward wound closure with regenerated silk fibroin and polylysine-modified chitosan composite bioadhesives as dressings. ACS Appl Bio Mater 3(11):7941–7951
53. Kang SK, Murphy RK, Hwang SW, Lee SM, Harburg DV, Krueger NA, Shin J, Gamble P, Cheng H, Yu S (2016) Bioresorbable silicon electronic sensors for the brain. Nature 530(7588):71–76
54. Tao H, Hwang SW, Marelli B, An B, Moreau JE, Yang M, Brenckle MA, Kim S, Kaplan DL, Rogers J (2014) Silk-based resorbable electronic devices for remotely controlled therapy and in vivo infection abatement. Proc Natl Acad Sci 111(49):17385–17389
55. Li C, Guo C, Fitzpatrick V, Ibrahim A, Zwierstra MJ, Hanna P, Lechtig A, Nazarian A, Lin SJ, Kaplan DL (2020) Design of biodegradable, implantable devices towards clinical translation. Nat Rev Mater 5:1–21
56. Kim DH, Kim YS, Amsden J, Panilaitis B, Kaplan DL, Omenetto FG, Zakin MR, Rogers JA (2009) Silicon electronics on silk as a path to bioresorbable, implantable devices. Appl Phys Lett 95(13):133701
57. Horan RL, Antle K, Collette AL, Wang Y, Huang J, Moreau JE, Volloch V, Kaplan DL, Altman GH (2005) In vitro degradation of silk fibroin. Biomaterials 26(17):3385–3393
58. Kim DH, Viventi J, Amsden JJ, Xiao J, Vigeland L, Kim YS, Blanco JA, Panilaitis B, Frechette ES, Contreras D (2010) Dissolvable films of silk fibroin for ultrathin conformal bio-integrated electronics. Nat Mater 9(6):511–517
59. Mannoor MS, Tao H, Clayton JD, Sengupta A, Kaplan DL, Naik RR, Verma N, Omenetto FG, McAlpine MC (2012) Graphene-based wireless bacteria detection on tooth enamel. Nat Commun 3:763
60. Pal RK, Farghaly AA, Collinson MM, Kundu SC, Yadavalli VK (2016) Photolithographic micropatterning of conducting polymers on flexible silk matrices. Adv Mater 28(7):1406–1412
61. Chen G, Matsuhisa N, Liu Z, Qi D, Cai P, Jiang Y, Wan C, Cui Y, Leow WR, Liu Z (2018) Plasticizing silk protein for on-skin stretchable electrodes. Adv Mater 30(21):1800129
62. Wang Q, Ling S, Liang X, Wang H, Lu H, Zhang Y (2019) Self-healable multifunctional electronic tattoos based on silk and graphene. Adv Funct Mater 29(16):1808695
63. Ensieh S, Hosseini SD, Priyanka G, Ravinder D (2021) Biodegradable materials for sustainable health monitoring devices. ACS Appl Bio Mater 4(1):163–194
64. Hemstreet JM (1982) Dielectric constant of cotton. J Electrost 13(3):345–353
65. Jayamani E, Hamdan S, Rahman MR, Bakri MB (2014) Comparative study of dielectric properties of hybrid natural fiber composites. Proc Eng 97:536–544
66. Loo F (2020) Asian plastic prices rebound but pandemic limits demand support for most markets. https://www.icis.com/explore/resources/news/2020/04/16/10496614/asian-plastic-pricesrebound-but-pandemic-limits-demand-support-for-most-markets
67. Fletcher C (2020) What happens to waste PPE during the coronavirus pandemic? The Conversation. https://theconversation.com/what-happens-to-waste-ppe-during-the-coronavirus-pandemic-137632
68. Karim MN, Afroj S, Rigout M, Yeates SG, Carr C (2015) Towards UV-curable inkjet printing of biodegradable poly(lactic acid) fabrics. J Mater Sci 50:4576–4585
69. Shahid-ul-Islam, Shahid M, Mohammad F (2013) Green chemistry approaches to develop antimicrobial textiles based on sustainable biopolymers—a review. Ind Eng Chem Res 52:5245–5260
70. Kan CW (2014) A novel green treatment for textiles: plasma treatment as a sustainable technology. CRC Press, Boca Raton
71. Karim N, Afroj S, Lloyd K, Oaten LC, Andreeva DV, Carr C, Farmery AD, Kim I-D, Novoselov KS (2020) Sustainable personal protective clothing for healthcare applications: a review. ACS Nano 14:12313–12340

The manufacturer's authorised representative in the EU is Springer Nature Customer Service Centre GmbH, Europaplatz 3, 69115 Heidelberg, Germany. If you have any concerns regarding our products, please contact ProductSafety@springernature.com

Printed and bound by CPI Group (UK) Ltd, Croydon, CR0 4YY
01/10/2026
02225857-0011